CAMBRIDGE LIBRARY COLLECTION

Books of enduring scholarly value

Technology

The focus of this series is engineering, broadly construed. It covers technological innovation from a range of periods and cultures, but centres on the technological achievements of the industrial era in the West, particularly in the nineteenth century, as understood by their contemporaries. Infrastructure is one major focus, covering the building of railways and canals, bridges and tunnels, land drainage, the laying of submarine cables, and the construction of docks and lighthouses. Other key topics include developments in industrial and manufacturing fields such as mining technology, the production of iron and steel, the use of steam power, and chemical processes such as photography and textile dyes.

Observations on Limes, Calcareous Cements, Mortars, Stuccos, and Concrete

An officer in the Royal Engineers, Sir Charles William Pasley (1780–1861) wrote on matters ranging from military sieges to architecture. In this substantial work, first published in 1838, he outlines the experimentally determined properties of various building materials, with a view to their practical application. Offering guidance on how to decide between different calcareous mortars and cements, Pasley discusses how to judge their comparative strengths. Heeding advice from the Institution of Civil Engineers, he made this work a broad overview, rather than simply focusing on his special area of interest: natural and artificial cements. His research on cements led to the large-scale manufacture of products such as Portland, patent lithic, and blue lias. Pasley discusses the research of other authors in the appendix. Also reissued in this series, in English translation, is Louis-Joseph Vicat's *Practical and Scientific Treatise on Calcareous Mortars and Cements, Artificial and Natural* (1837).

Cambridge University Press has long been a pioneer in the reissuing of out-of-print titles from its own backlist, producing digital reprints of books that are still sought after by scholars and students but could not be reprinted economically using traditional technology. The Cambridge Library Collection extends this activity to a wider range of books which are still of importance to researchers and professionals, either for the source material they contain, or as landmarks in the history of their academic discipline.

Drawing from the world-renowned collections in the Cambridge University Library and other partner libraries, and guided by the advice of experts in each subject area, Cambridge University Press is using state-of-the-art scanning machines in its own Printing House to capture the content of each book selected for inclusion. The files are processed to give a consistently clear, crisp image, and the books finished to the high quality standard for which the Press is recognised around the world. The latest print-on-demand technology ensures that the books will remain available indefinitely, and that orders for single or multiple copies can quickly be supplied.

The Cambridge Library Collection brings back to life books of enduring scholarly value (including out-of-copyright works originally issued by other publishers) across a wide range of disciplines in the humanities and social sciences and in science and technology.

Observations on Limes, Calcareous Cements, Mortars, Stuccos, and Concrete

And on Puzzolanas, Natural and Artificial

Charles William Pasley

Cambridge
University Press

CAMBRIDGE
UNIVERSITY PRESS

University Printing House, Cambridge, CB2 8BS, United Kingdom

Cambridge University Press is part of the University of Cambridge.
It furthers the University's mission by disseminating knowledge in the pursuit of education, learning and research at the highest international levels of excellence.

www.cambridge.org
Information on this title: www.cambridge.org/9781108070560

This edition first published 1838
This digitally printed version 2014

ISBN 978-1-108-07056-0 Paperback

OBSERVATIONS

ON

LIMES, CALCAREOUS CEMENTS,

MORTARS, STUCCOS, AND CONCRETE,

AND ON

PUZZOLANAS, NATURAL AND ARTIFICIAL;

TOGETHER WITH

RULES DEDUCED FROM NUMEROUS EXPERIMENTS

FOR MAKING

AN ARTIFICIAL WATER CEMENT,

EQUAL IN EFFICIENCY TO

THE BEST NATURAL CEMENTS OF ENGLAND,

IMPROPERLY TERMED ROMAN CEMENTS;

AND AN ABSTRACT OF THE OPINIONS OF FORMER AUTHORS ON THE SAME SUBJECTS.

BY C. W. PASLEY, C.B.

COLONEL IN THE CORPS OF ROYAL ENGINEERS,

F. R. S., &c. &c. &c.

LONDON:

JOHN WEALE, ARCHITECTURAL LIBRARY,

59, HIGH HOLBORN.

1838.

PREFACE.

THE first sheets of the present work were sent to press in May 1836, so that the unusual period of nearly two years and a half has elapsed between the commencement and the completion of printing it. This was owing to new questions concerning the properties of cements continually suggesting themselves, seldom occupying less than a fortnight and sometimes three months or even more, before they could be properly decided; besides which time was repeatedly lost by unforeseen failures in our apparatus, which required not only the defective-parts of the wood or iron work to be replaced, but sometimes also spoiled the cements, that were to be experimented upon, the replacing of which, if artificial, was a work of time. During this period also, as our views became more enlarged, the common modes of experimenting on the comparative strength of cements and mortars, which we had at first adopted, appeared to be unsatisfactory, so that in addition to these, we had recourse to a new method for ascertaining their comparative adhesiveness to bricks and stones, which is in fact their most important property, but to which little or no attention had been paid by former writers. This also was of course a cause of delay.

PREFACE.

As the whole of the opinions stated in this Treatise, were deduced from personal observation, or from the successive experiments alluded to, with the exception of those founded on or corroborated by some important facts extracted from a paper of Mr. Timperley, published in the first volume of the Transactions of the Institution of Civil Engineers; and as these experiments occupied so long a period of time, and the book was printed in portions, during the intervals between successive experiments, it cannot boast of such a methodical arrangement, as might have been adopted, had every experiment recorded in it been tried, before the first sheet was sent to press. Hence it was thought desirable, in order to compensate this want of arrangement by facility of reference, to make the Table of Contents more ample than usual, so as to form a sort of abridgement of the work itself, rather than a mere outline of the titles of the subjects discussed.

In this Treatise, the various properties of limes, of calcareous mortars and cements, as well as of puzzolanas, have been discussed in the amplest manner, and reduced to a system, no part of which, however, has been founded on mere hypothesis.

I trust, that a good deal of new information will be found in the following pages, upon a subject of immense importance to mankind, and that the principles developed in it, will prevent such crude notions respecting calcareous mortars and cements, from being adopted in future, as have occasionally caused the gradual and sometimes even the sudden ruin of the locks of canals, wharf and dock walls, and other pieces of brickwork or masonry, exposed to the action of water; for although sound principles have been acted upon by writers as well as practical men, how is the young student to decide, when he reads the discordant opinions of Semple, Smeaton and Higgins in this country, and of Belidor, Loriot, De la Faye, Vicat and Treussart in

France, all maintained with equal confidence; and when he finds no less discordant opinions maintained with no less confidence by practical men, trusting to what they call experience, which may be very limited, for unless a man of business be also a man of research, he cannot know what has been done in foreign countries, or even in distant parts of his own.

To do away this uncertainty, I have endeavoured to point out such modes of experimenting and of personal observation, as will enable an intelligent person, who has no experience of his own, to decide justly between the various sorts of calcareous mortars and cements, that may be held out by their respective partizans, as being the best, or if not absolutely the best, as being the most expedient in an economical point of view; which last consideration has caused the weaker water limes and sometimes even common lime, such as the chalk lime of the South of England, to be used for the brickwork or masonry of works of the above mentioned description, than which there cannot be a more mistaken economy, because the mortars of these limes are absolutely unfit to resist the action of water, and cannot insure the stability of walls built with them, and exposed to that action, for any length of time.

Of all the authors, who have investigated the properties of calcareous mortars and cement from time immemorial to the present day, our countryman Smeaton appears to me to have the greatest merit, for although he found out no new cement himself, he was the first who discovered in or soon after the year 1756, that the real cause of the water-setting properties of limes and cements, consisted in a combination of clay with the carbonate of lime; in consequence of having ascertained by a very simple sort of chemical analysis, that there was a proportion of the former ingredient in all the natural lime stones, which on being calcined, developed that highly important quality, without which walls exposed to water go to pieces, and those exposed to air and weather only, are comparatively of inferior strength. By this memorable discovery, Smeaton overset the prejudices of more than 2000 years, adopted by all former writers, from Vitruvius in ancient Rome, to Belidor in France, and Semple in this country, who agreed in maintaining, that the superiority of lime consisted in the hardness and whiteness of the stone, the former of which may or may not be accompanied by water-setting or powerfully cementing properties, and the latter of which is absolutely incompatible with them. The new principle laid down by Smeaton, the truth of which has recently been admitted by the most enlightened Chemists and Engineers of Europe, was the basis of the attempts, made by Dr. John at Berlin and by M. Vicat in France, to form an artificial water lime or hydraulic lime in 1818, and of mine to form an artificial water cement at Chatham in 1826, to which I was led by the perusal of Smeaton's observations, without knowing any thing of the previous labours of those gentlemen on the Continent, or of Mr. Frost, the acknowledged imitator of M. Vicat, in this country.

The natural cements of England, most improperly and absurdly termed Roman cement, though the discovery of their properties and use, is as much a modern discovery as the steam engine, and of more recent date, possess an infinite superiority over all other hydraulic mortars, not excepting puzzolana, which in combination with lime formed the strongest cement that the Romans were acquainted with. First discovered by Parker in 1796, our English cements are now prepared for use by a great number of manufacturers, some of whom may be depended upon as never offering a bad article for sale, and any individual who applies to them, will be sure to have it good; but in the public Departments, it is so common to advertise for tenders for building materials and to accept the lowest, that it tempts unprincipled men, who gain such contracts by offering to supply them at a lower price than would compensate the fair dealer, to endeavour to pass off not merely an inferior but an adulterated article upon the receiving Officers, of which I have known an instance at Chatham. This induced me to endeavour to lay down rules for judging of the quality of cement offered for sale, and for ascertaining whether it has been adulterated or not, by attending to which, the most unexperienced person may easily detect such frauds, in 24 hours or even less.

Instructions have also been given, by which Officers of Engineers or others, stationed in the Colonies, or employed in exploring some new country, will be enabled to judge, whether a calcareous stone, whose properties have not before been ascertained, is a water cement, a water lime, or a common lime.

Moreover, as it is very difficult to decide upon the value of different calcareous cements and mortars, by building portions of walls with them, which would not only require an inconvenient length of time to compare them, but would be liable to great uncertainty; rules have been laid down, by which their comparative strength may be judged of experimentally, in a short space of time such as ten days, with very little trouble and with the greatest accuracy; that is, if I am right in the opinion formed by me after the continual experiments of many years, namely that those cements and mortars, which set the soonest under water, and which are found to possess the greatest resistance and adhesiveness at the end of that period, will also prove the most durable, and will retain the same comparative superiority for any length of time. These rules, if correct, as I trust they are, will obviously be of the greatest importance, by enabling an Engineer to make a judicious choice between two sorts of cement or of mortar, both apparently good; but between which there may in reality be a considerable difference, which no other mode of comparison can detect so conveniently, or with so much accuracy.

The properties of cements first attracted my notice most particularly, so that when after many previous failures I succeeded in forming an efficient artificial cement on a great scale in 1830, I took very little notice of limes and none of puzzolana, in a Pamphlet printed by me in that same year, which contained my Observations on the Natural Water Cements of England, with a description of my Artificial Cement made in imitation of them. This pamphlet was sent by authority of the Master-General and Board of Ordnance to all the Royal Engineer Stations at home and abroad, and by the Court of Directors of the East India Company to all their Engineer Stations in India; and copies of it were presented by me to various Scientific Societies, as well as to several private friends. I discovered, however, on resuming the same subject six years afterwards, that my artificial water cement of 1830 was not the strongest, that could have been made of the same ingredients, as will be seen by a perusal of the present Treatise.

That the artificial cement, more recently made at the Royal Engineer Establishment in the years 1836, 1837 and 1838, is equal in efficiency to the best natural water cements of England, as set forth in the title page of this work, has been fully proved by numerous experiments, tried not only in presence of all the Royal and Honourable East India Company's Engineer Officers attending this Establishment, but also of Officers of other Regiments and Corps in the Garrison of Chatham, as well as of other Gentlemen, and of Clerks of Works, Builders and Artificers, whom I made a point of inviting to attend, as I wished to make my experiments as public as possible. The natural English cements were either supplied by Messrs. Francis and Co. of Vauxhall, or from her Majesty's Dockyard at Sheerness, both of excellent quality, and made, according to custom, of a mixture of calcined Harwich and Sheppy stones, which are better than either of those cements used singly. In such experiments, as were not tried for the purpose of ascertaining comparative strength, but for investigating the general properties of water cements, whenever the particular sort experimented upon is not specified, let it be understood, that Messrs. Francis's cement was used.

Other experiments recorded in the following Treatise have established an important fact before considered doubtful, namely that cement is no less suitable for masonry composed of the hardest and heaviest stones, than for brickwork; as was proved by the remarkable circumstance of a Bramley-fall stone breaking before the cement joint, which united it to another large stone of the same kind, gave way.

That a few courses of brickwork and cement bond, aided by hoop iron in the joints may supersede the necessity of chain timbers and bond timbers of all descriptions in the walls of buildings, as well as of wooden lintels over the openings of doors and windows and of bressummers over shop fronts, had been fully proved by the experimental brick beams of Mr. Brunel and of Messrs. Francis & Co, but as their experiments left the comparative importance of the cement and of the hoop-iron bond doubtful, I was induced to build three experimental brick beams, one with cement alone, another with cement and hoop-iron bond, and a third with Halling lime mortar and hoop iron bond, which on being all broken down fifty days afterwards, proved the vast importance of the hoop-iron bond, without which the brick and cement beam had comparatively very little resistance, and the brick and mortar beam would have had none. The practicability of executing domes or roofs with brickwork and cement, in forms absolutely impossible with the best lime mortar, was also proved by an interesting experiment.

PREFACE.

In 1836, when the first sheets of this work were sent to press, a sort of enthusiasm prevailed in favour of concrete, especially when moulded into artificial stone, which induced me to experiment upon the resistance of small square prisms of artificial stone, made of various limes and cements mixed with different proportions of sand and gravel, after allowing them sufficient time to set. The extreme weakness of these experimental prisms convinced me, that concrete was of very inferior importance, except when used in mass for foundations or for backing, to which Sir Robert Smirke, who first introduced the use of it in this country, very judiciously confined it. The unfavorable impression as to the value of artificial stone, produced by our experiments at Chatham has been confirmed on a great scale by the failure of the external surface of the concrete docks and wharf walls in her Majesty's Dock-yards at this place and at Woolwich; and even for foundations it requires to be used with judgment, and ought greatly to exceed in area the base of the walls to be supported by it; otherwise it may settle unequally and crack, as was the case in the experimental concrete casemate built at Woolwich, or it may sink to a considerable depth in soft soil, as was the case with the concrete wharf wall built by Mr. Ranger by direction of the then Admiralty Architect at the north extremity of Chatham Dock-yard, which has sunk about one foot deep in the mud, since it was first built, and which from the nature of its profile, combined with this deficiency of base, would have capsized and fallen into the Medway, if it had not been afterwards secured by iron bars and pieces of chain cable, acting as land ties, and fixed to piles driven at some distance in rear of it.*

In respect to water-limes, and water-mortars, or hydraulic

* This wall escaped my observation, until the whole of my Treatise was printed; otherwise I should have noticed it in the body of the work. The concrete casemate at Woolwich was described in Article 32 and its note, and its failure from unequal settlement, in the note to Article 33.

limes or hydraulic mortars as the French term them, I have tried fewer experiments, the results of which however have convinced me, that the opinion of Smeaton, since generally adopted, that the Aberthaw or blue lias is far the strongest of our English limes, is perfectly correct. Indeed it seems to possess rather greater strength than he gave it credit for, when he considered it necessary to mix it with puzzolana, for I have found, in confirmation of the efficiency of the system, since adopted for the masonry of several important works, and amongst others for that of the celebrated Menai bridge, that the mortar of this lime will set on being immersed, whereas the mortars of all the other English limes, that I have experimented upon remain in a state of pulp under water, unless improved by the addition of a proportion of puzzolana. But this substance, as imported from Italy, being too expensive to compete with the blue lias lime mortar; the only resource for making the other limes of England available for works exposed to the action of water, at a moderate expense, appears to be the use of artificial puzzolana, made by calcining the blue clay of the Medway or any other impalpable clay, whether alluvial or fossile. My own experiments induce me to put great confidence in this expedient, in which however, there is nothing original; because puzzolanas both natural and artificial have been in common use in the south of Europe since the time of the Romans, though the latter has been little used in this country, and the former I believe not at all, until Smeaton first introduced it about 80 years ago.

The subject of cements having frequently been discussed within the last 10 years at the Institution of Civil Engineers, where Mr. Frost and I, when present on those occasions, always explained our respective proceedings to the members, it was suggested to me that if the Treatise, which I then announced my intention of publishing, should take a comprehensive view of the whole subject of limes and mortars also, it would be more useful to practical men, than if confined to observations on natural and artificial water cements only, like my pamphlet of 1830. In compliance therefore with what appeared to be the wish of several members of that excellent and most liberal Institution, I have enlarged more upon limes, mortars and puzzolanas in the body of my Treatise than I had at first intended to do; and in deference also to their opinion, I have added in the Appendix an abstract of the writings and proceedings of the most eminent British and Foreign authors on the same subjects, that is of the more modern ones, for every author on limes and mortars, that I have met with, followed the authority of Vitruvius, until after the middle of the last century, when Smeaton came into the field, and pointed out the inaccuracy of his doctrines, in the same manner as Lord Bacon had set aside the implicit belief in the doctrines of Aristotle in matters of Philosophy.

Of the Foreign Writers, that I have met with, M. Vicat and his countryman General Treussart of the French Engineers are far the most distinguished, as they have both carried their experimental researches into the properties and use of hydraulic and common limes, and of the various mortars that may be made, by mixing them with sand alone, or with puzzolana or trass natural or artificial, with and without sand, in numerous combinations, to a far greater extent than any former writers or practical men had ever done; and therefore it must be acknowledged, how much not only their own countrymen but other nations are indebted to them, for the information contained in their respective works. But of the two, M. Vicat's writings appear to have made the greatest impression in France, probably from the circumstance of his having been prior to General Treussart, so that he was the first to draw general attention to the great importance of hydraulic limes and mortars in that country. They are both equally satisfactory in their experiments as to the hydraulic powers of different mortars, which they estimate by the same mode; but neither of them has attempted to ascertain the adhesiveness of their experimental mortars to stones or bricks, which as I said before is of much more importance than their resistance. Upon this last property, however, both M. Vicat and General Treussart have founded their opinions of the comparative strength of different mortars, possessed of the same water-setting powers. On perusing their proceedings with attention, M. Vicat's experiments on this point appear to me so very unsatisfactory, that I cannot help giving the preference to those of the General, which carry conviction along with them, being simple in their nature, and all described in the clearest manner, and perfectly congruous and consistent with each other. Whenever the opinions of these two gentlemen, or of any other authors native or foreign, noticed in the Appendix, differ essentially from mine, I have not failed to remark upon them; always stating however, whether I dissented in consequence of their being contradicted by the results of my own experiments, or merely as a matter of doubt, or of private opinion.

There seems reason to believe, that the proper mode of preparing and of using cement could not have been understood in France, when M. Vicat and General Treussart published, otherwise they never would have undervalued it, in the extraordinary manner they have both done, by declaring that they considered it much inferior to the hydraulic limes, which they experimented upon. This error, for I do not scruple to pronounce it such, can only have arisen from their having treated their cement and their limes in the same manner, which could not fail to spoil the former, as will be acknowledged by every English builder or workman, who has been in the habit of working with both. M. Vicat in particular goes so far in depreciation of cement, as to say, that although a vast quantity of it is now employed in and near London, it will gradually be disused as soon as the hydraulic limes shall be better known and appreciated in England. Nothing appears to me to be more unlikely than that this prediction shall ever be verified. The Aberthaw or blue lias lime, one of the strongest limes in the world, was and is well known in England, but as hydraulic mortar, it was immediately superseded by cement, for all private works of importance, as soon as the latter came into the London Market, although cement was then sold under Parker's patent, at a much higher price in comparison with lime, than has been customary since. Now as there was in those days a much greater difficulty in prevailing on the public to adopt any thing new, than there is at present, nothing can be more conclusive than the above fact; for it must be evident that private individuals, who had occasion to build a wharf on their premises, would never have put themselves to a great additional expense, by using this new cement in preference to the limes formerly in use, unless it had possessed a most decided superiority over the latter. In the public works of this country, on the contrary, whether executed by Government or by great commercial companies, cement has been less employed even for brickwork, by the Engineers intrusted with the direction of them: and in masonry it has never been used at all, in consequence of a strong prejudice prevailing against it, as being altogether unfit for connecting large stones, which I feel confident that our experiments at Chatham will tend to remove. In the mean time the wharf or dock walls, of the great public works alluded to, have either been built with the mortar of what was considered the best water lime, that could most conveniently be procured near the spot, or with mortar improved by the addition of puzzolana, in imitation of Smeaton's example, though not of the principle he acted upon, which was to use the best hydraulic mortar that he was able to form by a combination of the most suitable ingredients, known in his time. He found, that this consisted in a mixture of puzzolana, with blue lias lime; because the admirable properties of the English cements had not yet been discovered. If they had, I have no doubt that he would have used them for the construction of the Edystone Lighthouse, in preference to every other sort of mortar; and by so doing, he might have dispensed with some of the very ingenious but expensive expedients adopted by him for bonding all his stones together. To conclude, so far from assenting to M. Vicat's opinion, I am persuaded that the use of cement is not only increasing in England; but that it will in time supersede the most approved hydraulic mortars in other countries also, for works of importance exposed to the violent action of water.

C. W. PASLEY.

Royal Engineer Establishment,
Chatham, the 17th of September, 1838.

TABLE OF CONTENTS.

A*

CONTENTS.

* Provided that the precautions always used by skilful Architects or those afterwards recommended in Article 357 be adopted, in commencing the walls

CONTENTS.

Page

OF THE WATER CEMENTS OF ENGLAND ABSURDLY TERMED ROMAN CEMENTS.

* The downfal of these arches, which took place afterwards, is recorded in Article 319, and their construction more accurately described in Article CXLI of the Appendix.

CONTENTS.

* By experiments tried in 1836, it was found that this mixture did not produce so good an artificial cement, as another containing a smaller proportion of chalk, which was therefore adopted in preference.

* This conjecture was afterwards confirmed by burning and pulverizing the blue clay of the Medway, which made an excellent artificial puzzolana (See Articles 260, 261, and from 302 to 305 inclusive).

CONTENTS.

* The reader will find by after experiments, that the comparative adhesiveness of artificial cement and sand is much greater than these experiments make it (See Table X, Article 241).

Page

* This description, not being quite accurate, was afterwards corrected in Articles 318 and 360.

† But not without the aid of hoop iron, as was afterwards proved by the experiments recorded in Articles 322, 326 and 327.

CONTENTS.

* Some little inaccuracies in this article were corrected afterwards in Article 360.

† The construction of these semi-arches more accurately explained, and the circumstances of their downfal more fully stated, in the Appendix, Article CXLI.

CONTENTS OF THE APPENDIX.

CONTENTS.

* A continuation of this subject will be found in from Article CXXVII to CXXX inclusive, and afterwards in Article CXLII.

OBSERVATIONS,

&c.

PRELIMINARY OBSERVATIONS, AND GENERAL STATEMENT OF THE COURSE OF EXPERIMENTS PURSUED.

(1) In 1826, in consequence of having received an order the year before, from his Grace the Duke of Wellington, then Master-General of the Ordnance, that Practical Architecture was, in future, to form a part of the Course of Instruction for the Junior Officers of the Corps of Royal Engineers, attending the Establishment at Chatham, under my direction; I was induced to investigate the properties of water cements and limes, and adopted a very simple and expeditious mode of testing them; and afterwards I tried a great number of experiments, in hopes of obtaining an artificial cement from a mixture of chalk and clay. The clay experimented upon was an excellent brick-loam from Darland in this neighbourhood, which I was induced to use in preference to any other, in consequence of the celebrated Smeaton, in his researches into the qualities of lime stones, having declared that those adapted for building under water were all composed of a mixture of lime and of a clay suited for brickmaking,* so much so, that after dissolving the lime out of several of those stones, by an acid, he found the residue to be clay of such a quality, that on being burned, it was converted into a sound red brick, which experiment of Smeaton I had myself also verified. But after the efforts of several months, all my experiments with chalk and brick loam having completely failed, I gave up the pursuit, under the impression, however, that the thing was practicable, and that it would or might be done hereafter by a person more skilful in chemistry than myself.

* See Smeaton's Narrative of the Building, &c. of the Edystone Lighthouse, book iii. chap. 4.

(2) In 1828, my Friend and Brother Officer Major Reid having come to reside at Chatham for a few months, chiefly from the interest he took in the Architectural Course, then in progress, requested me to show him not only the mode I had adopted of testing the natural cement and lime stones, which had proved perfectly satisfactory, but also in what manner I had attempted to form an artificial cement. I was at first reluctant to repeat any of those experiments, which I told him could only lead to certain failure, but on his expressing a strong wish to see them notwithstanding, I complied with his request; and as my stock of brick-loam from Darland was at this time expended, and that place was nearly two miles distant, a different clay was used in lieu of it, which to our mutual surprise and satisfaction formed an excellent artificial water cement. This unexpected success was owing to my having given directions to a soldier, whom I employed in such experiments, to mix two parts of pulverized chalk and one part of clay together, without specifying what kind of clay, and he fortunately selected the blue clay of the Medway as being the nearest to the spot, and which on this account had been used by the Corps, to secure the powder-hoses of our experimental military mines, when prepared for explosion.

(3) Mere accident having thus led me to resume under more favourable circumstances an investigation, that I had entirely abandoned as hopeless, and having soon after discovered what I supposed to be the cause of the failure of the brown and the success of the blue clay, I proceeded to try many hundred experiments of a satisfactory nature; and have succeeded in rivalling the most powerful natural water cements by a great number of artificial compounds, which appear to possess the same properties, but are generally of a much lighter colour. I was engaged in this pursuit for more than three years; in which the winter months were chiefly devoted to experiments on a small scale, tried generally in crucibles in a common fire place, whilst the summer months were employed in burning my artificial cement in sufficient quantities for the practical purposes of building. My first efforts to make good cement on a great scale in the summer of 1829 failed, in consequence of some very extraordinary and unforeseen difficulties, that had not come to light in working on a smaller scale; but in the summer of 1830, these difficulties were surmounted, and my artificial cement has since been applied with success to all the practical purposes to which the natural

cements have as yet been used, of which conspicuous and satisfactory specimens may be seen at or near Brompton Barracks, Chatham. Two small brick tanks or reservoirs have been lined with it, both perfectly water tight, one in a garden attached to the Barracks, another at a little distance from them near some cottages occupied by married Soldiers of the Corps. It has been applied as stucco to the brickwork of all the gateways of the Barracks, and to the coating of two brick sentry boxes at the principal entrance. It has been run in moulds into the form of ornamental vases and chimney pots. It has stood the severest frosts; and in tenacity it was found rather superior to Harwich cement of good quality, as was proved by experiments, that will hereafter be noticed.

(4) At the period alluded to, I knew nothing of the researches of M. Vicat in France, who had previously succeeded in obtaining, by artificial means, limes having what the French term "hydraulic properties," which correspond with what are styled the water limes of this country. And it was not till several years afterwards, that I heard of a patent having been taken out in 1818 by Mr. Maurice St. Ledger of Camberwell, for making a lime with a mixture of pulverized chalk or other lime stone and clay: but I was well acquainted, and in occasional communication from time to time, with Mr. Frost, who had actually established a manufactory of artificial stone and cement near Northfleet before I commenced my operations. I shall hereafter briefly describe the proceedings of these gentlemen, and also of General Treussart in France, so far as they have come to my knowledge. In the mean time I shall explain my own, commencing with some preliminary observations, on the distinction between the various kinds of stones, that supply lime, plaster, or cement for building.

OF THE PURE LIME STONES, AND OF THE MORTAR THAT MAY BE MADE WITH THEM.

(5) The pure limes are those which consist entirely of carbonate of lime, that is of lime combined with carbonic acid gas or fixed air, in the proportion of about 5 parts by weight of the former to 4 of the latter; and they are all white, such as pure chalk, and the Carrara marble used by Statuaries, which are the purest lime stones in nature.

When a small piece of chalk or white marble is put into a glass of muriatic or nitric acid, diluted by about an equal quantity of water, the carbonic acid gas is expelled violently with great effervescence, and the lime is entirely dissolved in the liquor, which consequently retains its original colour and clearness. The same effect will be produced by these acids in their pure or concentrated state, but it is best to dilute them.

(6) When chalk or white marble are exposed to a strong red heat for a sufficient time, as in the process of being burned in a kiln, the carbonic acid gas is also driven off, and the stone is converted into *Quick Lime,* in which state it will have lost about four ninths of its weight, and if a fragment of it be now dropped into a glass of the diluted acids before mentioned, no effervescence will take place. The quick lime thus produced is capable of *slaking,* that is if a certain quantity of water be added to it, it will fall to pieces with great heat, assuming the form of a fine white powder, which after being sifted is mixed up with sand and an additional quantity of water, to form mortar for building.

Quick lime will also gradually slake by exposure to the atmosphere, especially in damp weather, and in this case the heat, which we may presume always attends the process of slaking, is not perceptible.

The pure or white chalk, which is the *Upper Chalk* of the Geologists, is always found mixed with flints, which are thrown aside in preparing it for the kiln, but occasionally there are also small portions of other siliceous matter imbedded in the same chalk, which cannot be detected until after it is burned, and are then known by their not slaking. These pieces technically termed *Core,* which will not pass through the sieve, when the lime is slaked for making mortar, are always thrown away. Portions of the pure chalk itself are also rejected as core, if they will not slake from having been underburned, which sometimes happens.

The powder formed by slaking quick lime with water is called *Hydrate of Lime,* as it consists at first of a mixture of solidified water and of quick lime. One cubic foot of chalk lime, burned in pieces not exceeding the size of a man's fist, weighs on an average 35 lbs. and is consequently the produce of about 63 lbs. of dry chalk.* When gradually slaked with the proper quantity of water, which occupies about two hours and a half, it will fill the space of one cubic foot and five sixth parts of a foot, and by the solidification

* One cubic foot of solid chalk perfectly dry weighs 95 lbs. but on breaking it into small pieces for burning, a cubic foot measure will only contain 63 lbs. of the same chalk of fair level measure.

of the water during this process, its weight will have increased from 35 to about 50 lbs, or in the proportion of 7 to 10. But when one cubic foot of pure quicklime is pounded dry without slaking, it will only occupy about two thirds of a cubic foot in the state of powder. Thus pure quicklime powder does not occupy much more than one third of the space of the slaked lime powder produced from the same quantity of chalk, their respective bulks being as 4 to 11.

(7) When pure lime after being burned is pounded and made up into a ball with water, it becomes very hot in a few minutes, and bursts with a kind of explosion into several pieces, which are sometimes thrown to a little distance. If such a ball be put into a cup full of water, it also generally bursts, but much less violently, and the pieces generally swell in the water. Sometimes it swells into a soft pulpy mass, of a much larger size, which eventually falls to pieces, being first diffused through and discolouring the water, which it renders warm, if the cup be small.

The smallest proportion of water capable of forming a ball with quicklime powder is one part of the former to three of the latter by measure. On increasing the comparative dose of water, by using one measure of this fluid to only two measures of quicklime, a much thinner paste will be produced, but this also will form a ball, which will fall to pieces like the former, though not so soon nor so violently.

(8) It was before stated, that well-burned lime fresh from the kiln contains no carbonic acid gas, and when first slaked it seems for the moment to be still free from it or nearly so; but it begins to imbibe this gas immediately afterwards, which may be known by its effervescing in the diluted acids, and in process of time all slaked lime whether mixed with sand or not, appears to recover the whole of the carbonic acid gas previously driven off in the kiln. Hence the lime contained in old mortar is reconverted into its original chemical state of carbonate of lime. Accordingly we found, on putting a piece of old mortar into a glass of diluted muriatic acid, that the lime dissolved with effervescence, leaving the sand at bottom; and also on burning a piece of old mortar in a brisk fire for three or four hours, the lime contained in it was evidently reconverted into quick lime, for on adding water to it, it slaked in the usual manner with great heat, and fell to pieces together with the sand which had been mixed with it, and on remixing them after this took place, they formed excellent fresh mortar, probably similar in quality to the original state in which the same bad been used more than half a century before.

(9) The sand used in making mortar should be *sharp*, that is angular not round, and *clean*, that is free from all earthy matter, or other than siliceous particles. Hence *Road Scrapings* always, as being a mixture of sand and mud, and *Pit Sand* generally, as being scarcely ever without a proportion of clay, should be washed before they are used, which is seldom necessary in river sand, this being cleaned by the force of the current which is the cause of its formation. None but clean sharp sand will ever form good mortar, and the intimate mixture of the sand and lime, which should be done with a moderate quantity of water, is of no less importance.

I have ascertained by repeated experiments, that one cubic foot of well burned chalk lime fresh from the kiln weighing 35 lbs as aforesaid, when well mixed with $3\frac{1}{2}$ cubic feet of good river sand, and about $1\frac{1}{5}$ cubic foot of water, produced about $3\frac{1}{4}$ cubic feet of as good mortar as this kind of lime is capable of forming.*

A smaller proportion of sand such as 2 parts to 1 of lime is however often used, which the workmen generally prefer, although it does not by any means make such good mortar, because it requires less time and labour in mixing, which saves trouble to the labourers, and it also suits the convenience of the masons and bricklayers better, being what is termed *tougher*, that is more easily worked. If on the other hand, the sand be increased to more than the above proportion of $3\frac{1}{2}$, it renders the mortar *too short*, that is not plastic enough for use, and causes it also to be too friable, for excess of sand prevents mortar from setting into a compact adhesive mass. In short there is a certain just proportion between these two ingredients, which produces the best mortar, which I should say *ought not to be less than* 3 *nor more than* $3\frac{1}{2}$ *parts of sand to* 1 *of lime*, that is when common chalk lime or other pure limes are used, for different limes require different proportions.

When the proportion of sand to lime is stated in the above manner, which is done by Architects as a part of their specification or general directions for the execution of

* Some readers may be surprised, that this mortar should occupy rather less space than the sand alone originally did, before the lime and water were added to it. The principal reason is that dry sand, and all dry loose materials generally, settle into a much smaller space when wetted. Hence the same quantity of sand measured dry, then moist, and afterwards wet, will occupy unequal spaces. The clean sharp river sand rather moist used by us, weighed about 87 lbs. per cubic foot. On gradually pouring water upon it in the measure it settled down from 12 to $9\frac{5}{8}$ inches in height, thus occupying only four fifths of the space, which it had before filled.

a building, it is always understood, when nothing is expressed to the contrary, that the parts stated are by fair level measure for the lime, and by stricken measure for the sand, and that the lime is to be measured in lumps, in the same state in which it comes from the kiln, without slaking or even breaking it into smaller pieces. It is of some importance that this shall be fully understood, because we have shown in Article 6, that quick lime from the kiln occupies when pounded a much smaller space, and when slaked into powder a much larger space, than it originally filled.

Without occasionally having recourse to actual measurements such as the above, it is impossible to form any just estimate of the proper proportions of lime and sand for mortar; but no such measurements are usual or necessary in practice, because the expert labourer employed in this operation, on receiving general directions to use as much sand as possible without making the mortar too short, will from habit serve out the proper proportions of lime and sand with all necessary accuracy, without measuring them.

THAT THE PURE LIMES, SUCH AS CHALK-LIME, &c, ARE GOOD FOR DRY SITUATIONS, BUT UNFIT FOR THE PURPOSES OF HYDRAULIC ARCHITECTURE, AND EVEN FOR THE EXTERNAL JOINTS OF BUILDINGS.

(10) The pure carbonates of lime form a good mortar for all dry situations, and for inside work, but not for building in damp or wet situations, in which they never *set*, as the process of induration is technically termed by workmen, but always remain in a soft pulpy state. If exposed to running water, or to the sea, as in the facing of wharf walls, &c, the lime is removed by degrees, partly dissolving and partly separating from all the external joints, leaving the sand unprotected, which is also washed away, and the destruction of the mortar will be followed by the downfal of the front of the wall, if built of bricks, or of rubble work composed of small irregular stones.

The unfitness of pure lime for the purposes of hydraulic architecture has been proved by several striking circumstances, that have come under my personal observation, of which I shall only mention a few. First, a great portion of the boundary wall of Rochester Castle having been completely undermined, nearly throughout its whole thickness, which was considerable, whilst the upper part of the same wall was left standing, I had always ascribed this remarkable breach to violence, considering it as having been the act of persons intending to destroy the wall for the sake of the stone; but on examining it more accurately, after I had begun to study the subject of limes and cements, I observed that the whole of this breached part was washed by the Medway at high water, and that all the mortar of a small portion of the back part of the foot of the wall still left standing was quite soft, but that towards the ordinary high water level it became a little harder, and that above that level it was perfectly sound and hard.* I observed the same process at the outer wall of Cockham Wood Fort, in this neighbourhood, which was also washed by the Medway, and of which the lower part had been undermined in the like manner, and the bricks of which it was composed carried away.

(11) I believe however that the above process is very tedious, and that it may require a great number of years for water to undermine a wall built of pure lime mortar. But I have ascertained that water has the same effect on the oldest mortar, if made of pure lime. For on taking a piece of the upper part of the same wall of Rochester Castle, in which the mortar was perfectly good and sound, and also a fragment of an excellent brick wall, within Chatham Lines, supposed to have been built in the time of the American war, and in which the mortar was also remarkably good, and putting them both into a vessel of water near a pump, and changing the water every day for about a year, the mortar of both was dissolved out of all the external joints, and several of the stones of the former, as well as of the bricks of the latter, became unconnected and fell to pieces.

(12) It is to be remarked, that this effect only takes place, when the water is constantly changed, because the proportion of lime, that can be dissolved in any given quantity of water is very small, and accordingly after having begun to act upon a fragment of masonry by the method before described of changing the water daily, and having thus dissolved part of the mortar in the external joints, on leaving the water afterwards unchanged for about six months, I found that no further effect had taken place. For as the water was saturated with as much lime as it was

* This breach can no longer be seen, having been repaired some years ago, when the ground outside of it, which was covered with several feet of water every tide and by mud at low water, was also raised higher than the level of spring tides, and protected by a wharf wall, so as to form an agreeable walk for the inhabitants of Rochester.

capable of dissolving, it lost the power of acting upon the remainder of the mortar, which therefore continued uninjured and perfectly sound. But the water to which masonry is exposed in nature is always changing by rain or filtration, even when stagnant, so that pure lime mortar should never be used in building under water, or in marshy situations.

(13) Pure lime is so little capable of resisting the action of water, that it is unfit even for the external joints of walls exposed to the common vicissitudes of the atmosphere. For by degrees the beating rains, to which the outside of such walls is subject, will gradually destroy the mortar of all those joints to a certain depth, as may be observed by inspecting old walls built with chalk lime mortar, which have not been meddled with for some years. Hence after a certain time, it becomes absolutely necessary to make good all those joints with fresh mortar, and this process technically termed *pointing*, if done with the same kind of lime, must frequently be renewed, and is so troublesome and expensive; that it should be avoided as much as possible, by pointing all the external joints of walls built with chalk lime mortar with cement, or with some superior kind of lime, as soon as it shall become necessary; which may not be the case until some years after the building is finished. I apprehend, that it should not be done from the first, which might appear the most natural and obvious arrangement, because mortar made of chalk lime being slower in setting and consequently longer compressible than any other, allows the upper part of a wall to sink by degrees to rather a lower level, which effect technically termed *settling*, although imperceptible to a common observer, is well known to persons of any experience in building, and the amount of which depends chiefly on the number of horizontal joints. Walls built with the water limes settle less, and those built with cement are entirely free from this action, because the cement used in the lower courses sets too soon, for the weight of brickwork or masonry added above to make any impression on the joints. Now though the difference of settlement, even between those extremes may be very small, it does not appear prudent to use more than one species of mortar in the same horizontal joints of a building, especially as it would give trouble to the workmen and occasion loss of time.

(14) Pure lime mortar has sometimes been used for the backing of wharf walls, the front or facing of which has

been protected by water cement, usually to the depth of about 18 inches or two bricks thick, from the outside of the wall. Even this system, though it does not like the former, described in Article 10, involve the entire ruin of the wall, is highly to be reprobated. The cement protects the pure lime mortar from the direct action of water in mass, but not against wet or damp, because the moisture penetrates through the pores of the brickwork and of the cement, and although not in sufficient quantity to dissolve the pure lime mortar, it effectually prevents it from setting, so that it always remains in a state of soft pulp, and is of no more use towards the consolidation of the wall, than so much moist clay.

(15) I have seen this circumstance exemplified, when part of the old wharf wall of Dover Harbour was pulled down some years ago, and also when part of the wharf wall of the old Gun Wharf at Chatham, faced with stone but backed with brickwork, has been laid open occasionally for temporary purposes, the backing mortar in both of which was in the state of pulp. The same effect was also observable in the wharf wall as well as in a basin wall of His Majesty's Dock-yard at the last-named place, which were both built entirely with bricks. In those walls the front of the brickwork consolidated by cement was perfectly strong and in one compact mass, whilst the backing mortar of the same walls was quite soft and moist, and in respect to the old wharf wall of the Dock-yard in particular, when part of it was ordered to be pulled down in the winter of 1834, the workmen were obliged to blast the brickwork of the front of it towards the river with gunpowder, this part having been laid in Sheppy cement, to the thickness of three feet; but after having got rid of this strong facing they removed the backing by spades, which they introduced horizontally into the soft pulpy joints of the mortar, and thus turned over each course of bricks with greater ease, than the operation of sod cutting in a meadow could have been performed by the same tool.

OF GYPSUM, ALABASTER, OR PLASTER-STONE.

(16) This stone which supplies the *Plaster* for cornices, ornamental ceilings, &c, and also for the casts or models of Statuaries, is very hard, and when pure of a handsome white colour not unlike white marble, and in this its natural state it is usually called *Alabaster*.

It is known not to be marble, by being insoluble in diluted acids, which have no action upon it, and also from the circumstance, that when burned in the manner proper for burning lime it will neither slake, nor will it even form plaster fit for use.

(17) The *Gypsum* fit for the purposes of building and of the Arts, is the *Granular or Massive Sulphate of Lime,* which when analyzed is said to contain about one third of its whole weight of lime, rather less than one half of its whole weight of sulphuric acid, and about one fifth of its whole weight of water.*

To test this stone chemically, pound a small quantity of it in a mortar, and pour the powder into a Florence flask half full of distilled water, and boil it for about a quarter of an hour over a spirit lamp. Filter the turbid mixture, through a paper filter, by means of a funnel, letting it drop into a glass. Pour the clear liquor into a couple of glasses, into one of which pour afterwards a few drops of the muriate or nitrate of Baryta, which is a test of sulphuric acid, and into the other a few drops of the oxalate of ammonia, which is the most delicate test of lime. Each of these will form a precipitate, the former indicating the presence of the sulphuric acid, and the latter of the lime.

(18) The quality of gypsum may be ascertained practically, in a very satisfactory manner, by the following simple process. Pulverize a quantity of the stone and place it in a shallow iron or tin dish, over a brisk fire, and stir it continually. In a short time, the water which was solidified in the gypsum will be driven off in steam, with so much violence, that the powder will be agitated in the vessel with all the appearance of boiling. When the disengagement of steam is no longer perceptible, hold a dry warm wine glass inverted over the powder, and if the inside of it should show no appearance of moisture, you may be assured that your gypsum has been burned enough, and that the powder now called plaster is in a state fit for the use of the Plasterer or Statuary. To judge of its quality mix a small quantity of it with water in the form of a ball, and it will set with moderate heat into a very hard fine white substance, and will even continue setting under water, but being partially soluble in that liquid in process of time, it is not applicable to the purposes of hydraulic architecture.

Another method of preparing gypsum for use is to break it into small pieces, and bake it in an oven, and afterwards grind it to powder. Both methods are used by

* See Brande's Manual of Chemistry, in which the proportions are stated as 26·5 of lime, 37·5 of sulphuric acid, and 17 of water. Total 81.

the British Manufacturers of this article, who generally use English stone, and yet the plaster made by them is commonly termed *Plaster of Paris,* near which city gypsum abounds, and from whence it probably was first imported into this country.

(19) Plaster exposed to air gradually imbibes moisture, which always injures and in process of time entirely spoils it, for in fact it reconverts it into the same chemical state, as in the original stone. Hence the powder prepared for sale is always preserved in air tight casks or in sacks. If spoiled, it may be recovered by burning it again.

OF THE WATER LIMES OR HYDRAULIC LIMES, HOLDING AN INTERMEDIATE PLACE BETWEEN THE PURE CARBONATES OF LIME AND THE WATER CEMENTS, AND OF THE MORTARS, CONCRETE AND ARTIFICIAL STONE, THAT MAY BE FORMED WITH THEM.

(20) The *Water Limes* or *Hydraulic Limes* are composed of carbonate of lime, generally mixed with silica, and alumina, and the oxide of iron, and occasionally with traces or very small proportions of other substances. Take away the carbonate of lime from this compound, and the remaining parts are the same as the component parts of the brown or blue clays of nature. Hence these lime stones are also termed *Argillaceous Lime Stones.* They all have more or less the property of setting in watery or damp situations, and therefore they are exceedingly useful for the purposes of hydraulic architecture, and nothing can be better for the foundations and backing of wharf walls; but even these limes when exposed to running water are partially soluble, and therefore it is desirable to protect the external joints of wharf walls, built with mortar made of these limes, by some superior kind of water cement. For this purpose *Puzzolana* from Italy, and *Trass* from the Rhine usually termed *Dutch Terrace* were formerly used. Recently, these two substances, both of which are considered of volcanic origin, have been superseded by the water cements of England, misnamed *Roman Cements,* for the purpose of mystery or for enhancing their value, by the ingenious patentee, who first discovered their strong hydraulic properties and introduced them into general use.

(21) The blue Lias lime stones are considered the strongest water limes of this country, and are found on opposite sides of the Bristol Channel near Watchet in Somersetshire

and Aberthaw in Glamorganshire, and also at Lyme Regis in Dorsetshire. The first of these, mixed with Puzzolana was used by Smeaton in building the Edystone Lighthouse. The Dorking or Merstham lime, and the Halling lime so termed from a village on the left bank of the Medway above Rochester, but which is also found near Burham on the opposite side of the same river, though not possessing such strong hydraulic properties, are also much esteemed; and these two limes, the former of which is considered rather the best, are more used in the metropolis than the blue Lias, probably from the greater proximity of the quarries where they are found, and from very little land carriage being required for either.

All the water lime stones are of a bluish grey or brown colour, which is communicated to them by the oxide of iron. They are usually termed *Stone Lime* by the Builders of the Metropolis, to distinguish them from common chalk lime, but so far improperly, that the Dorking lime stone is not much harder than chalk, and the Halling lime stone is actually a chalk and not harder than the pure chalk of the same neighbourhood, from which it is only distinguished in appearance by being a little darker.

In fact all the coloured chalks found in various parts of England, commonly termed *Grey Chalks,* which are the *Lower Chalks* of the Geologists, and generally free from flints, are possessed of hydraulic properties more or less powerful.

(22) When a fragment of any of the water lime stones is put into a phial of the diluted muriatic or nitric acids, the whole of the carbonate of lime contained in the stone is dissolved or driven off with effervescence, so that the stone falls to pieces, leaving the silica and alumina which are altogether insoluble in acids, together with the oxide of iron which is also insoluble under those circumstances, as a fine muddy sediment at the bottom of the phial.

(23) Most of the water-lime-stones when burned into quick lime by a heat sufficient to expel the carbonic acid gas, are of a yellowish or light brown colour, so that any calcareous stone, which after calcination assumes a kind of buff colour, may be presumed to have hydraulic properties.

All these limes slake with heat by the addition of water, and most of them also in process of time by exposure to air, but not so readily nor so soon as chalk lime or other pure limes.

(24) When the water limes after being burned into quick lime, are pounded and made up into balls with water, if one

of these balls be put into a basin of water, it will expand more or less and eventually fall to pieces; but those which contain the greatest proportion of carbonate of lime, and which therefore approach most nearly to the pure limes, will swell to greater size and fall to pieces in a shorter time, than those which contain less carbonate of lime, and which consequently approach nearer to the quality of a water cement. A ball of about an inch in diameter of the Halling lime for example will swell under water to twice that diameter or to eight times its original magnitude, in about an hour and a half, after which it will fall to pieces, and if the basin used be a very small one, this process will make the water quite warm. But the blue lias lime of Lyme Regis, which approaches very nearly to a water cement, when pounded and made into a moist ball sets very quickly, and if put into a basin of water soon afterwards, numerous cracks will probably be formed all over the surface, and eventually the ball may fall to pieces, but this effect may not take place, for several days or even weeks after such immersion, and in the mean time it swells very little, if at all. Smeaton remarks that the blue lias lime of Watchet slaked with so much difficulty, that when it was rather stale, the masons of that district used boiling water, which has a more powerful slaking effect than cold. If this failed, they considered their lime spoiled. In slaking any refractory lime like this, it is proper previously to break any very large lumps into smaller pieces, and to use a watering pot, and then to follow the common practice of covering it up with sand, which by confining the heat expedites the process desired; but for experimental purposes canvas is preferable to sand.

(25) We found by repeated experiments at Chatham, that one cubic foot of Halling lime weighed nearly the same when fresh from the kiln, and by the gradual addition of water that it dilated to the same increased bulk, in the state of quick lime powder, but when worked up into mortar not too short for use, that it would not bear quite so large a proportion of sand as the common chalk lime had done. This experiment leading to a result in opposition to a common opinion amongst the Builders of the Metropolis, which is that the Dorking and Halling limes, as being stronger limes, will, when made into mortar, bear more sand than common chalk lime; I was induced to examine the principle upon which they found this opinion, which on consideration appears to me to be erroneous; because these two limes and all the other hydraulic limes are undoubtedly in an in-

termediate state between pure lime which is the weakest, and the water cements which are the strongest, of all calcareous cements: and every one will acknowledge that the proportion of sand, which will make good mortar with chalk lime, would entirely ruin cement, which is scarcely capable of bearing one third of that quantity. Hence it follows that the hydraulic limes ought not to admit of so much sand as chalk, but that they will bear more than cement, without being injured.

Accordingly I conceive that *three cubic feet of sand to one of Dorking or of Halling lime, will be a good proportion for making mortar with those limes,* which approach very nearly to pure lime. The water required will be nearly one cubic foot, and the quantity of mortar produced will be about two cubic feet and nine tenths, being rather less than the original space occupied by the sand alone.

(26) But *the blue Lias Lime will not make good mortar if mixed with more than two cubic feet of sand to one of lime.* This opinion first formed by me from analogy, and in consideration of the blue lias approaching very nearly to a water cement, proved on due inquiry to be borne out by, and exactly conformable to, the practice of the masons of Lime Regis. But Captain Savage of the Royal Engineers, who was employed professionally some years ago in improving the *Cobb* or Pier of that little Seaport, which was done by tide work, and in which no other kind of lime was used, assured me that he found that a smaller proportion of sand than 2 to 1 made still better mortar. We have since ascertained by repeated experiments at Chatham, that 1 cubic foot of blue Lias lime from the kiln weighing 47 lbs., mixed with 2 cubic feet of sand, and about 3 quarters of a cubic foot of water, made mortar fit for use, but which could not have borne more sand without becoming too short. The average quantity produced was two cubic feet and one fifth, which contrary to the result obtained with the purer limes, occupied more space than the sand alone originally did.* We found also that blue Lias lime from the kiln, like all the other limes that we experimented upon, filled only

* Having ascertained by the usual test of the diluted muriatic acid (6), that the blue lias lime used in our first experiment was sufficiently burned, we tried a piece of it with cold water, which produced very little effect, and then another piece with boiling water, which slaked it almost immediately. Being aware therefore of the difficulty with which it slakes in cold water, we caused the cubic foot of blue lias lime used in this experiment to be previously pounded, but not very finely, in order to facilitate this process, notwithstanding which it still slaked very slowly, rendering this experiment much more tedious than any of our former ones. I have since ascertained from Captain Savage, that the difficulty must have arisen from our not having allowed sufficient time; for he assures me that blue lias lime made up into a heap, and treated as described at the end of Article 24, will always slake properly with cold water, if allowed to remain covered up for 18 hours.

about two thirds of its original measure, when reduced by pounding to the state of quick lime powder; but one cubic foot of blue Lias lime when slaked, only dilated into one cubic foot and a third of slaked lime powder, not including about one eighteenth part of a cubic foot of core, which we threw away. Hence it expands less by slaking than either chalk lime (6), or Halling lime (25).

(27) One of the most important uses of the water limes is in the formation of *Concrete* for foundations, &c, which if judiciously applied supersedes the very expensive troublesome and precarious process of piling and planking, even in the softest soils.* This is a recent improvement, first adopted by Sir Robert Smirke with success,† in the founda-

A piled foundation is peculiarly precarious, not only from the neglect or fraud of the persons employed, to which from their work being buried and from the value of wood, there is more than usual temptation; but even when perfectly well executed, from sometimes losing its perpendicularity in very soft soil, and from its being liable to rot in process of time, in all situations, excepting those which are constantly under water.

† In excavating for one of the piers of Waterloo bridge, the workmen had a good deal of difficulty, owing to the very compact state of the gravel forming the bed of the river, which every where else they had found perfectly loose. This effect had been produced by the accidental sinking of a barge load of lime over that spot some time before, which had cemented the loose gravel into a solid mass resembling the calcareous conglomerates of nature, which are gradually formed by a similar process. Mr. Rennie having mentioned this circumstance to Sir Robert (then Mr.) Smirke, the latter with great judgment availed himself of the hint, and subsequently used it in all his foundations, none of which have ever been known to fail. Part of the Penitentiary at Millbank, begun by another Architect in a different manner, before Sir Robert Smirke was employed there, was evidently giving way. The superior efficiency of concrete was also proved in a remarkable manner at the new Custom House, where the floor of the large apartment called the long room actually fell in, and the whole building was in danger, owing to the insufficient manner in which the piling had been originally executed, in a very difficult situation. At this period Sir Robert Smirke was consulted, who found it necessary to pull down a small part of the building but saved the rest of it by undersetting all the walls with concrete, to the average width of 12 feet, and to the depth of from 12 to 15 feet, that is until they found a natural bed of gravel, including one course of Yorkshire landing stones, and 12 courses of bricks laid in cement, having three or four offsets or footings between the Yorkshire landings resting on the concrete, and the base of the original walls. No other expedient could possibly have saved this fine edifice from entire demolition. It must be allowed that not only the ancient Romans, and after them the Moors, but even the Norman Barons of England in their feudal castles, used concrete, of which Kendal castle is one of the most striking examples; and more recently Belidor in his Architecture Hydraulique treats of Beton mortar, which is much the same, so that it is not absolutely new. In fact according to the old proverb, there is scarcely any thing new under the sun, but the merit of introducing this immense improvement systematically and generally into the modern practice of Architecture, is undoubtedly due to Sir Robert Smirke. Mr. Ranger has also great merit, the walls of a great storehouse in Chatham Dock yard having afterwards been underset by him, when the piled foundations were giving way, in a very simple and yet efficient manner, with concrete alone; and the formation of concrete into artificial stone being his invention, and a very ingenious and useful one.

tion of the Penitentiary at Millbank, where the soil consisting chiefly of peat moss was soft to a very great depth. His example has since been followed by other Architects as well as by Civil Engineers and Builders, who have used concrete not only like him for foundations, but for the backing of wharf walls and docks, and more recently for entire walls and buildings.

(28) Concrete is formed by mixing lime, coarse gravel, and sand together, with a moderate quantity of water, which is usually done on a large square board, having a margin raised a little above it on three sides only. The lime used for this purpose has usually been reduced to fine powder by pounding or grinding it, whilst fresh from the kiln; and it is generally considered of so much importance not to slake it until ready for use, that it has been customary to mix it with the gravel and sand in a dry state for a little while, before the water was added; after which the whole of these ingredients have been intimately mixed, with as much expedition as possible, by employing two labourers to work together at each of the mixing boards, which being always placed as near to the spot previously prepared for the foundation as possible, the concrete is either thrown down at once, or wheeled a little way and dropped from a temporary scaffold with movable planks, which is usually necessary, into the excavation, where it is spread and levelled, and trodden down or sometimes rammed by other labourers below. It is considered useful to throw it from a height, as each shovelful or barrowful compresses the mass already below by falling upon it. Concrete is sometimes only laid in trenches cut for the foundations of the several walls of the building, and of such a width as to enable it to occupy a much larger base than the walls themselves, without which it would be of little or no use: but in the foundations of more important edifices, especially if the soil be very soft, it is more usually spread over and somewhat beyond the whole area on every side, in successive layers of about a foot in thickness. The most experienced Builders are of opinion, that its depth should never be less than four feet and a half, and that it need seldom exceed 6 or 7 feet, under ordinary circumstances.* Concrete made in this manner, according to the system first introduced by Sir Robert Smirke, throws out a moderate heat on the slaking of the lime, and soon begins to set,

* But if a solid stratum, whether horizontal or inclined, can be found by excavating somewhat deeper, the thickness of the concrete should be increased, so as every where to rest upon it.

forming in time a kind of artificial rock, supporting the whole weight of the walls of a great building, upon so large a base, as to prevent those unequal settlements of the various parts, in proportion to their unequal weights, which first cause unseemly cracks in the walls, and in time the ruin of the building. As soon as a foundation of this kind is finished, the heaviest building may immediately be commenced upon it, and however quickly the walls may be carried up, the concrete will have set sufficiently to bear their whole weight, for no failure has ever occurred in thus using it, although I am inclined to believe that many years may elapse before the great masses of concrete, buried in such foundations as those of the Penitentiary, which extend beyond the whole area of the building, and are said to be 11 or 12 feet deep, can attain their maximum of induration.

(29) When concrete, which was only used for foundations by Sir Robert Smirke, in the manner that has been described, was afterwards applied by other Architects or Builders to the formation of walls, which I believe was first done at Brighton, and of which the Sea wall East Cliff affords the most remarkable example, it became necessary to use a mould or hollow wooden profile suited to the proposed form of the wall, into which the mixture was thrown and allowed to set. The simplest mode of accomplishing this object, is that which has been long used in Spain, and in part of the South of France, for constructing houses, frequently of more than one story high, with rammed earth, which is called *Tapia* in the former, and *Pisé* in the latter country. Two sets of boards are ledged together, each set forming an oblong of any convenient length, but of moderate height, which are connected near the top and bottom by iron bolts to keep them at their proper interval apart. After the foundation is finished by cutting a trench, and throwing concrete into it in the usual manner, the mould thus formed is placed upon it, on the same level as the natural ground, and filled with concrete, and when that has set sufficiently, it is moved to another situation, which may be done by drawing out the iron bolts, after unscrewing the nuts by which they are confined; and thus the whole of the first course of concrete, of a height corresponding to that of the mould, is finished over the whole base of the wall or walls. The moulds, for more than one would be necessary, are then put together on a higher level, and another course

of concrete is formed by the same process, and so on, until the whole height of walling necessary is completed in regular courses, upon each of which the moulds rest successively as the work rises.*

In the Sea wall at Brighton, boarding was used in the front only, because the back of the wall either rested on the natural chalk cliff, or was made good with chalk rubbish, which superseded the use of woodwork on that side. The concrete used in this wall was formed of a mixture of grey chalk lime obtained from a distance, with the coarse shingle and sand found on the beach at the foot of the cliff; for the base of this wall is every where above the level of high water, except in a small portion of its length, where the lower part of it has very judiciously been protected against the action of the Sea, by permanent woodwork fixed in front of it. Thus the Sea wall of Brighton, except in this small portion only, has no pretension to be considered as a wharf wall.†

(30) Still more recently Mr. Ranger has extended the same process to the formation of artificial stone, for which he has taken out a patent, and which under many circumstances may be considered a very great improvement, although it only differs from the concrete before used, in being mixed with boiling water instead of cold, and in being formed in moulds of moderate size, instead of great masses.‡ Numerous examples of this mode of building may now be seen, of which two houses (Nos. 16 and 17) on the north side of Pall Mall, which are faced with Mr. Ranger's Artificial Stone, and a School in the parish of Lee, very near Blackheath, are the most convenient for inspection of any in or near the Metropolis.§

The moulds used by Mr. Ranger in forming his artificial stone, have a flat board for the bottom, which is some-

* The extraction of the iron bolts after the mould is removed, leaves holes extending entirely through the concrete. In fixing the mould on a higher level, the lower bolts are passed through the same holes, from which the upper bolts were previously taken out.

† The Sea wall at Brighton, which I inspected whilst in progress, was intended to be every where 2 feet 6 inches thick at top, with a slope of one third, and the Foreman assured me, that in one part, it would be no less than 60 feet high, so that its thickness at the base there would be 22 feet 6 inches. Its height varied, being much less in other parts. The boarding used on one side only, was 20 feet long and 4 feet 9 inches high.

‡ The system of using boiling water, though not generally adopted except by Mr. Ranger, has long been known to promote the rapid slaking of lime. (See Article 24).

§ The walls of this School are entirely built of concrete, chiefly made of gravel dug on the spot, which being incorporated with stiff brown clay was well washed previously to using it, and mixed with Thames sand, without which it would have been much too coarse.

what larger than the base of the proposed stone. The short upright sides of the mould are fitted in close between the two other upright sides, whose ends project a little way beyond the former, and are then secured by iron cramps, which being properly wedged up keep all tight. A couple of oak trenails pass through both sides of the mould towards the ends. A number of these moulds being laid out in a line, the mixing boards are placed near them, in which the sifted quick lime powder and the proper proportion of gravel and sand are mixed with boiling water, with all expedition; and the mixture is thrown immediately into one of the moulds, in which it is continually rammed as it rises, until the mould is quite full, when a smooth surface is made upon it afterwards by a plasterer with his float, with fine mortar of the same kind of lime; and this may be done also on any other of the sides of the artificial stone, that may require it, that is on such as are to appear in the front of a wall. Mr. Ranger has not only formed his artificial stones of moderate size like common building stones, but has made very large massy blocks for ornamental cornices, in which the mouldings have been finished by the plasterer, after the mass of each block has been cast in the mould. The use of boiling water causes the lime to slake much more rapidly and with greater heat than cold water would do, and is so far an advantage, that it enables the men to work with fewer moulds in making artificial stones; but as my own experiments have not convinced me, that rapidity of slaking or setting either makes cement or mortar of the same quality better eventually, than if allowed to slake or set more slowly, I do not see that there can be sufficient advantage to compensate the expense of boiling the water, which Mr. Ranger also does, in preparing large masses of concrete for foundations, &c. When the artificial stones made in this manner have set sufficiently, for which about half an hour is considered a reasonable time in working with Halling lime, the moulds are taken to pieces, first drawing out the trenails, which leaves a couple of cylindrical holes in each of the artificial stones, for the purpose of afterwards hoisting them up to their proper position on a wall, without injuring them. The wedges are then knocked out and the cramps disengaged, which allows the upright sides of each mould to be removed, after which the artificial stones are disengaged from the bottom boards of their respective moulds and laid on skids to dry, until they set more thoroughly, and after the experience of some years, it is now considered desirable, that they should remain two or three months in this state, before they are used.

(31) Sir Robert Smirke informed me some years ago, he himself being then the only person who had made use of concrete to any extent, that he would not scruple to erect the heaviest Dock for shipbuilding upon a concrete foundation, instead of the very expensive but certainly excellent foundations of piles and other woodwork, and of bricks laid in cement, which had been at that time used for the fine new granite Docks at Sheerness and Chatham, built by the celebrated Mr. Rennie. At that time I believe no one contemplated the still bolder measure, since adopted by Mr. Taylor the present Architect of the Admiralty, who gave in a plan a few years ago for building Docks of the largest size, and also wharf walls both at Woolwich and Chatham, entirely with Mr. Ranger's concrete and artificial stone, which having been approved by the Lords Commissioners of the Admiralty, was ordered to be carried into effect. From the numerous experiments which I had previously tried on limes and cements, it was my opinion at the time, that these Docks and Wharfs after being finished would gradually give way at the surface, unless protected against the action of water in mass, as well as against those collisions to which such walls are liable, by a facing of stone or of brickwork laid in cement. This arrangement has subsequently been adopted by Mr. Taylor himself for these Docks, which have been faced with granite, but not for the Woolwich wharf wall, where the facing of artificial stone still remains exposed to the Thames every tide, and where time will show, whether my apprehensions of its insufficiency are well founded or not.*

(32) Upon the whole, it may be allowed that the use of concrete and artificial stone, if kept within proper bounds, that is so as not to attempt to supersede stone and brick-

* Let it be fully understood, that my opinion was only expressed to private friends, but neither made known to the public nor to any official person. In fact I was particularly glad that the experiment of concrete docks and wharfs was tried by Mr. Taylor, which I considered at the time not only a very bold, but so far a safe one, inasmuch as I had not the smallest doubt of the efficiency of this substance for the great mass of the work, and if it should fail in the external surface, it appeared to me that this part of the concrete might be cut away, and a facing of granite added and bonded into it, without difficulty and with very little more expense than if this arrangement had been adopted from the first. I believe that the chief difficulties attending the execution of the concrete Docks at Woolwich and Chatham were in some measure independent of the material used, owing to land springs, &c.

work entirely, is a very great improvement. When I first saw the Sea wall of Brighton in progress some years ago, I could not help being struck by the facility, which the like application of concrete would afford for fortifying sandy islands, where neither stone nor earth fit for brickmaking are to be found, and which are often important military posts. Adopting this arrangement, it would not be necessary to import any thing but quick lime in barrels or sacks for building the scarp and counterscarp revetments, and even for casemates, as was proved last year at Woolwich, by an experiment tried by order of the Master General of the Board of Ordnance, at the suggestion of Lieut.-Colonel Fanshawe of the Royal Engineers, who proposed that a vault of 18 feet span, 5 feet rise and 4 feet thick at the haunches, being of the dimensions suitable for a military casemate,* should be built with Mr. Ranger's concrete, on the marsh adjoining to the Arsenal at that place, and exposed to the fire of heavy mortars and guns, which was done in little more than two months after it was finished; and contrary to previous expectation, it was less injured by the vertical fire of 13 inch shells, each loaded so as to weigh 200 lbs. than by the direct fire of 24 pound shot. Notwithstanding this favorable result, I would recommend brickwork laid in cement always to be used for casemates, where it can be had, in preference to concrete, for the former not only sets immediately, but if afterwards covered with a coating of pure cement, brick casemates afford a dry wholesome shelter for troops, whereas the concrete would always be damp.

(33) One very striking advantage of Mr. Ranger's artificial stones is the greater conveniency of applying concrete to the walls of a common building in this form, than in mass; but it appears to me that they also possess another advantage, less obvious but not less important. If laid out for two or three months, before they are used, the great surface which these stones present to the air, combined with their small bulk, enables them to set much more thoroughly, than an equal quantity of concrete in mass could possibly do in the same time, for I believe that

* The vault was 18 feet long and 6 feet thick immediately over the crown of the arch, where it was formed with a ridge, in the usual manner. The piers which supported it were 8 feet wide and about 4½ feet high to the spring of the arch, with foundations 10 feet wide, having an offset of 1 foot on each side. It was commenced in the beginning of February and finished on the 17th of March, and the centers struck four days afterwards. The piers were built in successive courses of about 1 foot thick, and the arch was begun from both sides, in the form of voussoirs extending its whole length, the eleventh of which completed it, by forming the key stone.

the induration not only of mortars and concrete, but even of cement, is gradual, taking place first at the surface, and from thence proceeding regularly towards the center.*

(34) Neither gravel without sand, nor sand without gravel can form good concrete. The large pebbles composing the former, if mixed with quicklime powder and water would only be cemented together by lime paste, or as it is technically termed *Lime Putty*, filling up the large interstices between them, which is known to be the weakest form of lime. On the other hand, fine sand alone would produce nothing better than a mass of common mortar, which has very little strength in itself, but is excellent for cementing larger materials. The proportions of the gravel and sand used are of little importance, provided that the former be rather large, and the sand sufficient to fill up the interstices in it, for which purpose a mixture of coarse and of fine sand is better than one sort only. But no such mixtures are necessary, in using the sandy gravel of many parts of the Thames, where it is found in the state most suitable for making good concrete, and is employed both for this purpose, and as ballast for shipping.

(35) It being known that clean gravel and sand when put dry into any measure, will almost immediately settle to a lower level on the addition of a certain quantity of water, and it being a matter easily proved that no ramming can possibly compress them afterwards, it appears to me, that the proportion of lime used in concrete should be just sufficient to combine with the gravel and sand in this compact state; that is after the violent action of slaking shall

* In the experimental casemate at Woolwich, mentioned in the preceding article, the concrete at the surface was every where quite hard, but several smooth cylindrical holes, about 6 inches in diameter, formed in one of the piers or side walls to the depth of about 3½ feet, by the penetration of the 24 pound shot fired into it, afforded me an opportunity of observing that the lime, not only at that depth but even much nearer to the surface, was quite soft and moist. The 24 pound shot shattered the hard surface first, producing a breach resembling the crater of a small mine 2 or 3 feet in diameter and about a foot deep, beyond which in continuation of their course, they bored the regular cylindrical holes before mentioned. The 13 inch shells did not penetrate much more than one foot into the top of the arch, producing holes nearly an impression of their own form, without shattering the concrete much, but by their concussion they also produced some fine cracks in the arch. This however had been previously injured by the unequal settlement of the piers, one of which had sunk nearly a foot, and the other about 4 inches less, in the very soft marshy soil on which they had been founded, the nature of which may be understood, from the circumstance that the 13 inch shells, which happened to miss the casemate, sank 15 or 16 feet below the surface.

have subsided,* which causes a temporary expansion, that is counteracted whilst in operation, by the usual process of treading down or ramming concrete.

(36) In their specifications for concrete foundations, it is usual for Architects to direct, that it shall be made by mixing so many parts, usually not less than 6 nor more than 8 of gravel and sand, with 1 part of ground or pounded and sifted quicklime;† but it was before stated in Articles 6, 24 and 25, that quicklime powder occupies less space, than the same quantity of lime does when received from the kiln. Hence the above two proportions, if stated as usually done in the specifications of the same Architects for making mortar, would only be equivalent, the first to 4 measures of gravel and sand, mixed with 1 measure of lime from the kiln, and the second to 5⅓ measures of gravel and sand, mixed with 1 measure of lime from the kiln; but even when thus explained and corrected, the same quantity of lime is mixed with so much more siliceous matter in making concrete, than in common mortar, that when both are good of their kind, it might be supposed on a hasty view of the subject, that the same lime had a more powerful cementing action in concrete than in mortar. This however is not the case. When the ingredients for making concrete are mixed together, the lime combines with the fine sand only, forming a mortar which cements the large gravel and coarser sand together, in the same manner in which the like mixture of fine lime and sand, previously made up into

* Formerly great importance was attached not only to the proper incorporation or mixture of all the ingredients, but also to the continual beating of mortar, which was carried to such an extent, that Smeaton reckons it a fair day's work of a labourer, to mix and beat up 2 bushels of quicklime powder with 1 bushel of Dutch Tarras, vulgarly termed Terrace, which probably would not make more than 1½ cubit foot of mortar. This practice was not only recommended by him, but by Dr. Higgins, who also investigated the subject of calcareous cements with great industry and sagacity. In the present day the beating of mortar has become obsolete, at least I have never seen it used in any of the great works, which I have inspected whilst in progress in or near the metropolis, for many years past; and with great respect for the opinion of those two able men, I doubt much whether the advantages were sufficient to counterbalance the extra labour involved in this troublesome and very tedious process. The most esteemed mortar now is that made by the pugmill, which mixes it extremely well and compresses it also, as it forces out the ingredients which enter at the top of the mill, through a small hole at the bottom of it; and this is done so expeditiously, that the quantity of mortar, which might be a day's work of a labourer, according to the beating system of Smeaton and Higgins, would be mixed in two or three minutes by the pugmill.

† In grinding quick lime, the under burned parts of the core are retained, which are rejected as prejudicial in making mortar. But when previously reduced to fine powder as in making concrete, a little core can do no harm.

mortar, cements the stones of rubble masonry built of small materials. Hence in reality, the lime used in concrete does not combine efficiently with more sand than it does in mortar; and its cementing properties and power in both are precisely alike.

(37) For example, in the concrete used in the Sea wall at Brighton before mentioned, six parts of large shingle and sifted sand were mixed with one part of grey chalk lime from the kiln. But in making it, the ingredients were not all mixed together at once, according to the usual custom. Three parts of sifted sand and one part of lime were actually made into mortar by labourers first, after which this mortar and the coarse beach shingle were thrown, in equal quantities, namely by one barrowful of each at a time, into a pugmill worked by a horse, which in mixing them into concrete produced a more intimate union of all the ingredients, than could have been effected by manual labour, it being well known to practical men, that no common workman will ever mix mortar so well or so quickly, as can be done by this very useful and simple machine. The concrete of the Sea wall of Brighton was therefore formed with lime previously slaked, on the same principle as in making common mortar, instead of working with quicklime, according to the system used by Sir Robert Smirke, and refined upon by Mr. Ranger.*

(38) But whatever proportion of the ingredients may be stated by an Architect in his specification, the same practice prevails in making concrete, as in mortar (9). No precise measurement ever takes place, the matter being left to the sagacity of the labourers employed, who produce mixtures of uniform quality with extraordinary accuracy. Those in the employment of Mr. Ranger, in working with Halling lime, make up one barrowful of sandy gravel into concrete at a time, which they mix with two small shovelfuls of sifted quicklime powder, the proportion being apparently about 8 measures of the former to 1 of the latter, in combining which they use a small pailful of boiling water.

(39) It may be observed, that as different limes require different proportions of sand in making mortar, from the circumstance, that those which have the most powerful hydraulic properties are not capable of combining with so

* It appears worth while to record this difference, because in working with limes that slake easily, such as the Dorking and Halling limes, if the concrete formed both ways should be equally good in the end, it may be convenient to dispense with grinding them.

much sand as common lime,* it follows that different limes must also be used in different proportions in making concrete. For this reason if it were attempted to make concrete, or to make artificial stone, with the same proportion of gravel and sand and of blue lias lime, which is the most suitable in working with Halling lime, I conceive that the concrete thus formed would be much too friable. It is essential that this should be understood, lest the best lime should lose its character by being injudiciously used.

GROUTING DEFINED. IN WHAT RESPECT IT DIFFERS FROM CONCRETE. ITS GREAT UTILITY.

(40) When a thick wall of rubble masonry, chiefly of small and irregular materials, is built in successive courses, with all the stones at first laid dry, excepting the outside ones of each course, or when long irregular stones in the form of an arch are laid dry over a wooden centering, and quicklime slaked with excess of water, either with or without sand, is poured in a liquid state, into the mass of dry materials composing every such course or arch, the work is said to be *grouted;* and this process which is of great antiquity is so analogous to the mode of forming concrete foundations above described, that these were often called grouted foundations, when first introduced. It is proper however to make a distinction between the terms *Grout* and *Concrete,* because the best Architects are now in the habit of causing brick walls to be grouted, in a manner which from the uniform dimensions of bricks, and the extreme regularity with which they are laid, has no resemblance either to the ancient grouting or modern concrete. This is done sometimes in every course, sometimes in every third or fourth course only, of a brick building. A bed of mortar being laid on the uppermost course of the finished part of the wall, and the outside bricks only being placed, so as to leave a kind of channel in the middle, water is thrown upon the wall, and the mortar and water are mixed together by raking backwards and forwards with a kind of long handled iron hoe called *a Larry,* which converts the mortar into grout, making it so liquid that it runs down into all the vertical joints of the course below it, and completely fills them.† This system of grouting is very judicious, because it prevents any of those joints from being left wholly or

* See Articles 9, 25 and 26.

† This must be done in thin walls by pouring the grout out of pails.

partially dry, which is too often done by bricklayers working in haste, and which if not observed at the time cannot be detected afterwards, without pulling down a part of the wall. Indeed it requires great attention, even for a very good bricklayer to fill or as it is technically termed *to flush* all his vertical joints with common mortar alone;* but no part of these joints can escape from the grout or liquid mortar, which penetrates into every crevice. This system was followed by Sir Robert Smirke in building the new British Museum, in which every course of the brickwork was grouted, wherever the walls were two bricks and a half thick or upwards, but in the thinner walls of the same building, which however are comparatively few in number, every third or fourth course only was grouted. Upon this subject, I may be permitted to remark, that unless every course be grouted, it appears to me that there is a risk of the grouting not penetrating lower than the single course immediately under it, for the beds of plastic mortar in the next courses below that, have sufficient consistency to intercept the grouting, unless those beds themselves should have been imperfectly laid, which seldom or never happens, even when middling or indifferent bricklayers are employed. For this reason, one can scarcely expect sound brickwork, unless every course be grouted, especially in thick walls, although the more general custom is to work with mortar only. When one of the massy walls of the new British Museum, after being grouted in the manner before described was cut through for some temporary purpose, it was remarked that the brickwork resisted the tools of the workmen quite as much, and appeared equally firm in the joints, as if the latter had been filled with plastic mortar instead of grouting.†

The same risk of part of the vertical joints being left dry may occur also in masonry, and I am not aware that there is any method of guarding against it more effectual than to grout each course.

* Any person, who has examined old brick buildings, whilst being pulled down, must have observed, how many of the vertical joints are entirely or partially dry. I believe that this has often been the cause of disastrous fires. I once knew a skirting board in a room at Brompton set on fire from this cause, in consequence of a family in the next house having made a small oven on the other side of the same party wall.

† This circumstance is an argument against the necessity of beating mortar, according to the system formerly in repute. (See the Note to Article 35.)

OF MAGNESIAN LIME STONES.

(41) There are no magnesian lime stones, that I am aware of, in the South of England, and none are imported into London for building purposes. The small specimens that I have been able to procure appear to me to possess hydraulic properties, and I hear that such is their character in those parts of the North of England where they abound; but as all those specimens were coloured, generally with a yellowish tint, and only partially soluble in the diluted acids, it follows that they must have some additional ingredient in their composition, besides the carbonates of lime and of magnesia, the latter of which when pure is also white, and entirely soluble in acids like the former, but with less effervescence. From this circumstance, as well as from some experiments with the pure carbonates of lime and of magnesia tried by me, that will afterwards be recorded, I have doubts whether any magnesian lime stone composed of these two carbonates only, if such should be found in nature, will either make a very good water lime or water cement.

OF THE WATER CEMENTS OF ENGLAND, ABSURDLY TERMED ROMAN CEMENTS, AND OF THE MODE OF TESTING THEM, AND OTHER NATURAL CEMENT STONES. THEIR PECULIAR PROPERTIES AND ADVANTAGES. THAT THEY SHOULD NEVER BE USED FOR CONCRETE.

(42) The term *Water Cement* holds good in a double sense, implying not only one which sets under water, but one which being of a calcareous nature is prepared for use by mixing it with water.*

The best water cements of England are the Sheppy and the Harwich cement, which are most used about London, and the Yorkshire cement is also said to be good, but of this I speak from report chiefly. They should always be designated in this manner by the places from whence they

* In contradistinction to *Oleaginous Cements* such as *Mastich*, which is a mixture of litharge stone dust and linseed oil, and to *Resinous Cements*, such as the *Mineral Fusible Cement*, proposed by Mr. Fitz Lowitz, and experimented upon by Lieutenant-Colonel Sir J. M. Frederick (then Captain) Smith of the Royal Engineers in 1829, who reported that when prepared for use by melting it in a caldron over a fire, and mixing it with twice its own weight of sand, that it appeared stronger than the Sheppy or Harwich Cements, though the latter was used pure. But it was then about 50 per cent dearer than the natural cements, and in most situations its inflammability would be a strong objection against the nse of it, for I found that it would burn even in the flame of a candle.

have been obtained, and not by the absurd title of Roman Cement, which serves only to mislead the ignorant, and being used alike for each, and all, and for every mixture of those cements, prevent those who are better informed from knowing what sort of cement any manufacturer of that article is offering for sale, unless they make minute inquiries, or personal observations, at his cement kilns.

These cements differ chemically from the water limes, only in containing a smaller proportion of carbonate of lime, and a larger proportion of silica and alumina in their composition, than the latter. Practically they differ from lime in several remarkable particulars.

First. The water cements will not slake at all, after being burned, boiling water having very little action upon them, and cold water none at all,* unless they be previously pulverized, by pounding or grinding them, which is therefore always done in preparing them for use.

Secondly. The fine calcined powder thus obtained when made up into a ball with water, does not swell or go to pieces, whether in air or if put immediately afterwards into a basin of water, but sets or hardens without any perceptible alteration of its bulk, under water, in an infinitely shorter time, than the best limes do in air. The first of those properties was no doubt the cause, which prevented men in this country from discovering the nature and use of the cement stones on the South East coast of England, until within the last forty years, when the light of Chemistry began to be more generally applied to the practical arts.† The second property is that which

* I have found by repeated experiments, that when cement from the kiln is sprinkled with boiling water, the pieces usually crack or split, and much more heat is produced than is due to the temperature of the water, so that all the symptoms of incipient slaking are observable; but this process seems suddenly to stop short, for the cement does not fall down into a fine powder, and thus the heat goes off without the fragments losing their solid form, at least within the space of two days, after which period not anticipating any further change I have always thrown them away.

† The properties of the Sheppy cement were first discovered by Mr. Parker, who took out a patent for the sale of it under the title of Roman Cement, but which from him as the discoverer was at first often termed Parker's Cement. Afterwards Mr. Frost, whose more recent labours I have noticed in Article 6, discovered that the stone on the beach near Harwich belonging to the Ordnance Department had similar properties. The Yorkshire Cement is sometimes termed Atkinson's Cement, from the name of the Architect who introduced it in London, or Mulgrave Cement, from the Earl of Mulgrave. It appears to me that the names of all water cements should be derived from the localities where they are obtained, which is always done in respect to limes. The use of such cements, a recent discovery first made in this country, was entirely unknown, not only to the Romans, but to all other nations, until Mr. Parker's time.

renders them particularly valuable, under situations and in circumstances, where the best limes would fail.*

Thirdly. Cement is always weakened by the addition of sand, whereas every kind of lime is improved by it, so that in all cases lime mortar made with the proper proportion of sand, is much stronger than lime putty.

Fourthly. When cement powder, whether pure or with the addition of sand, is mixed with water, for the purposes of Practical Architecture, it must always be used before it becomes warm, after which process it must never be disturbed, so that any portion of it that could not be used immediately must be thrown away; whereas mortars made of all limes, and more especially of the pure limes, may be disturbed, and if not used immediately they may even be remixed by adding more water, without materially injuring them.†

(43) To analyze a cement stone, or any other compound mineral accurately, so as to obtain the precise proportional weight of each of its component parts, is not only one of the most difficult operations of practical Chemistry, but demands much more time than Architects, Engineers, or Builders in full employment can spare. I beg therefore to suggest the following simple and satisfactory mode of proceeding, in testing a stone supposed to be a water cement, which ought of course to be of a bluish grey, or brown, or of some darkish colour, as white indicates pure lime stone or gypsum.

From their containing silica and alumina, which are the

* A joint of brickwork laid in pure cement becomes harder in half an hour, than the best mortar made of Dorking or Halling lime would be in six months or even more. Recently on inspecting a fine brick building in Southwark, now ready for receiving the roof, I observed that all the joints of the vaults and of the flat arches over windows, which were laid in cement mixed with sand in equal parts, were perfectly hard, resisting the edge of a shilling, whilst the joints of the mortar which was no less good of its kind, being made of the best Halling lime mixed with clean river sand, were so soft, even in the lowest and first finished parts of the walls, that the edge of a shilling raked out any part of them with almost as much ease, as if they had been composed of sand alone.

† Notwithstanding this distinction between cement and limes, the properties of all calcareous stones appear to me to be so analogous, that I cannot help thinking that it is possible, that cement mixed with a moderate quantity of sand, and made up with water into a heap, and left for some hours or even for some days, might bear remixing, and that in process of time, it might even set, like a heap of remixed lime mortar. But as the chief or only advantage of cement over such limes as the blue lias or Aberthaw is its instantaneous setting, and as I conceive that remixed cement will probably set if at all more slowly than lime mortar, the rule never to disturb cement that has been used, and never to use cement that has been long mixed, should always hold good in practice.

component parts of clay, the natural cement stones, on being touched by the tongue, indicate sensibly the presence of clay, which is also detected by the smell after wetting them. They only dissolve partially in diluted acids, leaving a more copious sediment, than any of the lime stones.

(44) The first test therefore is to pour a little diluted muriatic or nitric acid into a wine glass, and drop a fragment of the stone into it. If no action take place, it is neither a lime stone nor a cement, but if it effervesce and fall to pieces with a muddy sediment, it is probably a water cement.

(45) To ascertain this point, break the stone if necessary into compact fragments not exceeding one inch and a half in thickness, and put two or three of these into a common fire place, first heating them gradually, that they may not burst into too many small pieces, and keep them exposed to a full red heat for about three hours.

At the end of this time, take one of your specimens out of the fire, and put a small fragment of it into a glass of diluted muriatic or nitric acid. When the stone is burned enough, that is when all the carbonic acid gas has been driven off by the heat, no effervescence will take place in the acid. A moderate effervescence will show that it is a little underburned, but violent effervescence will show that it is very imperfectly burned, and it should therefore be put into the fire again.

At the same time great care must be taken not to overburn the cement, which is always injured, if not entirely spoiled, by excess of heat. When moderately overburned its colour is changed to a much darker tint, than ought to belong to it, which may easily be judged of after a few trials, and which is a proof of incipient vitrification; for by excess of heat all the natural cement stones may be fused into a dark coloured glassy substance, resembling the volcanic product called Obsidian.

If therefore the calcined cement stone do not effervesce with acids, and if at the same time its proper colour be not changed to a darker one on taking it out of the fire, these circumstances prove that it is properly burned.

(46) In this case, you must take your specimens and pound them in a mortar until they are reduced to the state of impalpable powder. To ascertain whether the calcined cement be sufficiently pounded, take a pinch of it between your finger and thumb and rub it backwards and forwards.

If it feel gritty to the touch, so that after having rubbed away the finer parts a number of small solid particles remain, let the whole of the powder be pounded finer, until on repeating the same process, no such grittiness is perceptible and no such particles remain. This is of the greatest importance, for I have ascertained that the best cements both natural and artificial are very much deteriorated, if imperfectly ground or pounded after calcination.

(47) The next process is to take a small quantity of the fine powder, and mix it up with a moderate quantity of water by means of a spatula or strong knife, upon a slate or porcelain slab, and knead it into the form of a ball by turning it between the palms of the hands. It will then soon become warm, and if it be a good water cement, it will not only set or harden in the heating, but if put into a basin of water, it will continue hard and even become harder in the basin. A ball of good cement, will set under water, even if put in quite soft and before it becomes warm, and in this case it makes the water dirty by throwing off a little scum from its own surface. It is better therefore to wait, till after having attained its greatest warmth it shall have begun to cool a little, before you put it under water.

The setting of such cement balls may be much retarded, even to double the usual period or more, if you mix the same calcined powder with a larger quantity of water, so as to form a thin or liquid instead of a stiffish paste, and in this case the heat thrown out will be diminished in the inverse proportion of the time required in setting: but eventually the slow setting ball becomes quite as hard as the others. I have ascertained by experiment, that one fourth part of water to one part of cement powder by measure is necessary for making a stiffish mixture, whilst one half measure of water to one measure of cement powder produces a very thin one. Between these extremes the proper proportion of water appears to lie. Less than one fourth would not suffice to make up the whole of the mixture into a paste, whilst more than one half would be inconvenient or injurious; for cement powder may be spoiled, or as it were drowned, if stirred up in a basin or bucket with excess of water; but this can never occur in practice, nor is any measurement of the proportional quantities of cement powder and of water either usual or necessary; for the expert bricklayer's or plasterer's labourer always mixes those ingredients together to the proper average degree of consistency, with very great accuracy by the eye alone.

During the process of setting, experimental cement balls always throw out vapour, which may generally be seen rising from them, as long as the heat continues, and which moistens the surface of any dry substance, on which they may be laid.

(48) The Sheppy cement stone is usually found in boulders or round nodules, and when dry is of a light brown colour. The Harwich cement stone is obtained in larger masses, and is of a dark bluish brown colour. When properly burned the former is of a light brown, the latter of a nut brown colour. When the calcined powder of each is mixed with water for use, both mixtures are brown, but the former is of a much lighter tint than the latter, which approaches to blackness.

(49) Since cement from the kiln will not slake even with water, much less by the action of air alone, it might be preserved in this state for a long time in a dry room, as I have proved by experiment; but calcined cement being of no use until it is pulverized, this is always done at the mill of the Manufacturer, to save the necessity of every purchaser providing himself with an apparatus for grinding it. This process, which enables water to act upon cement, makes it also subject to the action of the atmosphere.

(50) After calcined cement powder has been mixed up with water for use, it recovers in time the whole of the carbonic acid gas driven off by the fire. Hence the cement used for the purposes of Practical Architecture like the lime in mortar, is brought back but much more quickly, than the latter, to the same chemical state, as in the original stone. Hence also a fragment of old cement from a building will partially dissolve in the diluted acids leaving a residue, and on reburning old cement and pulverizing it, the calcined powder thus obtained will be quite as good as that from the natural stone.

(51) Cement powder exposed to air, especially if damp, also gradually recovers the carbonic acid gas previously driven off from it; and whilst this process is in progress, it is injured and eventually spoiled. For this reason the cement powder prepared for sale is always preserved in casks or sacks.

When cement powder damaged by exposure to air, or from any other cause, has become too *stale* for use, it may be converted into good cement by reburning it, the practicability of which may be ascertained by burning some stale cement powder in a crucible in a common fire place. But if any considerable quantity of damaged cement should require to be restored in this manner, as it could not conveniently be burned in a kiln whilst in the state of powder, it would be necessary first to mix it up with water into balls or lumps of a convenient size, which if allowed sufficient time will probably set so far as not to fall to pieces. If however the cement powder should be too stale for this purpose, let a small proportion of fine clay, not exceeding one tenth part by measure be added to it, in making these lumps, which will cause them to hold together in the kiln, without materially injuring the quality of the cement. It would not be worth while to take so much trouble in this country, but the knowledge of this expedient may perhaps be useful in some of the British Colonies where cement is only obtained by importing it from England.

THAT THE MOST POWERFUL WATER CEMENTS ARE, LIKE LIMES, PARTIALLY DECOMPOSED BY AND SOLUBLE IN WATER, WHEN FIRST PREPARED FOR USE. THAT THIS ACTION IS TOO INSIGNIFICANT TO AFFECT THE SOLIDITY OF THE WORKS IN WHICH THESE CEMENTS ARE USED, AND THAT AFTER A CERTAIN TIME IT CEASES. CHEMICAL TEST FOR ASCERTAINING THIS POINT.

(52) When a cement ball made as above described is put immediately under water, it always throws off a little scum as was before mentioned, which partly falls to the bottom of the basin as a sediment, and partly rises to the surface in a very thin film, thus proving that the cement is partially decomposed by the separation of a part of the carbonate of lime, on being first immersed. If after a day or two the ball be taken out of the dirty water, and put into clean, the same effect will take place, but in a slighter degree, and on continuing to change the water from time to time, the films successively formed on the surface will become smaller and smaller, and at last imperceptible to the eye. When no apparent action of this kind takes place, wash your ball well in distilled water, and scrub it with a clean tooth or nail brush, in order to get rid of any impurity on the surface of the ball, and when it is perfectly clean put it into a basin of distilled water, and let it remain there about a week. Then pour some water out of the basin into a glass, and let a few drops of the oxalate of ammonia fall into it. If the water should still remain clear, it shows that the cement ball has now become proof

against the action of water; but if a milky precipitate should take place, it shows that the unfavourable action of the water on the cement, at first clearly indicated by the scum thrown off from the surface, has not entirely ceased, for the presence of this precipitate proves that the water has dissolved a small portion of the lime belonging to the cement. Some recent experiments induce me to suspect, that balls made of the oldest and best natural cements, if kept in distilled water much longer than a week, would give out a slight precipitate. But the quantity of matter, which they must undoubtedly lose in order to produce this effect, is far too insignificant to be worthy of notice. Indeed without so very delicate a chemical test as the above, it would elude our observation. If you cannot obtain distilled water, clean rain water will answer the purpose, but no other should be used, because the water of wells and springs and still more that of rivers is always impure, and often contains lime in solution, which would of course defeat the object of this experiment.

It may require some weeks, before the cement balls will cease to yield a copious precipitate with the oxalate of ammonia, though they will have set at the surface so as to resist the action of the nail, in a much shorter period. Before this degree of induration takes place, they will generally stain the back of the hand or a piece of white paper, if rubbed against them after being wetted.

(53) Cement balls if kept in the air after being made, throw out an efflorescence at first, which when seen in the sun or by candlelight is in the form of very brilliant crystals, but these if washed off after a certain time are never renewed. The same efflorescence is observable in the fronts of buildings, when first stuccod with cement.

(54) That the partial decomposition by, and solubility in water, of cement, especially when first prepared for use, is too insignificant to injure the solidity of wharf walls, and other works of Hydraulic Architecture, in which it is employed, may be admitted, not only from the circumstance, that I have never observed any apparent diminution of the size of the experimental balls made by me from the Sheppy or Harwich cement stones; for the marks of the printer's types, with which I usually impressed them whilst moist, have never been erazed, although not penetrating more than about the twentieth of an inch, nor have even the lines caused by the seams of the hand in kneading them into balls been entirely obliterated; but also from the well known fact, that in

wharf walls built with bricks and cement, no pointing of the external joints has been ever required, which would undoubtedly have been the case, had any perceptible decay of those joints occurred.

(55) In short, whilst the water cements and water limes are similar in their component parts, and general properties, the great distinction between them as applied to the purposes of hydraulic Architecture is, that the setting process at the surface comes so much sooner to full perfection in the former than in the latter, that the water, which after this period ceases to act perceptibly upon either, has not time to injure the external joints of the cement, but is enabled in the mean while to make a considerable impression upon those of the lime.

The greater solubility even of the water limes than of cement was proved to my satisfaction, by the following circumstance. The new vaults in part of the London Docks built, under the superintendence of Mr. Palmer then Engineer of that company, of brickwork laid in Dorking lime mortar, were left during the winter of 1827 exposed to the action of rain penetrating from above, as the warehouses by which they were intended to be covered were not then begun. Consequently I observed, that this mortar although very good of its kind was giving way, for the lime in many of the joints had dissolved into stalactites, which hung from various parts of the vaulted roofs.* In the Thames Tunnel on the contrary, no part of the cement in which the brickwork is laid, has ever given way in this manner.

(56) We found by experiment at Chatham, that two thirds of a cubic foot of calcined Sheppy cement powder, which is equivalent to one cubic foot of cement from the kiln, would not bear much more than one cubic foot and one third part of a foot of sand, without evidently becoming too short for building purposes. This is equivalent to a mixture of two measures of sand to one of cement powder. But experience has shown that even this proportion of sand is too great in practice, for the builders of the Metropolis, who have used immense quantities of the Sheppy and Harwich cements for many years, agree that more than 5 parts of

* The same lime would not have given way in this manner in a concrete foundation, or in the backing of a wharf wall, because these rest upon and are inclosed by solid materials on all sides, so that there is no vacant space, as in a vault, for the lime to fall into, or escape by. Consequently as it cannot get away from its original position in combination with the sand and gravel with which it has been mixed, it sets in process of time without dissolving.

sand to 4 of cement powder, or $1\frac{1}{2}$ measure of the former to one measure of the latter, injures the cement, by retarding its setting and rendering it too friable, whether used as mortar for walls, or as stucco for the fronts of houses; but they consider that equal parts of sand and of cement powder, involving a smaller proportion of the former ingredient, are still better.

But not to lose sight of the just comparison between cement and lime, these proportions when stated in the same manner as we did in treating of lime, imply that whilst *one measure of cement from the kiln will not bear more than two thirds or at the utmost five sixths of a measure of sand without injury,* one measure of the various sorts of lime from the kiln, according to its quality, will bear two or three measures of sand or even more. (See Articles 9, 25 and 26.)

(57) There are two properties of cement, which ought to be thoroughly understood.

First. It only sets rapidly when made up into small balls or in very thin joints. In large masses or in thick joints, the rapid induration takes place near the surface only, from whence it extends towards the center so very slowly, that the cement there may remain in an imperfect state for a very long time. This property it has in common with lime mortars and concrete, which when in mass set more slowly at the center than at the surface, in the like proportion (33).

Secondly. As was before remarked in Article 42, cement is always weakened by sand, no matter how small the proportion of that ingredient may be, so that if both materials were equally cheap, it would be best to dispense with sand altogether in using cement as mortar for building walls, but not in using it as stucco for plastering the fronts of houses.

Numerous experiments have convinced me of the truth of these maxims, which any of my readers may easily verify. In respect to the latter in particular, take a small quantity of the best cement powder, mix it with three or four times its bulk of fine sand and make it up into a ball with water, and you will find that instead of setting it, will either remain quite friable or crumble to pieces, both under water and in air.

Upon the whole cement sets most quickly, and unites itself most powerfully to bricks or stones, when it is perfectly pure or unmixed with sand, provided only, that the joints be thin, I should say not exceeding half an inch in thickness. For this reason, in forming cement into chimney pots, copings, &c, where the general thickness much exceeds the above dimension, and consequently where pure cement alone would not make sound work, instead of frittering away its strength with sand, I would recommend fragments of broken tiles or gravel, to be mixed with it, the interstices of which are such as to allow the pure cement which fills them, sufficient body to attain a due degree of strength, without being quite so large as to retard its setting, and thereby cause weakness in the central parts of those spaces.

(58) This opinion is confirmed by the experience of the Thames Tunnel, where the justly celebrated Mr. Brunel, who commenced differently, has finally adopted the system of using nothing but cement for all his brickwork, which he mixes with sand in equal measures for the foundation and lower part, and with half that proportion of sand for the piers, which support his arches; but from the spring of the arches upwards, as being the most important part of the work, he uses only pure cement. Indeed if the use of this admirable material had not been discovered, the execution of the Thames Tunnel would have been impracticable,* for if it had been attempted with the best mortar, the pressure of earth would have crushed some parts of the brickwork before the mortar got consolidated, and in other parts the lime would have been washed out of the joints, as was the case in a new basin in Chatham Dock-yard, where several landsprings washed the lime entirely out of those portions of the concrete foundation, which lay in their course, until channels were formed for carrying off each of those little rivulets by iron pipes leading to a culvert.

* Under the protection of that most ingenious piece of mechanism, the massy iron shield, the brickwork of the Thames Tunnel is built in successive portions or ribs of 9 inches in length, but where the soil is bad in portions of half that length only, which are connected by cement alone, the bricks of one rib not being toothed or bonded at all into those of the next. This defective bond, adopted from necessity, does not impair the strength of the work, because the judicious system followed renders it impossible that any new rib can separate from the brickwork previously finished, they being screwed up close together by means of the shield, for a space of time longer than is necessary for the intermediate vertical joints of cement to set, after which these joints become stronger than the bricks themselves, as I had an opportunity of observing in the beginning of the present year, 1836, when the brickwork laid in cement, with which the head of the Tunnel had been blocked up during the temporary suspension of the work, was cut away in order to continue the excavation. In this operation, which was exceedingly laborious, the solid bricks themselves were more frequently fractured than the cement joints.

(59) But the remarkable strength of pure cement is still more proved and in a wonderful manner, by the two extraordinary semi-arches built for experiment by Mr. Brunel near the entrance of the Tunnel, which spread out from the same common central pier, like two immense branches from contrary sides of the trunk of a tree, one of which is 60 feet and the other about 37 feet long, the rise of the former being 10½ feet, and of the latter 10 feet; and this last is loaded at its extremity with a weight of 62700 lbs, suspended below it, as shown in the annexed figure. It is well

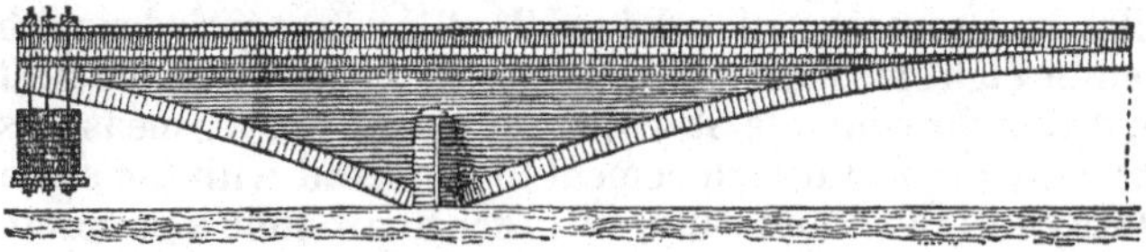

known that cement alone will not hold together above a certain limited number of bricks thus unsupported, and therefore much must be ascribed to the pieces of hoop iron introduced horizontally in the joints of the brickwork, but unless cement had the property of setting almost instantaneously, and thus combining brickwork properly supported into one solid mass, having the joints and bricks of equal strength, the hoop iron could not possibly communicate such extraordinary stability to the same kind of brickwork hanging in the air. This ingenious arrangement of Mr. Brunel will probably be found hereafter of great value in practical Architecture.*

(60) The rapidity with which cement sets and dries in air, neither admitting of settlements nor producing damp in walls built with it, is no less advantageous in domestic Architecture, by allowing an alteration or repair in a build-

* Mr. Brunel proposes its application to the construction of large brick arches, without the necessity of previously constructing a substantial and expensive wooden bridge, technically termed *the Centering*, to support such arches until they are keyed in or completely finished; instead of which he uses only a small light scaffold for the bricklayers, which is suspended from the finished part of the arch, and continually moved forward as the work proceeds. It occurs to me that this ingenious invention of Mr. Brunel may be applied also to the construction of Tunnels in loose soil, by carrying on the brickwork in oblique portions suited to the natural slope of the earth, instead of continuing it always vertically, which requires strong woodwork, for supporting the loose earth, until each new portion of brickwork is finished.

ing to be made in the least possible time, and with the least possible inconvenience or injury to the health of the persons occupying it. By the use of cement for example, the temporary houses of Lords and Commons were finished in a dry wholesome state ready for use, in the short space of three months after they had been destroyed by fire in 1834, though the work was executed in a very unfavourable season.

(61) From the circumstance that all limes expand after being mixed with water, and in this process attach themselves firmly to sand which increases their strength, whereas cement does not expand when mixed with water, and is always weakened by sand; it follows that whilst one measure of calcined Halling lime powder with seven or eight measures of gravel and sand may be the best proportion for concrete, it will require nearly three measures of Sheppy or of Harwich cement to be mixed with the same quantity of gravel and sand, in order to produce a concrete equally sound.

Hence cement should never be used for making concrete. If used for that purpose in the same proportion that would be proper in working with lime, the cement concrete thus formed will never set at all, either under water or in air. If used in a larger proportion, in order to get rid of this defect, it would involve a greater expenditure of a much more costly material without producing a better article than the common concretes, formed with the proper kinds of lime.

In building with cement, there is however no objection to cement grouting, which when pure cement is used, may be made by mixing up the calcined powder with water only, but in building with cement and sand, the latter ingredient also may be added to form a part of the grout.

(62) In building brickwork with cement, it is more usual, and excepting under circumstances of peculiar difficulty, it is certainly better to mix the cement with a moderate quantity of sand, than to use it pure, for this mixture diminishes the expense of the work, and though certainly weaker than pure cement, it is strong enough, inasmuch as it produces joints equal in strength to the bricks themselves.

(63) Pure cement has always been used as the finishing coat for the lining of tanks, and the coating of casemates, and we know that in both cases it has proved effectual, preventing leakage in the former, and rendering the latter perfectly dry. I am not aware that cement mixed with sand

has the same property, at least I can certify that it is not water tight, in the manner in which it is usually laid on by common plasterers. But for stucco this mixture is not only cheaper but better than pure cement, as being less liable to crack in setting, especially in hot weather when exposed to the sun. All the cement stucco, however, that I have seen, in spite of the admixture of sand, cracks more or less in process of time, which I conceive is partly caused by its not being sufficiently compressed by the workmen in laying it on, which they never will attend to, except in working at Tanks or Casemates, where the same careless manner, in which they execute their cement stucco, would inevitably occasion leaks, and defeat the object in view; for I believe that even pure cement is not watertight, if negligently laid on. I conceive that from this cause chiefly, the cement fronts of the new buildings in and near the metropolis, and indeed in every part of the kingdom are completely reticulated by an immense number of fine cracks, which are usually superficial or *hair cracks*, but which elude observation from these fronts being afterwards washed over or painted with some light colour to take away the sombre hue of the natural cements. But unless the cement stucco should peel off from the face of the wall, which sometimes happens when very unskilfully applied, these hair cracks on its own surface do no harm.

OF SLOW SETTING CEMENT STONES, SUCH AS THE BLACK ROCK OF QUEBEC. THAT THEY MAY BE USEFUL UNDER PARTICULAR CIRCUMSTANCES.

(64) The Sheppy and Harwich cements when made into moist balls begin to set almost immediately, whether in air or under water, and in the course of half an hour or sometimes much less, they emit that clear ringing sound, which is a proof of hardness, on being thrown against the side of a basin; and no water cements inferior to these would be accepted in the English markets. But it is possible that a less powerful water cement might have its value in other countries, if found abundantly in any district, where superior water cements, or the ingredients for imitating them by artificial means, were not to be conveniently obtained.

For example, a number of specimens of what is called the black rock of Quebec were recommended as being good water cements by Captain Baddeley of the Royal Engineers to the Master General and Board of Ordnance, by whom I was ordered to examine and report upon them. Accustomed only to the powerful water cements of England, I should have rejected the whole of them at once, if the respect, which I considered due to the opinion of another Officer, had not induced me to give more time and care to their examination, than I had ever done to any former specimens of natural stone supposed to have cement properties. The quantity of carbonate of lime contained in most of them was evidently too small, since on being immersed in the diluted muriatic acid, they did not always fall to pieces, but retained their original form as solid fragments of stone, after all the lime had been dissolved and the carbonic acid gas driven off with effervescence. And on being burned in the fire, and afterwards pounded and made up into moist balls in the usual manner, they threw out very little heat, showing scarcely any symptom of setting, and remaining as a soft pulpy mass under water. But on giving them longer time, by allowing them to remain several days in the air, I found that they had acquired a stony hardness, and on then putting them into a basin of water, it had little or no effect upon them, and finally after having kept them one or two months longer, and immersed them when perfectly clean in distilled water, for several days, I found that the oxalate of ammonia produced no effect upon this water, which afforded the most satisfactory test of their quality. Indeed they yielded less to this test, than our best English water cements had done under the like circumstances. I recommended therefore, that the specimens reported by me as being the best, for some were very indifferent, should be tried on a greater scale at Quebec, by burning a sufficient quantity in a kiln, to be used in the first Government work, in which waterproof cement might be required, but not in tide work, for which I considered that the slowness of their setting rendered them less suitable.*

* It is rather curious, that Captain Baddeley who suspected the black rock of Quebec to be a water cement from its chemical properties, could not succeed in making it available as such, until he read my observations on water cements printed by authority for the use of the Royal Engineer Department in 1830, whilst on the other hand his recommendation of this stone led me to the conclusion, that very slow setting water cements, such as I would before have condemned, are not without their value under particular circumstances. On receiving a copy of my report, he wrote that he had ascertained that this cement becomes waterproof in about six hours after being mixed, instead of requiring some days, as I supposed. Not having tried the precise period, in which this effect took place, and having no more specimens of the stone to experiment with, I cannot verify this fact personally; but I have no doubt of the correctness of Captain Baddeley's statement. Probably the use of boiling water might render the Quebec cement fit for tide work, or even in working with cold, that the temporary action of water upon the external joints might not be sufficient to injure the work materially.

Having now explained the properties of the water limes and cements of nature, I shall proceed as I before promised (3), to describe in sufficient detail the course of experiments upon artificial water cements, which occupied a portion of my time for several successive years.

OF THE SYSTEM PURSUED IN THE EXPERIMENTS FIRST TRIED ON A SMALL SCALE IN CRUCIBLES, IN 1829 AND 1830, IN ORDER TO FORM ARTIFICIAL WATER CEMENTS, BY VARIOUS COMPOUNDS.

(65) After having succeeded by accident in making a good water cement with chalk and blue clay, as stated in Article 2, and having given a good deal of time to the chemical analysis of the natural cement stones, which I found attended with the difficulties before alluded to (43), and which led to no satisfactory result; I determined to relinquish this unpromising investigation, and to adopt the system of working synthetically, by combining together all the substances likely to form an artificial cement, first by twos, and afterwards by threes and fours, in various proportions, determined generally by measure not by weight. All the natural stones, and other dry substances experimented upon were measured in the state of impalpable powder to which they were previously reduced by pounding them in a mortar. The natural clays and some other substances were measured in a moist state. The powders were dropped or thrown into the measure with sufficient force to fill every part of it, and were stricken by scraping off all that stood above the rim; but they were never rammed or pressed down. Each of the measures used was open at bottom as well as top, and laid on a slab whilst being filled, and the clay or moist substance was not only put into it by one of the open ends, but by the other also, after inverting the measure, without which we found that vacant spaces were always left. When thus put in and stricken at both ends of the measure, it was forced out by a wooden plug fitting the opening, but of about double its height.

After measuring the several ingredients, we mixed them intimately together, adding a moderate quantity of water, with a spatula upon a slate or porcelain slab; after which we made them into balls of about one inch in diameter, and dried them for a little while before the fire, until they would not stick to one another, after which we put as many of them into a Cornish crucible, as it would hold, and burned them for about three hours in a common fire kept up to a full red heat.

We found that the ingredients cannot be pounded too fine or mixed too well, and that water is absolutely necessary. If imperfectly pulverized or mixed carelessly, the most suitable ingredients may fail, and if mixed dry and burned in a crucible, they will also fail.

(66) To prevent our crucibles from breaking by sudden changes of temperature, we heated them gradually before we exposed them to the full degree of heat necessary, and after taking our calcined cement balls out of them, we put them into the fire again, and by diminishing the heat allowed them to cool gradually. We always purchased covers with our crucibles, and when they broke we replaced them by iron ones.

(67) To judge whether our artificial cement mixtures were burned enough, we followed the same rule, as in burning the natural cements, explained in Article 45. Taking our crucible out of the fire, we dropped a fragment of one of the balls contained in it, into a glass of diluted muriatic acid. If it effervesced violently, which proved our mixture to be underburned, we put it into the fire again: but if it did not effervesce, and was not burned darker than its original colour, we considered it burned just enough and not more than ought to be.

(68) We therefore next proceeded to pound our calcined cement balls in a mortar, until we reduced them to the state of an impalpable powder, leaving no gritty particles when rubbed between the finger and thumb, after which we made our calcined cement powder into moist balls with water, proceeding exactly in the same manner as in trying the quality of the natural water cements. (See Articles 46 and 47.)

(69). All our experimental balls of calcined cement powder, whether natural or artificial, were marked with Printers' types, whilst in a soft state, before they heated and set, the letters and numeral figures impressed upon them, showing by appropriate abbreviations, the component parts and proportions of each ingredient, and the day of the month and year, when they were made. Without this precaution the record book of our experiments, which was kept in the form of a journal would have been of little use. The types necessary for marking each ball, must always be arranged beforehand, and used without delay, because the natural cement stones of England, and the best artificial compounds when burned in a common fire-place set so quickly, that there is no time to spare. We also marked the raw mixtures previously to burning them, when several sorts were experimented upon at the same time.

(70) As all our experimental balls of calcined cement, with some few exceptions which shall be noticed hereafter, were put into a basin of water soon after they were made and marked, those which failed in setting permanently, did so under various circumstances. Some of them fell to pieces in a state of powder at the bottom of the basin, without any apparent separation of the lime, or action upon it by the water. Others fell to pieces after swelling to a much greater size. Some appeared to set in the most promising manner, but afterwards cracked, and eventually split to pieces, or become soft throughout. These results took place almost immediately in some mixtures, but not until after several days or even weeks in others, and in all cases, films of lime formed at the surface of the water, unless the defective mixtures were thrown away immediately. Some mixtures hardened at the surface only, proving soft inside on being broken, after having been made several weeks. We therefore adopted the rule of breaking all our artificial cement balls with a hammer and chisel, in a month or two after they were made, and if we did not then find the inside of a stony hardness, we considered the cement an imperfect one.

In order to obtain this result of perfect hardness throughout, it is however absolutely necessary that the experimental balls of calcined cement shall not exceed one inch in diameter, because as was before observed (57), the induration of the central parts of larger masses of cement is tardy, and analogous to the gradual induration of common mortar.

THAT ON REPEATING THE SAME EXPERIMENTS IN 1836, IT WAS FOUND MOST CONVENIENT NOT TO USE CRUCIBLES.

(71) Having presented numerous specimens of the artificial cements made by me in 1829 and 1830, to the Royal Institution, to the Institution of Civil Engineers, and to some other Scientific Societies, without keeping duplicates of all, I was induced at the beginning of the present year to repeat most of those experiments, not only to obtain new specimens of those that were deficient, but to verify generally the most important results; and having in these more

recent experiments dispensed with crucibles, in all those mixtures which were of a plastic nature, and therefore not liable to crumble to pieces in the fire, which I had seldom done in former years; I consider this upon the whole to be not only the most economical but the best mode of proceeding, and I therefore recommend it to any of my readers, who may be disposed to try similar experiments. In using this method it is best to make up the raw cement mixtures into balls of about two inches and a half in diameter, with a moderate quantity of water. If they should split a little in the fire which sometimes happens, it is of no importance; and if the surface should be blackened with smoke, let it be scraped off before the ball is pounded, and no harm will be done, but this circumstance only occurs in a dull fire, which ought not to be used.*

OF THE INGREDIENTS WHICH WERE USED IN OUR EXPERIMENTS.

(72) The ingredients used in our experiments were as follows.

First, Carbonate of Lime. For this, which is the most important ingredient in all water cements, whether natural or artificial, we used pounded chalk.

Secondly, Silica. For this we used pounded flints, previously heating them red hot and then quenching them in water, according to the custom of porcelain manufacturers, without which they cannot conveniently be pulverized.

Thirdly, Pure Alumina. This was partly purchased in London, and partly made for us by a chemist in this neighbourhood.†

Fourthly, Silica and Alumina. For this compound we sometimes used Dorsetshire pipe clay, the purest natural combination of these two ingredients; sometimes a mixture of pounded flints and pure alumina.

Fifthly, The Protoxide of Iron. For this we used the scales struck off from their iron by the anchorsmiths in His Majesty's Dock-yard at this place, which we reduced to fine powder by pounding them in a mortar.

* Rather more care is necessary in burning the natural cement stones or any other hard stones without crucibles than with them, owing to the tendency of these stones to burst into small pieces in the fire; but this is easily prevented after a little practice in such experiments. (See Article 45.)

† Mr. Tatum of Rochester, who informs me that for this purpose he dissolved a quantity of alum or sulphate of alumina in distilled water, and added subcarbonate of soda to the solution until the acid was neutralized, after which he obtained the pure alumina by repeatedly washing and then drying the precipitate with heat.

Sixthly, The other Metallic Oxides, Carbonates, &c. We purchased the peroxide of iron and the other metallic oxides, as well as the calcined magnesia and the carbonates of magnesia, soda, potassa, &c, used in our experiments in the state in which they are usually sold by the chemists, and of the best quality.

Not only the clays as before mentioned but the pure alumina, and the carbonate of magnesia were measured moist. The pulverized chalk, the other carbonates and all the metallic oxides were measured dry.

OF THE MIXTURES, WHICH SUCCEEDED OR FAILED IN MAKING GOOD ARTIFICIAL CEMENTS.

(73) *Carbonate of Lime and the Metallic Oxides.* A calcined mixture of pounded chalk with the protoxide of iron generally fell to pieces, but in some few cases it hardened a little under water, but never so as to form a cement, for it was easily broken by the fingers, and always dirtied the hands. A mixture of pounded chalk with the peroxide of iron also generally fell to pieces, but occasionally formed a soft mass that never hardened under water. A mixture of pounded chalk with the oxides of manganese would neither harden nor adhere together, but always fell to pieces under water.*

(74) *Carbonate of Lime and Silica.* This mixture did not succeed in forming a water cement, above once in ten trials in former years, but it succeeded frequently on repeating the same experiment in the present year, although the cement formed seldom set into a very sound substance, under water. I can only account for the difference in these results, by supposing that the flints formerly used may have been badly pounded, which defect is sure to produce failure or imperfection in all such experiments, but which I know was not the case recently. In this mixture, the chalk and silica previously to adding the water being both in the state of dry powder, it was most convenient to determine their properties by weight, and we found that equal weights of these two ingredients produced the most satisfactory result. On afterwards ascertaining the proportions of the same quantity of each ingredient by measure, we found that it was nearly equivalent to a mixture of 2 measures of

* All the mixtures mentioned are supposed to have been calcined or properly burned in the fire unless the word *raw* be prefixed, which denotes an *unburned* mixture.

dry chalk powder with 1 measure of flint-paste, by which I mean the same flint powder mixed up with water into the consistency of moist clay.*

(75) *Carbonate of Lime and Alumina.* This mixture always succeeded this year in forming a water cement, of a white or greenish colour, but not of a very hard sound nature; and of all the proportions tried, a mixture of equal measures of pulverized chalk and of alumina-paste proved the most satisfactory.† It had almost always failed in former years, from our having used only about half of the above proportion of alumina.

(76) *Carbonate of Lime and Carbonate of Magnesia.* When pulverized chalk and the moist carbonate of magnesia are mixed together and burned, in any proportion of the latter ingredient not containing less by measure than one half of the former, they form an imperfect cement which if kept a day or two previously to immersion will not always fall to pieces in a basin of water, but seldom or never sets well either in air or under water. On using a greater proportion of magnesia, by combining equal measures of chalk powder and of magnesia paste, the mixture will not always fall to pieces on immediate immersion, and it even sets sometimes into a very sound close grained cement.

(77) Upon the whole, all the double compounds formed, by mixing the carbonate of lime with one other ingredient only, proved rather unsatisfactory, as the two best of them, namely the mixture of chalk with silica, and the mixture of chalk with alumina, did not produce a very good water cement, and the mixture of chalk with magnesia did not always succeed, whilst the others invariably failed.

(78) *Carbonate of Lime, Silica, and Alumina.* Four measures of pounded chalk, 1 measure of flint-paste, and 2 measures of alumina-paste, being a compound of the two mixtures described in Articles 74 and 75, always succeeded

* Mr. Frost, whose operations were before noticed in Article 4, had actually formed an artificial stone of this mixture, before I commenced my course of experiments.

Dry chalk powder averages in weight about 175 Troy grains per cubic inch. With 10 cubic inches of chalk powder mix about $2\frac{1}{4}$ cubic inches of water, and the chalk paste thus formed will occupy only about 5 cubic inches.

Dry flint or silica powder averages about 280 grains per cubic inch. With 10 cubic inches of this powder mix about $3\frac{1}{2}$ cubic inches of water, and the flint paste thus formed will compress into about $7\frac{1}{2}$ cubic inches.

† Dry alumina powder averages about 80 grains per cubic inch. With 10 inches of this powder mix about 5 cubic inches of water, and the alumina paste thus formed will compress into about $6\frac{1}{4}$ cubic inches.

in setting under water, and seemed rather better than either of these mixtures singly.* But still more satisfactory results were obtained from a mixture of pulverized chalk and pipe clay, and the best proportion for uniting these ingredients appeared to be 5 measures of the former to 2 of the latter, the chalk being measured in the state of powder, and the pipe clay in the state of a stiffish paste, as it would be used by the pipemaker.

(79) *Carbonate of Lime, Silica, and Alumina, with the Protoxide of Iron.* To each of the two mixtures described in the preceding Article, we added various proportions of iron from about one tenth to one hundredth part of the whole compound, all of which proved satisfactory, but more especially the latter, that is the mixture of 5 measures of chalk, 2 measures of pipe clay and about one half or one third part of a measure of the protoxide of iron, which produced excellent water cements of a light bluish grey colour, apparently not inferior to any of the water cements of nature; but in consequence of the peculiar hardness of those mixtures after calcination, we found that the pounding of them required very great care and more than usual labour, which persons when first employed in such experiments are sure to neglect, and which if neglected, become as I must again repeat, a certain source of failure or imperfection.†

OF THE ESSENTIAL COMPONENT PARTS OF AN ARTIFICIAL WATER CEMENT. THAT THEY MUST BE IN THE FINEST POSSIBLE STATE OF IMPALPABLE POWDER OR OF PASTE.

(80) Knowing that the natural water cements are chiefly composed of the carbonate of lime, silica, alumina, and the oxides of iron, I had now by working synthetically with all these ingredients obtained an artificial water cement apparently no less perfect; and it appeared that next to the carbonate of lime or chalk, which is indispensable, silica and alumina were the most important ingredients; for these made a water cement without the iron, whereas the iron never succeeded without them, notwithstanding which, it was of great use, since it caused them to set more quickly and at

* From the nature of these two mixtures, there can be no doubt, that any combination of the two will form a water cement, though I only tried them in the proportion stated, which implies 2 parts of chalk with an equal weight of silica powder, and 2 parts of chalk with an equal measure of alumina paste, all mixed together with water.

† I found that my assistants if left to themselves always failed from this cause, in producing a good water cement with the hard calcined mixtures, even after they had become expert in working with others of a less refractory character.

the same time communicated to them a superior degree of compactness and hardness.

I must here observe most decidedly, that no artificial mixture however judicious will make a water cement, unless the solid ingredients be reduced to the state of impalpable powder, and unless the moist ones also be in the finest possible state, previously to being burned.

EXPERIMENTS CONTINUED.

(81) *Carbonate of Lime, Silica, and Alumina, with other Metallic Oxides.* We next tried the same mixture of chalk and pipe clay with other metallic oxides. The addition of the peroxide of iron occasionally formed a good cement, but the success of this mixture is very precarious. The addition of the oxides of manganese or of red lead always formed good cements with the same ingredients.

(82) *Carbonate of Lime, Silica and Alumina, with the Carbonate of Magnesia.* On mixing 5 measures of pounded chalk, 2 measures of pipe clay, and from 1 to 5 measures of the carbonate of magnesia together, they formed an excellent artificial water cement, and the same satisfactory result was obtained, on using a somewhat smaller proportion of magnesia. This cement, when moderately burned, was of a handsome white colour, but when burned rather more it assumed a darkish slate colour; and we ascertained by some recent experiments, that it was equally good in both those states. Hence I suspect that cement a little overburned may not be materially injured, provided that no part of it is vitrified; but in former years having always thrown away those mixtures, which were burned to a darker colour than usual, I cannot speak with certainty as to this point.

To return to the carbonate of magnesia, it thus produces as an ingredient in a water cement, the same favourable effect as the protoxide of iron, or perhaps it may be pronounced superior to it or to any other metallic oxide, inasmuch as the carbonate of magnesia will sometimes combine with the carbonate of lime alone into a good water cement (76), which none of the metallic oxides are capable of effecting.

(83) *Carbonate of Lime, Silica, Alumina, and the Metallic Oxides, as obtained by a mixture of Chalk and the Clays of nature.*

First, Chalk and blue Alluvial Clay. This mixture of pounded chalk and of blue clay from the Medway, although the first of my artificial cement mixtures, which succeeded, is placed here amongst the quadruple compounds, after several of the others. I need scarcely say that the chalk supplies the carbonate of lime, whilst the blue clay furnishes the silica, alumina and the protoxide of iron.

Having tried various proportions of these ingredients, I finally adopted 5 measures of chalk to 2 measures of moist clay, as being that which produced a cement most similar to the Sheppy cement in its qualities.

Considering the clay as unity, the chalk thus used is 2½ times the measure of that ingredient. Three measures of chalk to one of clay will also make a water-setting cement, but not so good as the former proportion. A greater excess of chalk such as 3½, 4 or 5 to 1, spoils the mixture as a water-cement, but converts it into a water-lime. A smaller proportion of chalk on the contrary such as 2 to 1, makes a very good water cement, and even 1½ measure of chalk to 1 of clay makes a good cement, but one that sets with moderate heat and very slowly, for according as the proportion of chalk is diminished, the heat in setting is diminished also, whilst the period required for this process is prolonged.

The artificial cement balls formed of chalk and blue clay are of a lightish brown colour, but when put into a basin of water, the lower surface if in close contact with the basin is sometimes green, and on breaking one of these balls the inside also is sometimes green, but after the fractured part has been exposed to air, it assumes the light brown tint of the original surface.

When tested by the oxalate of ammonia in the manner explained in Article 52, the artificial cement made with 2½ measures of chalk and 1 of blue clay appears to be acted upon by water, in the same degree nearly as the Sheppy or Harwich cements.

(84) *Secondly, Chalk and the brown alluvial Clay or surface Clay of Rivers.*

The alluvial clay, which abounds in estuaries near the mouths of rivers, consists of soil washed away from the upper country after rains, which being then held in suspension in the stream according to the chemical phrase, is deposited on the beds of those rivers, near the sea, where the current is usually more moderate. In its original state it is of the natural brown or reddish brown colour of the soil, and the surface or upper part of it, which is exposed to the air at low water, always retains the same colour, for about an inch in depth, but the remainder of it, which is inaccessible to free air, acquires a dark blue colour approaching to blackness. Thus the great mass of alluvial clay is blue, whilst the surface only is brown.*

Having tried a mixture of pounded chalk and of the brown alluvial clay or surface clay of the Medway, it made apparently an excellent cement, which set under water immediately, but after a few days, it began to crack at the surface, and eventually fell to pieces in about a week.

THAT BLUE CLAY, OR ALLUVIAL CLAY FRESH FROM A RIVER, IS AN EXCELLENT INGREDIENT FOR A WATER CEMENT, BUT THAT IT LOSES THIS PROPERTY, IF KEPT UNTIL IT BECOMES STALE.

(85) *Chalk and stale Alluvial Clay.* The blue alluvial clay, treated of in the preceding Article, when taken out of a river and exposed to air, loses its dark blue colour, which such exposure gradually changes to the same light brown tint, that distinguishes the surface clay, after which it also loses its powerful cement properties, and may then be termed *stale,* in contradistinction to its original state of *fresh alluvial clay.*

Wishing to ascertain how much of this clay consisted of solid matter, I weighed it repeatedly when fresh from the river, and finding that its average weight was 87 lbs. per cubic foot, I dried it carefully on iron plates without burning it, and ascertained that it lost eleven twentieth parts of its weight nearly; so that more than one half of the blue clay of the Medway must consist of fluid or volatile matter. The very offensive smell emitted during the process of drying it, led me to suppose that fresh alluvial clay must contain a much greater quantity of hydrogen gas, than I had before suspected, which proved to be the case in an experiment afterwards tried by me, in which I was kindly assisted by Mr. Griffith,† and I believe that the escape of this gas, which is of course facilitated by exposure to air, is the chief cause of the change of colour from blue to brown, during the transition of the alluvial clay from fresh to

* The same effect is observable in drains after violent rains. The fluid mud washed away from the roads is of a brown colour, whilst running in gutters exposed to air, but after falling into covered drains, where it is no longer accessible to the free action of the atmosphere, it acquires the same dark blue colour belonging to alluvial clay, as may be observed from time to time when the sesspools of such drains are cleared out.

† Who had formerly been the assistant of Dr. Faraday, in the Chemical Department of the Royal Institution, and was then delivering a Course of Lectures on Chemistry at the Philosophical and Literary Institution of Chatham and Rochester. Mr. Griffith's Report on the blue clay of the Medway is annexed in the Appendix.

stale, which robs it of its cement properties. But as this change is gradual, it is proper to explain under what circumstances it takes place, and how one may judge whether its injurious effects have come into full operation.

(86) When blue clay fresh from the river is mixed with chalk and made up either into balls for burning in the fire, or into larger lumps or cubes for burning in a kiln, the colour of the clay predominates so much over that of the chalk, that these balls or lumps are still of a darkish blue or slate colour, but the air very soon changes the outside of the mixture to a light drab or dirty white colour; and by degrees this change extends from the surface inwards, until it takes place throughout the whole mass, which it does sooner or later in proportion to the smaller or greater size of the ball or lump, and at this period of total change of colour, the mixture will have entirely lost its powerful cement properties. In small experimental balls of about an inch in diameter, exposure to air for about 24 hours only in a dry room, previously to burning, will injure the mixture, and 48 hours will spoil it; whereas in balls or cubes of about 2½ inches diameter or side, exposure to air for ten or twelve days will not materially injure it.

As a change of colour at or near the surface only is of little importance, if you break one of the balls or lumps of this mixture prepared for burning, and examine the colour of the inside, you may always judge, whether it still retains or has lost its power of forming a good water cement.

(87) *Chalk, stale Alluvial Clay and Coal Dust, or other Combustible Material.*

It having been suggested to me by an emiment Chemist, to whom I mentioned the success of the fresh and the failure of the stale alluvial clay, that it probably was occasioned by the iron in the clay having gradually changed from the state of protoxide to that of peroxide,* and that the addition of pulverized charcoal or coal dust, to the raw cement mixture injured by air, might by promoting a contrary change in the iron during the process of calcination, correct this defect; I proceeded according to his advice, after having mixed chalk and stale alluvial clay together in the usual proportion, to add various proportions of charcoal powder and coal dust, and when the latter was not less than one fifteenth part by measure of the whole compound, it generally produced the favourable effect anticipated by the

* This was long before the presence of hydrogen gas, in any considerable quantity, as a component part of the blue clay was suspected by me.

distinguished individual alluded to, and I found that the same result was obtained by adding any other combustible matter, such as rosin, soot, sawdust, or oils and tars of various sorts, to the raw mixture of chalk and stale alluvial clay.

(88) *Chalk, stale alluvial Clay, and the Protoxide of Iron. Chalk, stale alluvial Clay, and the Carbonate of Magnesia.*

On adding the protoxide of iron, or the carbonate of magnesia to the raw cement mixture of chalk and stale alluvial clay, these ingredients also improved the mixture, and appeared to restore its original virtue to the clay.

(89) *Chalk, blue or fresh alluvial Clay and Coal Dust.* Under an impression, that blue alluvial clay was the best ingredient that could be used for forming an artificial water cement, in combination with chalk, and having ascertained that when this mixture became stale and gradually lost its virtue, it might be restored by remixing it with some of the substances mentioned in the two preceding Articles, it appeared desirable to ascertain whether any of those ingredients would act as a preventive to the same evil, which by saving the necessity of a second mixing, would obviously be better than to allow the raw cement to spoil first, and to cure it afterwards. But it seemed evident that iron would be of no use as a preventive, because the same exposure to air, which injured the alluvial clay would also spoil the iron, by changing it from the protoxide into the peroxide; and most of the other substances, especially the carbonate of magnesia, were too expensive. It therefore occurred to me, that a proportion of fine coal dust, if mixed with the proper proportions of chalk and of blue clay in the first instance, although it would not prevent the clay from being injured by exposure to air, yet that it would, during the process of calcination, remedy this evil, and produce as good cement, as if the clay had not been allowed to become stale; and it also seemed probable, that the proportion of coal dust used in the original raw cement mixture might not be altogether wasted, but might assist in the burning of the cement. Under these impressions, I was of opinion for more than a year, during the course of my crucible experiments, that the best artificial cement mixture consisted of 5 measures of pounded chalk, 2 of blue clay, and half a measure of coal dust, and in working on a greater scale I frequently adopted these proportions; but afterwards, on allowing a mixture of this kind to dry for a

year, we found that the air had spoiled it also, and that it required an additional proportion of coal dust, much greater than the first, to restore the mixture thus deteriorated; so that in this case, the coals mixed with the chalk and clay at two successive periods, together with the coals required in burning the mixture, more than doubled the expenditure of coals, that would have sufficed for the latter purpose, if the mixture of chalk and blue clay alone had been put into the fire soon after it was first made up, without allowing time for the air to spoil it.

(90) *Chalk and washed alluvial Clay.* It occurred to me at one time that the mixture of chalk and blue clay, if washed repeatedly by stirring it up with a great excess of water, and then allowing the mixture to subside to the bottom, and draining off the clear liquor from the top, might thereby be greatly improved and yield a better cement after calcination. This process however produced the same injurious effect upon the clay as exposure to air, and I believe in both cases, that it results not only from a change in the state of oxidation of the iron in the clay, as stated in one of the preceding articles (87), but also from the escape of the hydrogen gas contained in the clay being facilitated by its diffusion through a great quantity of water, during the process of washing, which causes the same quantity of fresh alluvial clay to become stale, much sooner than exposure to air alone would do (85).

(91) *Chalk and brown Pit Clay.*

By the term *Pit Clay,* in contradistinction to the *Alluvial Clay* of rivers or lakes, I mean clay dug out of the ground, not including the surface, which even when free from vegetation is never used for pottery or brick-making.

It was before mentioned, that our first efforts to make a water cement by a mixture of chalk with an excellent brown brick loam entirely failed (1), whilst the mixture of chalk and of blue alluvial clay succeeded (2). On endeavouring to analyze those earths, but without reference to the quantity of gas that might be contained in either, to which I did not turn my attention till long afterwards, I found that they both contained silica, alumina and the oxide of iron; and the only apparent difference, which I could at that time discover, was after repeated washings, when I found that the blue clay contained a very small portion of fine white sand, whereas the brown clay contained a considerable proportion of coarse gritty sand. I therefore pounded the brown clay in a mortar, until I made it as fine as the blue, after which, on mixing it with pulverized chalk, it formed a good cement. The same result occurred, on separating the coarse sand from the brown loam without pounding it. Thus an important rule in artificial cement-making came to light, that only the finest clays of nature are fit for that purpose, agreeably to the principle, that various substances are capable of chemical combination, when reduced to an impalpable powder, but in no other state.

(92) On experimenting afterwards with other brown or reddish brown clays of a sufficiently fine quality, from the high ground between Upnor and Hoo, they all formed good water cements, with chalk, in the usual proportion of 5 measures of the latter to two of the former, and these pit clays have the advantage of not losing their cement properties by exposure to air so soon as the alluvial clay, for whilst a month or six weeks will ruin the raw cement cubes made of the latter, it may require two or three times as long to ruin the former. Indeed the fine brown clays are so much less liable to injury from this cause, that I entertained great doubts whether the air affected them at all, until I found that the same mixture of chalk and of fine brown pit clay, from which I had made a considerable quantity of excellent cement, entirely failed seven months afterwards, on my burning some cubes of the same mixture, which I had preserved for this purpose. In one respect the brown clay cement mixtures are much inferior to the blue, inasmuch as the former burn into a much harder substance and require more labour in pulverizing them afterwards than the latter, which is a very great disadvantage both in a natural and in an artificial cement.

There being no blue pit clays in the neighbourhood of Chatham, though they abound in other parts of England, I have not tried any of them, but I have no doubt of their efficiency; and upon the whole it appears to me, that any fine clay, whether alluvial or not, will make an excellent water cement in combination with chalk, provided that it shall not have been subject to the free action of the atmosphere, either from its original position being at the surface, or from subsequent exposure.

(93) *Chalk and Tile Dust.* Tiles being usually made of the fine brown pit clay, which I found to be a good ingredient for an artificial cement, I tried various proportions of pounded chalk and pounded tile dust, mixed together with water and then burned in a crucible in the usual

manner, all of which failed with only one exception, being greasy to the touch and either going to pieces or not setting under water.

(94) *Chalk and Slate Dust.* As roofing slate according to the Mineralogists, is merely clay in a hardened state, both containing the same component parts,* I tried various proportions of chalk and of pulverized slate, treated as in the preceding Article, which formed an imperfect cement, that after being kept under water for some weeks, crumbled to pieces on being taken out.

(95) *Chalk and Fuller's Earth.* The same reason induced me to try the Fuller's earth of Reigate, which is reported also by the Mineralogists to consist of the same component parts as slate nearly.† But the mixture of chalk with this earth entirely failed.

Hence as slates, which are without plasticity, and tiles, which have lost that property in the fire, always failed, whilst the component parts of the same substances, namely silica, alumina and oxide of iron, as contained in the clays of nature always succeeded, in forming an artificial cement in combination with chalk; I inferred that plasticity is essential to an artificial cement, especially as the want of this property seemed also to be the chief cause of the failure of the Fuller's earth, which when made up into a form with water is not plastic, but splits in drying in the air, and when burned in the fire with every precaution, instead of forming a sound brick, breaks into a number of small fragments in the state of porous unsound slag or scoriæ. But on washing the Fuller's earth repeatedly, in order to get rid of the small portion of salt, also said to form a part of that substance, we found that it became more plastic, and in this state it frequently succeeded in forming a water cement, but inferior to that of the plastic clays.‡ The same cause may perhaps account for the superiority of pipe clay, which possesses great plasticity, over the mixture of pounded flints and prepared alumina, which had none.

* According to Kirwan it contains silica 38, alumina 26, magnesia 8, lime 4, and peroxide of iron 14 parts in 100. See Jameson's Mineralogy.

† According to Klaproth, it contains silica 53, alumina 10, magnesia 1·25, lime 0·5, muriate of soda 0·1, potassa a trace of, oxide of iron 9·75, and water 24 parts in 100. See also Jameson's Mineralogy.

‡ This experiment proves the utility of the common practice of passing the earths used in brick-making through the wash-mill. The blue clay of the Medway burns into a hard sound brick, but it is too fine or as brickmakers term it *too strong*, for the common purposes of brickmaking, for which they prefer a *milder*, that is a coarser clay.

THAT HARD LIME STONE, IF USED AS AN INGREDIENT FOR A WATER CEMENT, MUST BE BURNED TWICE, FIRST IN ITS NATURAL STATE, AND AFTERWARDS SLAKED AND MIXED WITH CLAY.

(96) The only very soft lime stones that I am acquainted with are chalk and a yellow magnesian lime stone, of which I procured a specimen from the Geological Society, and which appeared rather softer than chalk. In preparing these as ingredients for an artificial cement, the obvious mode of proceeding is to grind them into fine powder; but if we suppose it necessary to make an artificial cement in some country, where none but hard lime stones are to be procured, which I believe must have been the case in those districts of France where M. Vicat was employed, the grinding becomes such a troublesome and expensive process, that the simplest mode is first to burn those hard limestones and slake them, in order to reduce them to a more tractable form, and then to mix them with clay and burn them a second time, to obtain the artificial cement desired. This was the system adopted by M. Vicat, no doubt from necessity, and which I thought it desirable to enter into, as a part of my own experiments, although I should certainly not recommend it in this country, in which good water limes are too common, to render it worth while to make artificial ones, by the process indicated in Article 83, and too valuable to use them as ingredients for an artificial water-cement, which may be obtained equally good and much more economically by the simpler method explained in the same Article.

In order to ascertain the best mode of forming an artificial cement, with any hard stone consisting of pure carbonate of lime, such as white marble, and clay, we used the quicklime of chalk which is the same as the quicklime of white marble; but as chalk broken into irregular lumps could not be measured with any degree of accuracy, we had recourse to weight, for determining the proper proportions of the lime and clay in the peculiar experiments now under consideration.

After due investigation, we found that any given weight of well burned chalk lime, and consequently of any other pure quicklime fresh from the kiln, combined with twice its own weight of blue clay fresh from the river will form an excellent water cement, observing however, that the quicklime after being weighed must be slaked with excess of water into a thinnish paste, and allowed to remain in that state about 24 hours, before it is mixed with the clay. These proportions by weight are as nearly as possible equivalent to a mixture of 5 measures of chalk powder to 2¼ measures of blue clay, absolute identity of proportion not being necessary, as was explained in Article 83, and it being always desirable, to avoid fractions, whether measure or weight be used for proportioning the ingredients of an artificial cement.*

(97) *The slaked Water Limes, or slaked Hydraulic Limes, and Clay.*

To make a water cement, by a mixture of the slaked water-limes and clay, the same process must be followed, by burning the lime first, weighing it in the state of quicklime fresh from the kiln, slaking it with excess of water for 24 hours, and then mixing it with blue clay fresh from the river.

But in respect to the proper proportional weights, as all the water limes contain a certain proportion of clay in their composition, whilst the pure limes contain none, a given weight of the former when burned into quick lime, must not be mixed with double its own weight of blue clay, which is proper in working with chalk or other pure lime, but with a smaller proportion.

Thus for example we found that any given weight of the blue lias quicklime, when mixed with an equal weight of blue clay, the former fresh from the kiln, the latter fresh from the river made a good water cement.

To produce the same effect, with other hydraulic limes, each of these respectively must be mixed with a different proportion of clay, according to its own peculiar quality. Thus for example, a given weight of Dorking lime fresh from the kiln would require more than its own weight, but less than twice its own weight, of blue clay, to make a good water cement, this lime being in an intermediate state between the blue lias lime, to which the former, and chalk lime, to which the latter of those proportions is suitable.

(98) *Lime and Clay burned separately.* Having burned chalk lime and clay separately, and mixed them together into balls with water, after previously pounding them, I never could succeed in obtaining a good artificial cement by this method, as the balls always went to pieces on immediate immersion.

* 555 grains of quicklime are the produce of 1000 grains of dry chalk which in the state of powder will fill 5 measures, of which 1110 grains of blue clay, being double of the first named weight, will fill 2¼.

Yet as the expedient of mixing pounded tile dust with lime, has been used in the Mediterranean for the lining of cisterns, whenever puzzolana could not be procured, even since the time of the Romans, I have no doubt, that the mixture of slaked quicklime and of calcined clay, which has been adopted and strongly recommended by General Treussart in France may be very useful for many purposes in Hydraulic Architecture, although much inferior not only to the natural, but to most of the artificial cements produced in the course of my experiments.

(99) *Of the additional substances by which the best natural Water Cements may be spoiled.*

It having occurred to me, that it might be desirable to ascertain, what ingredients would injure or ruin the best natural cements, as it appeared probable that the same might produce a like unfavourable effect upon artificial mixtures used as substitutes for them, the following experiments were tried.

Sheppy Cement mixed with Charcoal or Linseed Oil.

Having pulverized the Sheppy cement and mixed it up into balls with a proportion of pounded charcoal or of linseed oil, these ingredients did it no injury. In fact as they were burned out during calcination, they only served as fewel to promote this process.

(100) But we found that oil or grease of any kind has a most injurious effect on cement powder, if mixed with it, not before, but after it has been burned; for they converted it into a pulpy mass, that would not set at all.

Having mixed the calcined Sheppy cement powder with salt water, and also immersed the balls thus formed in salt water soon afterwards, they appeared to set nearly as well as if fresh water had been used, but threw out a much greater efflorescence from the surface.

(101) *The Sheppy Cement and the Carbonate of Potassa.*
The Sheppy Cement and the Carbonate of Soda.
The Sheppy Cement and the Muriate of Soda.

Having pulverized the Sheppy cement stone, and mixed it successively with the above-mentioned alkaline salts in various proportions, and burned each mixture, we found that the cement could bear a proportion of nearly one third without being much deteriorated, but not more. Hence as salt is certainly unfavourable, I would never mix cement with salt water, except in case of absolute necessity.

(102) *Chalk, stale Alluvial Clay, and the Alkaline Salts.* Before I became aware of the great quantity of hydrogen gas contained in fresh alluvial clay, having as-

certained during my repeated washings of the blue clay before alluded to (90) that the clear water drawn off contained no iron, but was impregnated with alkaline matter, I was induced to mix well washed and consequently stale alluvial clay with chalk in the usual proportion, of 5 measures of the latter to 2 of the former, adding small proportions of the carbonates of soda and of potassa, and of the muriate of soda, all of which appeared at first to produce a most favourable effect, but eventually failed.

(103) *Chalk, Pipe Clay and the Alkaline Salts.* To pursue the same investigation, previously to the failure of the last mentioned mixtures, I added a proportion of the same alkaline salts to the usual cement mixture of 5 parts of chalk and 2 of pipe clay, and all of them yielded artificial cements that set immediately under water in a most promising manner; but in the course of a period, not exceeding from six to ten or twelve days, most of them cracked at the surface, and the inside of them proved unsound. Some few, especially those in which the salt was not more than one fiftieth part of the whole compound remained good, and proved in breaking them to be hard inside, but these mixtures are too precarious to be depended upon.

THAT SALT IS VERY PREJUDICIAL TO BRICK EARTH.

(104) A curious circumstance was observed in the experiments now alluded to. Our former mixtures of chalk and pipe clay had always burned into a kind of brick much harder than any of our other calcined compounds. The smallest addition of the muriate of soda to the same mixture completely deprived it of its plasticity, and caused it to burn into a soft friable mass, capable of being rubbed to pieces by the fingers.

This experiment and those relating to Fuller's earth recorded in Article 95, combine in proving, how very injurious a proportion of salt must be to the earth used in brickmaking.

(105) *The Sheppy Cement & the Carbonate of Magnesia.* The carbonate of magnesia being the only substance that did not injure the Sheppy cement, when mixed with it in a greater proportion than one third, and treated as in Articles 99 and 101, we increased the dose of magnesia gradually, and finding that an equal quantity of magnesia did not injure the cement, we tried two measures and afterwards three measures of magnesia to one of cement; and even this great excess of the former ingredient did not deprive the latter of its virtue as a water cement.

THAT MAGNESIA IS A WATER CEMENT.

(106) From this unexpected circumstance, I was induced to try magnesia alone, unmixed with any other substance, and having burned the carbonate of magnesia, and made the calcined powder into a ball with water, it gave out no perceptible heat, but at the end of twelve hours, when it had become hard, I put it into a basin of water and it continued to set in that fluid. On mixing Henry's calcined magnesia, which of course required no burning, with water, and forming it into balls the same effect took place. Thus the curious fact developed itself, that pure magnesia is in itself a water cement, but a slow setting one, for it did not succeed on immediate immersion.

THAT THE ARTIFICIAL CEMENTS ABOVE DESCRIBED ARE CAPABLE OF RESISTING THE SEVEREST FROSTS.

(107) In the very severe winter of 1829-30, having previously prepared a great number of the experimental cement balls described in the foregoing Articles, I exposed them on the top of a porch for about a fortnight in January 1830, and as a good deal of snow fell which would have protected them if left there unmoved, I took them down every morning, laid them before a fire until they were quite warm, immersed them in a basin of water for about half an hour and then replaced them on the porch, previously sweeping off the snow if any had fallen during the preceding period. All the artificial cement balls, which had been approved by us before, stood this severe test quite as well as the Harwich and Sheppy cement balls, which I also exposed in the same manner. In fact the frost did no injury to any of them.

I have however observed, that frost injures the best cement, whether natural or artificial, if laid on as stucco in an injudicious manner, which if the proper mode of applying it be understood any workman can effectually guard against. This point will be explained hereafter.

OF THE EXPERIMENTS ON A GREATER SCALE WITH KILN BURNED ARTIFICIAL CEMENT. THAT OUR FIRST ATTEMPT IN 1829 FAILED, OWING TO THE RAW MIXTURE HAVING BEEN SPOILED BY WASHING IT.

(108) Having towards the close of the year 1828, and in the beginning of 1829, tried as many experiments on a small scale, as I then considered necessary, I determined to prepare a considerable quantity of artificial cement

composed of chalk and blue clay, with a view of applying it on a larger scale, to those purposes for which the natural cements have been used in Architecture.

The chalk, after having been broken small and dried in the air, was pounded in small quantities at a time, in iron troughs that had belonged to a forge, by iron rammers made for the purpose, and was passed through sieves with brass wires, having 25 meshes to the inch, being the finest used in the Ordnance gunpowder works. A large mass of dry pulverized chalk being thus provided, 5 cubic feet of it were laid on a wooden platform, and made into a paste, with a moderate quantity of water, after which 2 cubic feet of the blue clay were added, and the whole intimately mixed together on the same platform by shovels. When a sufficient quantity was prepared, the mixture was next moulded in the same manner as common bricks, excepting that water was used instead of fine sand to prevent adhesion. After these bricks of raw cement, which were twelve inches long became drier, they were cut into five equal parts, each forming a cube of rather less than $2\frac{1}{2}$ inches side, this being the average size of the lumps into which chalk is usually broken, before it is burned, in the common open lime kilns in Kent.*

(109) Having prepared a considerable quantity of these raw cement cubes, on burning a few of them in a fire place before I sent them to the kiln, the calcined cement seemed rather inferior to that obtained from my former experiments, and on examining the remainder of it, it appeared that the workmen had not taken sufficient pains in making and mixing it, the cubes being coarse grained and porous, and having the dark colour of the inside mottled with specks of white chalk. I therefore threw the whole of this mixture away, and to remedy these imperfections, I determined to adopt the process of washing, which is known to break down all clays, and which is therefore always used in the preparation of mixed brick earths previously to moulding. I caused the blue clay to be squeezed through a coarse sieve, to remove pieces of vegetable matter, with which it was occasionally adulterated, and ordered the chalk after having been pounded and sifted as before, to be ground still

* I made my moulds exactly 12 inches long, and $2\frac{4}{10}$ inches wide by $2\frac{4}{10}$ inches deep, in order that 25 bricks or 125 cubes should be exactly equal to one cubic foot. Thus by merely counting the number of bricks, we could ascertain the quantity of raw cement made, without the trouble of measuring it.

finer in an iron mill that had been used for grinding linseed. After the ingredients thus prepared had been well mixed in the due proportion, the compound was put into a large tub, with about twice its bulk of water and the whole was well stirred, after which the mixture was allowed to subside and the clear water from the top was drawn off by a syphon. More water was then added, and the same process repeated for about ten days. After the water was last drawn off, care was taken to remix the ingredients well, lest the heavier parts of the original mixture from having been so often held in suspension and deposited again as a sediment, should have got to the bottom. The mixture was then moulded and cut into small cubes, in the manner before described, which being well beaten with little wooden bats, whilst moist, were thereby reduced to somewhat less than their original size. These cubes of raw cement, when dry, were extremely sound and of homogeneous colour and texture, so that I expected that they would now make a perfect cement, instead of which, I was surprised and disappointed to find, on again burning a few for trial, before I sent the whole quantity to the kiln, that the calcined cement obtained from them would not harden at all under water, but remained in a soft pulpy state altogether unfit for the purpose in view.

In consequence of this circumstance, which was before mentioned in Article 90, I gave up all further proceedings for that year, it being rather late in the season, and renewed my crucible experiments in hopes of discovering the cause of this unexpected failure, together with the means of avoiding it for the future, which occupied me for the remainder of that year and part of the next; but having already recorded the whole of my experiments on a small scale whether tried before or after the period alluded to, it is unnecessary to repeat them here.

SECOND FAILURE IN 1830, OWING TO THE RAW CEMENT MIXTURE HAVING BEEN SPOILED BY TOO LONG EXPOSURE TO AIR.

(110) About the beginning of March, 1830, having finished the preliminary experiments alluded to, I reverted to my original intention of preparing a sufficient quantity of artificial cement to operate with on a large scale, and as an addition of coal dust had in some of my crucible experiments (87) cured the raw cement mixtures that

had been spoiled by excess of washing, it appeared reasonable to suppose, that this ingredient introduced into the original mixture might prevent that evil. I therefore prepared a considerable quantity of raw cement, composed of a mixture of 5 measures of chalk, 2 measures of blue clay, and half a measure of fine coal dust. The dust was obtained by pounding small coals in the iron trough, and was mixed with the pounded chalk dry, and the two were ground together in the oil mill before mentioned. They were then mixed up with water on the wooden platform, and lastly the clay was added, which had only been squeezed through a sieve without washing, and the whole of the ingredients were intimately mixed by the shovel.

Some of the cubes thus prepared, being burned for trial in a common fire place, yielded a very good cement; but instead of sending the remainder of my raw cement immediately to the kiln, I thought it expedient first to ascertain by experiment, how long the calcined cement would keep if exposed to the air of a dry apartment, before it lost its virtue, and whether it would keep better in calcined cubes or in powder. In experimenting upon this subject, I made up successive portions of my calcined cement powder into balls with water once a week, and found that as it gradually became staler, these balls were less perfect at each trial: but they were not so much injured during the first two or three weeks as to be unfit for use. Afterwards they deteriorated more rapidly, till they became at last completely unserviceable, and would not set under water at all.* The same calcined mixture remained much longer perfect in the form of cubes, but there are strong objections to this mode of preserving cement (49).

Two months elapsed before this experiment, which I did not wish to hurry, was brought to a conclusion, after which, reserving only a few more cubes, I sent the remainder of the raw cement, made at the commencement of this period and amounting to more than 20 cubic feet, to a lime kiln in Chatham, where a very intelligent foreman, after spoiling a few of the cubes put in for trial, burned the remainder of it remarkably well, by using a

* The calcined cement powder experimented upon was kept in a shallow dish. Cement powder kept in an air-tight cask or other deep vessel open at top, would be much longer in spoiling, because the upper surface for a couple of inches down, though injured itself, would protect the remainder of the mass below from the action of the air.

smaller proportion of coals, than he had been in the habit of using for burning chalk. The fewel and the cement cubes were laid according to custom in alternate layers, in this kiln, which was of the common form of an inverted cone open at top: but as the quantity sent by me was not sufficient to fill the whole kiln, the cement cubes were placed in the center of it, with chalk above and below them.

(111) Having satisfied myself that the cement thus made was burned to a nicety, I caused it to be pounded in the iron trough, and sifted with the fine sieve before described (108), and having put it into air tight casks for use, I was surprised and disappointed, on trying it in the usual manner, to find that the experimental balls made with it, fell to pieces under water, either immediately, or at the end of a few days, after hardening at the surface. On burning the few cubes of the same raw cement, that still remained, in a common fire place, in order to ascertain whether the burning in the lime kiln might not be prejudicial, the same effect followed. Hence it became evident, that the mixture of chalk and blue clay, which made an excellent artificial cement, when burned immediately, became unfit for that purpose, if exposed to air for two months previously to this process. Subsequent observations and experiments on a small scale, the results of which have already been recorded, threw more light upon the subject, and proved that the length of period necessary for changing this mixture from good to bad, depended upon this size of the lumps into which it is made, very small ones being ruined in a day or two, and large ones preserving their cement properties much longer, in proportion to their magnitude, as explained in Article 86.

Having thus spoiled my first kiln-burned artificial cement, I prepared about 25 cubic feet more of the same mixture, which I sent to the same lime kiln at Chatham, ten days after it had been mixed. Unfortunately the proprietor of the kiln had in the mean time engaged a new foreman, who obstinately persisted in applying the fewel to the raw cement cubes, in the same way that he had been accustomed to, in the burning of common chalk lime, in consequence of which all the artificial cement put into his hands was completely vitrified, so that we were obliged to throw it away.

FIRST SATISFACTORY EXPERIMENTS WITH KILN-BURNED ARTIFICIAL CEMENT IN 1830. CONTINUATION OF SIMILAR EXPERIMENTS, WITH EQUAL SUCCESS, IN 1831 AND 1832. AVERAGE TIMES OF PREPARING AND BURNING OUR MIXTURES, &c.

(112) I therefore found it necessary to take the whole process into my own hands, by employing Private James Menzies, an intelligent soldier, who had been my assistant in all the preceding experiments, to manage in future not only the mixing and moulding, but also the burning of my artificial cement; and Mr. Nash a coal merchant of Gillingham in this neighbourhood, who had himself tried many experiments on cements, but on a different principle from mine, obligingly allowed me the use of a small lime kiln about 4 feet in diameter at top and 6 feet deep, which he had chiefly built for experimental purposes. In this little kiln, than which nothing could have answered better, we burned, at four successive periods of the same year, about 140 cubic feet of raw cement. In the first of these batches of artificial cement, we used the same mixture as before of 5 measures of chalk, 2 measures of blue clay and half a measure of coal dust; and in burning it, after putting in shavings and wood at the bottom of the kiln, we laid half a bushel of coals over the wood, then four bushels of the raw cubes, after which another layer of half a bushel of coals, then four bushels of cubes as before, and thus we continued applying the coals and cement cubes in alternate layers, until the kiln was filled, using one measure of coals to eight measures of raw cement, the former being broken rather small, so that no piece of coal used exceeded an inch in thickness; and both being thrown loosely into the baskets with which we measured them.

The cement thus obtained was certainly underburned, as it effervesced in acids more than was desirable, but finding that it hardened under water, I determined to use it, although I considered it from the above circumstance to be of an inferior quality. I therefore built a small brick tank in the garden attached to the Field Officer's quarter occupied by me in Brompton Barracks, Chatham, and plastered the inside of it, with the cement, that has just been described, which was applied pure about three quarters of an inch thick, and was well worked in, and compressed, and brought to a very smooth surface by the trowel. The tank was then covered over by a flat brick arch laid in cement, with a man hole in the center; and a small brick drain was formed to convey the rain water to it from a roof, and another to carry off the waste water, if it should rise nearly to the spring of the arch, which drains also were laid in the same cement. In the rough brickwork of the body of the tank, instead of common lime mortar, I used the inferior cement previously burned in the kiln at Chatham (111) mixed with sand, and which although quite unfit for the more delicate purposes of a water cement, seemed to be in the same state nearly as the blue Lias or Dorking limes, and therefore superior to common lime for all hydraulic purposes. A fortnight after the plastering of the tank was finished, I caused it to be filled with water, which was done nearly six years ago, and it still remains perfectly water-tight.

(113) In the second batch of cement, burned at Gillingham, we used the same mixture of chalk, blue clay and coal dust, as before; but increased the proportion of coals or fewel, by using one measure of coals to seven only of the raw cement cubes. Afterwards we found, that about one of the former to seven and a half of the latter was sufficient for this peculiar mixture.

In the third and fourth batches of raw cement prepared for burning at the same kiln, we dispensed with the coal dust altogether, using 5 measures of chalk to 2 of blue clay; and we merely pounded and sifted the chalk, without grinding the powder afterwards in the mill; and in consequence of there being no fewel combined with the raw cement in this mixture, we used one measure of coals to five measures of the raw cement cubes, in burning them, which proportion we always adhered to afterwards, as the best for this mixture; but if we could have kept the little kiln constantly burning, as is often the case in lime works, by adding more fewel and cubes at top from time to time, in proportion to the quantities drawn from below, a smaller proportion of coals than one fifth would have sufficed, after the first portion of calcined cement should have been drawn; and it may be observed also, that whether kept constantly burning or not, a larger kiln would probably have required less fewel under the like circumstances.

The two last mentioned batches of artificial cement proved to be the best of any of the kiln burned cements, that we had yet made; and the previous process of mixing was much simpler and cheaper, than when coal dust was used; notwithstanding which the results of some crucible experiments tried the same year, induced me afterwards to return to the use of coal dust again, which I did not finally abandon, until more conclusive evidence convinced me that it was not advisable to use it at all.* (See Article 89).

(114) In 1831 we burned seven batches, and in 1832 five batches more, averaging about 35 cubic feet of artificial cement each, the whole of which has been applied to practical purposes. The mode of proceeding finally adopted was this. In preparing our artificial cement for the kiln, we had the chalk powder all ready for mixing, before we brought up the clay from the river, in order that the latter might be used quite fresh. Generally from 10 to 12 and never exceeding 14 days were allowed to elapse, between the moulding and the burning of the artificial cement, for the purpose of drying the cubes,† which were not injured by this delay, though from the principle before explained, if they had been made much smaller, as for example only one inch each side, they would have been completely spoiled in a much shorter period (86).

After lighting the kiln, we generally allowed 24 hours to elapse, before it was drawn, or sometimes more, in order to give our cubes time to cool, for although they were sufficiently burned in a shorter time, we found that they were liable to break or crumble to pieces when handled, if taken out of the kiln warm.

(115) It is proper to observe that one of the batches of artificial cement burned in 1831 proved defective, inasmuch as it would not set under water, and when afterwards used as stucco, it cracked much more than any of our former specimens had done. This was the first batch of artificial cement made without the presence of my intelligent assistant James Menzies, he having been ill at the time; which circumstance was very discouraging to me, as success seemed to depend upon one individual; for the men who made it, declared that they had followed the same proportions and mode of proceeding, that he had done, in every respect. Afterwards on interrogating them more strictly,

* As I had only three or four soldiers employed, in preparing my artificial cement, who could not work constantly, owing to the military duties of which they had to take their share, it required about a month to prepare each portion of about 35 cubic feet for the kiln. Hence we were only able to prepare six batches of artificial cement, including the two burned in Chatham, in 1830, before the early frosts of that year set in.

† We dried one batch of our raw cement in a couple of days by laying the cubes on iron plates over a fire.

the cause of failure was discovered. The man who served out the pulverized chalk had compressed it in the measure as closely as he could, so that he had forced a much greater quantity into the same space than had ever been used before: and thus as the blue clay in its moist state is incompressible, the true proportion used in this mixture was $7\frac{1}{2}$ measures of chalk powder to 2 of blue clay, which is proper for a water lime, but not for a water cement (83), and to which if we had known it at the time, we would have added more sand as being a lime stucco (56).

(116) In the first batch of cement made in 1832, we used a fine brown pit clay from the vicinity of Upnor, which had previously been employed by Mr. Frost, in his artificial cement works. As this was much drier and more compact than the blue alluvial clay, we mixed it with water, till it was brought to the same moist state as the latter, after which we mixed it with chalk in the same proportion. This also made a good artificial cement, but had the disadvantage before alluded to of being a great deal harder than the blue clay mixture after calcination (92), and the stucco made from it also cracked more than that formed of the latter. Hence we did not use the brown clay a second time, it being evidently inferior to the blue.

Besides finishing the garden tank before mentioned, and stuccoing and coping a dwarf brick wall by which it was inclosed, the remainder of our kiln burned artificial cement of 1830, and the whole of that burned in 1831 and 1832, were applied to practical purposes at or near Brompton Barracks, as was before mentioned (3).

(117) Finding at the end of the first year, that the stucco applied to the gateways, though generally of a handsome light brown colour, was not of the same tint throughout, some parts having become darker than others, we washed over the whole of our stucco with fluid cement, in order to obtain a more uniform colour, which expedient we adopted only from economy, as we considered oil colours if not applied too soon, to be decidedly preferable for all such stuccos, whether made of natural or of artificial cement.

REMARKS ON THE KILN-BURNED ARTIFICIAL CEMENT, USED FOR THE PURPOSES OF PRACTICAL ARCHITECTURE IN 1830, AND THE TWO FOLLOWING YEARS.

(118) From four to six years having now elapsed, since our artificial cement was applied to all the purposes of Practical Architecture on a sufficiently great scale, we are

now enabled to judge of its quality, which appears to rival that of the best cements of nature. The tank first made in the garden at Brompton Barracks has remained perfectly water-tight as was before mentioned, and that near the married soldiers' cottages is in the same state, but the mode of filling the latter has proved unsatisfactory.* In plastering both of these tanks, pure cement only was used; and wherever we formed cornices or other ornamental mouldings, or rusticated joints, pure cement was also used at the outer surface, to the depth of a quarter of an inch or sometimes more; because we found that cement mixed with sand was not capable of producing such *sharp arrises* technically speaking, or in other words such well defined angles, as we desired.

Most of our cement chimney pots were also composed of pure cement at the surface, for although formed of small fragments of tiles and cement powder mixed with water, well stirred up together and poured into the mould, we found that the pieces of tile usually got towards the center, being seldom within a quarter of an inch of any of the sides of the mould; whence the fluid cement, though it effectually pervaded the whole space, was in greater mass at the sides than any where else. Our cement copings, likewise formed in a mould, being made of pure cement and fragments of tiles, not always cast fluid, but sometimes worked together by the trowel and by the hands, had also the top and sides of each piece of coping formed of pure cement, to the depth or thickness of about half an inch from the surface, or sometimes more. Those artificial coping stones were all set in pure cement.

* The rain water from the adjacent road first flowed into a deep sesspool, from whence it passed over by a small covered drain, and fell into a filtering bed of gravel and fine sand, from which it ascended to a second drain of the same description, and then fell into the tank or reservoir. Owing to the exposed state of the road, this water was so very dirty that it soon choaked up the filtering bed. This arrangement was copied from the great military tank constructed by the late Major-General Forde, R.E. on the Western heights Dover, where it answers admirably, from being on a greater scale and supplied with cleaner water from those heights. Having since examined the extensive filtering apparatus of the West London Water Works near Chelsea, constructed by Mr. Simpson the Engineer of that Company, in which the water of the Thames, which after rains is exceedingly dirty, is filtered through a bed of sand and gravel by descent, falling from thence into apertures in small brick arches, from whence it issues remarkably clear, I am of opinion that the same arrangement would have succeeded with our small tank, as filtering by descent seems to be better suited for muddy water, than by ascent. Even at Chelsea, the several filtering beds require to be cleaned from time to time by scraping off the upper surface of the fine sand, as soon as it becomes covered by and impregnated with a coat of mud, after the removal of which fresh sand is added.

(119) In respect to our artificial cement stucco as applied to plane surfaces, having little experience in the use of cement at that time, I thought it desirable to try several sorts; and accordingly in some of the gateways, I used two measures of cement powder to one of sand; in others equal measures of each; in others I directed one measure and a half of sand, to be mixed with one of cement; and in others two or three measures of sand to one of cement. The sand was all of good quality and sharp, but in some gateways we used it all very fine, in others all coarse, and in others coarse below and finer at the surface. In some few cases we laid on the stucco in two coats, sometimes scoring the surface of the first coat, and sometimes not; but generally we adopted what we found afterwards to be the most judicious system, of applying the stucco in one coat, not absolutely laid on all at once, but taking great care, after having applied a certain thickness over the brickwork, that the second or outer portion, whether of the same or of a finer quality, should be laid on before the first had begun to set.

(120) Having recently cut out pieces of the cement copings before mentioned (118), and pieces of stucco, $2\frac{1}{2}$ inches square, from each of the gateways down to the brick work, we found that the pure cement from the copings, in some parts nearly an inch in thickness, was exceedingly hard and sound. In respect to the stuccos of the gateways, varying from about 5 eighths to one inch in thickness, we found that those, which had been formed of two measures of cement to one measure of fine sharp sand, were very good, being every where of the same hardness and compactness nearly as Bath stone; and a portion of stucco formed with equal measures of cement and fine sand was also good and sound, yet not quite equal to the former; but those stuccos, in which the proportion of fine sand by measure exceeded the cement, were hard at the surface only, all below being softer and weaker, where the fine sand was in moderate excess, and nearly as dry as dust, where it was in great excess.* But the best stucco of all, was that which I know to have been formed of equal measures of coarse sand and of cement; and indeed all the stucco in which coarse sand was used proved to be good; but not having been present

* All stuccos, whether formed of a mixture of lime, or of cement, with sand and water, being worked up hard at the surface, it is impossible to judge of their quality, without cutting solid pieces entirely out in this manner, so as to ascertain, whether they are equally hard throughout their whole thickness.

myself at the stuccoing of those gateways, where I had directed coarse sand to be used, in the proportion of two or three times the quantity of the cement powder with which it was to be mixed, I have great doubts whether the workmen did not adhere to the usual practice of plasterers, and use much less of the former ingredient than I had ordered; for I found in most of those men the greatest reluctance to deviate from their former habits, and in cutting into the portions of stucco now alluded to, the coarse sand did not appear any where to exceed the cement by more than one half; and in one part in particular, where I had expressly ordered coarse sand to be used, we found nothing but very fine sand.

(121) In respect however to two gateways, where I had myself superintended the work in person, and for which I can therefore vouch, we found that the stucco of one, composed of 2 measures of cement to one of fine sand, which had been worked in and compressed with great care by one of the plasterers at my desire, was much better than the same kind of stucco, which had been applied by another plasterer, whom I allowed to work in the same hasty manner that he had previously been accustomed to, before he entered the Corps. The former stucco was of equal hardness and compactness throughout, and showed no cracks for nearly a year after it was applied, and on examining these cracks lately at the end of five years, they are merely hair cracks, not extending more than the 50th part of an inch, and I have even doubts whether they penetrate much deeper than the cement wash used for colouring the stucco (117).*

Some very small portions of our artificial cement were injured by frost, owing to our inexperience in the mode of applying it, but we soon discovered the peculiar circumstances, under which this takes place, which are easily guarded against, and which I shall now proceed to explain.

* The gateway at the N. W. angle of the Barrack Square leading from the Officers' quarters to the North Stable yard, and the outer gateway of that yard, were stuccod with 2 measures of cement powder mixed with 1 measure of fine sand. The gateway at the N. E. angle of the Barrack Square was stuccod with equal measures of cement and of coarse sand. The Eastern wing wall outside of the Principal Gate leading into Brompton at the S. W. angle of the Square, was stuccod with a mixture of very fine sand and cement, the sand being in excess. The convex back wall of the entrance leading from that gate to the South stable yard was stuccod with $1\frac{1}{2}$ measure of sand to 1 measure of the inferior cement mentioned in Articles 110 & 111. These two last mentioned are the only bad or indifferent stuccos out of the whole. All the others are good and some excellent. The artificial cement made of the brown pit clay from Upnor was used in stuccoing the rustic piers of the garden gate behind the center of the Officers' quarters, but not for the ornamental cercophagi over those piers, which were made of the blue alluvial clay, and which are the finest specimens of our artificial cements.

RULES FOR PREVENTING CEMENT IN OUTSIDE WORK FROM DETACHING ITSELF, ESPECIALLY AFTER FROSTS.

(122) New brickwork intended to be coated with cement stucco is left rough on purpose, but when old brickwork is to be stuccod, the mortar must be raked out of the joints for about an inch deep, and roughened, and the bricks scored by a light axe, and the surface of the wall should be perfectly clean (100), for which reason if blackened by city-smoke, or otherwise greasy, it should be scraped as well as scored, and well moistened or rather wetted with water, before the stucco is applied. It is also desirable that more labour should be given to this process, than is usual in England, where the plasterer instead of compressing the whole of the stucco, lays on the body of it in haste and very loosely, and then waits till it is on the point of setting, after which he works up the surface only, with a smaller proportion of finer sand, and thus produces a hard but thin crust over a coarse porous substance.

(123) The only precaution necessary to prevent cement whether pure or mixed with sand, from being injured by frost, is never to allow any part of it to set completely before a second coat is applied, for in this case the joint between two such coats is always a weak part, and if water penetrates into it, will be sure to give way if attacked soon after by frost. In stucco applied to the face of a wall, this defect may not come into notice, because the rain generally falls off without penetrating quite so deep as the vertical joint between the outer and inner coat, and at top such joints are always protected by a projecting cornice or coping, as was the case in our gateways. Hence we did not discover the weakness of any of those joints, until we cut into them at the end of five or six years, as was before mentioned (120), when the two coats immediately separated as soon as the chisel reached the inner one, showing that they had not had any effectual adhesion; but in respect to part of our cement coping, not previously made in pieces in a mould, but formed by the plasterer on the wall itself with fragments of brick, and with cement stucco laid on in several coats, the weakness of the joints between those successive coats became apparent in a few weeks after the work was executed, which was rather late in the year; for the rain water having penetrated into those joints which were near the surface, and freezing afterwards, broke off the upper coats, which were very thin, one after another.

The same effect occurred in respect to a small ornamental column, from the base of which a torus or semicircular moulding detached itself, and also in respect to the first gateway of the Barracks stuccod by us, from which part of an ornamental fillet separated, after the first hard frosts of the same season. The torus and fillet in both cases, projected from the surface of the first coat of stucco, and the joints connecting them to it were nearly vertical and exposed at top, so that the water naturally descended into those joints where it lodged itself, and when frozen broke off those little ornaments, but only partially. In short, though several severe winters have occurred since, no injury has been done by the frost to any of our artificial cement, except under those circumstances of its being applied in two coats, with the joint injudiciously formed and exposed. In those cases, we observed that the corresponding surfaces of the first coat, which was still adhering to the wall, and of the second coat, which had detached itself, were perfect, for the plasterer had in all of them scored the former well, expecting that this precaution would make the second adhere to it, and the marks upon the two surfaces which had been in contact, concave on the one and convex on the other, corresponded as exactly, as the coin does to the die, by which it has been stamped. Thus cement sets too quickly to admit of its being laid on in two coats, like common plasterer's work of lime mortar. If applied in two thicknesses, the second must be laid on almost immediately, and long before the first has begun to set, so that properly speaking these two thicknesses form only one coat.

(124) In respect to our cement chimney pots cast in moulds as before-mentioned, and to such of our artificial coping stones as they may be termed, which were formed in moulds, no part of the cement used in either having been allowed to set partially, and consequently there being no internal joint, or weak part, they have all stood the severest winters to which they have since been exposed without the smallest injury.*

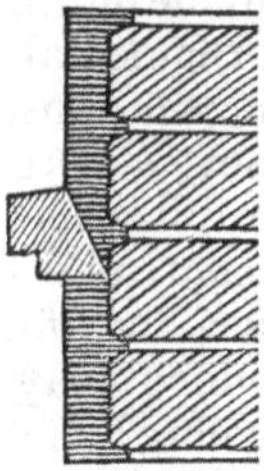

(125) The fillet before-mentioned, which partially detached itself had been added to the surface of our cement stucco, after previously cutting a triangular groove to receive it, which having the base of the triangle at bottom, as shown in the first of the annexed

* Some of the cement chimney pots have since been removed, on account of the aperture which was only 7 inches square in the clear at top, being rather too small for the fire places, especially those used as kitchens. We were not aware, when we made our mould for casting them, that this size would prove objectionable. Nine inches would have been sufficient.

figures, allowed the water to descend into the joint. If this triangular section had been reversed by having the base uppermost, this effect could scarcely have taken place.

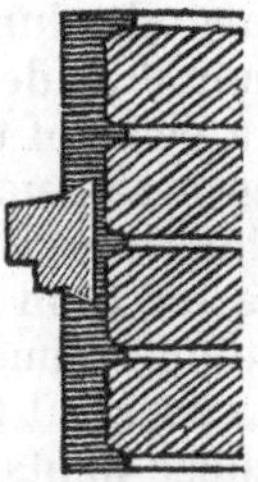

To remedy it I caused the whole of this fillet to be cut off, and the triangular groove enlarged until it assumed a dove-tailed form, as shown in the second figure annexed, thus obtaining what plasterers technically term *a key*, which has the same power as a dovetail to prevent separation.

(126) On inquiry, the best plasterers have since assured me, that the joint between two coats of cement is always weak, and that the same little failures which I experienced would have taken place with the best natural cements, and therefore they always avoid using two coats of cement work if possible, but should that be absolutely necessary, they take extraordinary precautions to obtain that sort of key, which will prevent the second from separating from the first.

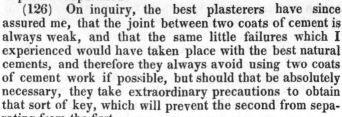

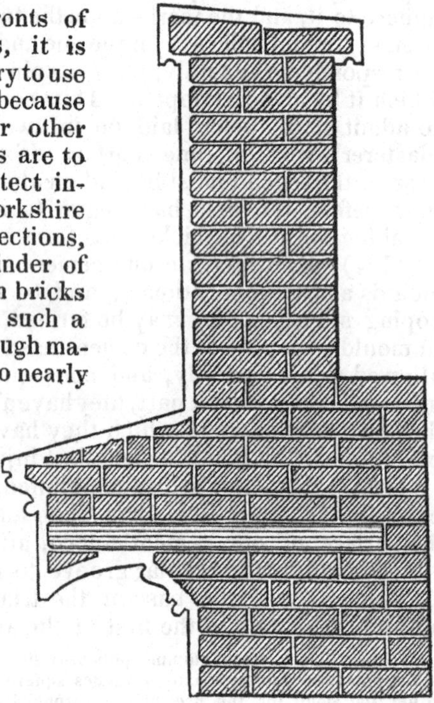

In the cement fronts of new brick buildings, it is scarcely ever necessary to use two coats of cement, because wherever cornices or other ornamental mouldings are to be formed, the architect introduces pieces of Yorkshire stone for all bold projections, making up the remainder of those projections with bricks cut if necessary, in such a manner, that these rough materials approximate so nearly to the outline of the ornamental mouldings proposed, that only a thin external application of cement is necessary for the finishing, which may every where be laid on in one single coat, according to rule. The annexed figure shows the facility with which this object may be effected, being the section of a cornice and parapet, such as might be used for a first-rate house in London. One course of Yorkshire stone, having small pieces of cut bricks supposed to be attached to its lower surface by pure cement, supports the whole projecting part of the cornice, the rest of which is formed of common bricks, whilst the coping of the parapet above is formed partly of bricks on edge and partly of bricks laid flat or in their usual position. The joints of the brickwork are shown, whilst the coating of cement afterwards to be applied, nowhere much exceeding an inch in thickness, has its outline also distinctly defined, and is left unshaded

(127) Where woodwork is used to obtain such projections, which is not so good an arrangement as the former, but often practised in common dwelling houses as being cheaper, it is proper to fix battens to the surface of the wood, and over them laths, in order to obtain a proper key for the cement.

(128) In forming circular projections of cement of moderate relief over brickwork, it is usual to drive large nails, and to connect them by pieces of spun yarn or rope yarn passing round and between them in zigzags, the whole of which being covered by and imbedded in the cement afterwards applied, prevents it from detaching itself. In this manner we prepared for the semicircular architraves over the doors of our brick sentry boxes near the principal gate of Brompton Barracks, driving long nails into the joints of the brickwork before it was stuccod, which having spun yarn passed round them afterwards, in the manner described, have effectually secured those mouldings, which project further than the rest of the stucco, and were laid on as a second coat.

(129) Circular cement mouldings, not attached to a plane surface, such as the bases and capitals of columns, are often cast in moulds, in two or more pieces, and fixed by pure cement, round the rough shaft previously formed of brickwork. When we consider the prodigious strength given to projecting cement work by Mr. Brunel's ingenious arrangement of hoop iron, before described (59), there could be no difficulty in forming the boldest projections with bricks and cement alone, aided by iron, supposing that Yorkshire stones, or other stones of a similar description, or rough slates of inferior value, suitable for this purpose, could not be procured. For it is well

known that a row of single bricks may be made to stand out from the plane surface of a vertical wall by cement alone, and although such rows of bricks are so far weak, that if you go on increasing their number and consequently the length of the projection, beyond a certain limit, they will break off near the wall by their own weight, yet the length to which they may be carried out without this risk, far exceeds that of the Architectural Projections usual for the largest Edifices. Hence as I said before, the aid of iron would render projecting bricks laid in pure cement perfectly safe and efficient, that is under the rather improbable hypothesis, also before stated, of a deficiency of more appropriate materials.

(130) Moreover bricks connected by pure cement may previously be formed into masses, or artificial stones, in order to save time in placing them in tide-work or in drains; and for the latter purpose, they may have the arch-like form, which expedient has recently been used in Chatham Dock-yard; but as the various modes of using cements in general have now been noticed, as far as appeared necessary, I shall enlarge no further upon this subject, but return to our own proceedings.

EXPERIMENTS PROVING THE STRENGTH OF OUR ARTIFICIAL KILN-BURNED CEMENT.

(131) The Master-General and Board of Ordnance having directed the arched roofs of some of the casemates in Chatham Lines to be plastered with Harwich cement, in 1831, Captain Streatfield the Engineer Officer in charge of that work, borrowed some of our artificial cement from curiosity, in order to compare it with the Harwich cement about to be used there, which he considered to be of good quality. In trying the strength of each, a brick placed horizontally in the direction of its length, was attached by means of pure cement applied to one side of it, to the vertical surface of the wall of one of the casemates, the cement being well compressed in applying the brick, which was steadied for about 2 or 3 minutes by holding the point of the trowel firmly against it. The bricklayer then left it, until he had mixed up as much pure cement with water as was necessary for the next joint, after which he applied another brick; and the cement set so quickly that the time required for placing each brick did not average more than 6 or 7 minutes; and thus he proceeded, adding one brick to another until the mass gave way. Not being pre-

sent myself at all these experiments, I was informed by the Foreman of Bricklayers that he considered my Artificial Cement to be rather stronger than the Harwich Cement, with which it had been thus compared; for both of them had supported ten or eleven bricks before the joints gave way, but that the artificial cement had supported the greater number more frequently. I was myself present at three experiments only. In the first, my artificial cement supported eleven bricks, but the joints gave way on the application of the twelfth. In the second, eight bricks were attached to the wall as before, after which 24 hours were allowed to elapse, and then four more were added, which composed a solid mass weighing 73¼ lbs, projecting 2 feet 11¼ inches from the wall and nearly 9 inches in width, and 4½ inches in depth, as shown in the upper projecting row of bricks in the next figure, which is marked as *standing out flat*. On adding a thirteenth brick the mass broke down.

In the third experiment, eight bricks were attached to the wall one after another in the expeditious manner that has been described, after which one brick more was added daily, under an expectation that this system might render it possible to attach a greater number than before. This mass also broke down like the former, on the addition of a thirteenth brick. But none of these masses ever broke into more than two or three parts, and in the last experiment a portion of brick belonging to the original wall was torn out, instead of the cement joint given way.

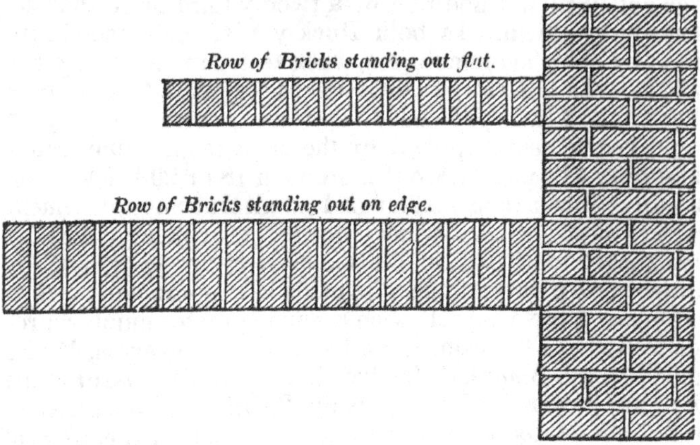

(132) I was not aware at that time, that the more usual method, in trying the strength of cement by attaching

bricks to the upright surface of a wall, was to place them vertically in the direction of their length, forming a mass nearly 4½ inches in width and 9 inches in depth as represented in the lower projecting row of bricks in the same figure, marked as *standing out on edge*. But supposing an equal number of bricks to be cemented together in both, in which case the quantity of solid matter projecting from the wall is equal, the second form having double the depth must be capable of supporting a much greater weight than the former: on the same principle as a piece of four inch plank, if placed flat and used as a beam by fixing it at one or both ends, will bend and break under a weight which it could support with perfect safety, if placed on edge.

The Master Bricklayer in Chatham Dock-yard informs me, that he attached 14 bricks to a wall in this second manner, by cement which he considered to be of good quality, in which experiment he allowed a quarter of an hour for the placing of each brick, and the setting of the cement joint, but that on adding a fifteenth brick the mass broke down. I suspect that this cement must have been rather stale, because Mr. John Hopkins the Master Bricklayer of Woolwich Dock-yard, to whom I made a reference, on hearing that he had tried similar experiments, informed me that in the course of several trials he once succeeded in sticking out 22 bricks from a wall by cement from the same manufacturers, which set so quickly that he was only obliged to allow 5 or 6 minutes in the placing of each brick, and it was not until the addition of a twenty third brick that the mass broke down. In both Dock-yards, the cement supplied by other manufacturers, when tested in the same manner, proved inferior to the weakest of these two. Upon the whole I am persuaded, that the average quality of approved cement supplied by the most respectable manufacturers, will not stick out more than 18 or 20 bricks from a wall in this manner, although I admit that Sheppy cement of very superior quality and burned to a nicety, may by using particular care in the mode of applying it, and by allowing of plenty of time, be made to support more. For example, I have myself seen a much greater number projecting thus, at the cement works of Messrs. Francis, White and Co. near Vauxhall Bridge, Lambeth, who assured me that they have repeatedly stuck out 29 bricks in this manner; and it was stated a few years ago, at one of the meetings of the Institution of Civil Engineers, by Members of that Society, that so many as 30 bricks had been stuck out in one day, and 33 bricks in 33 days, in the same manner.

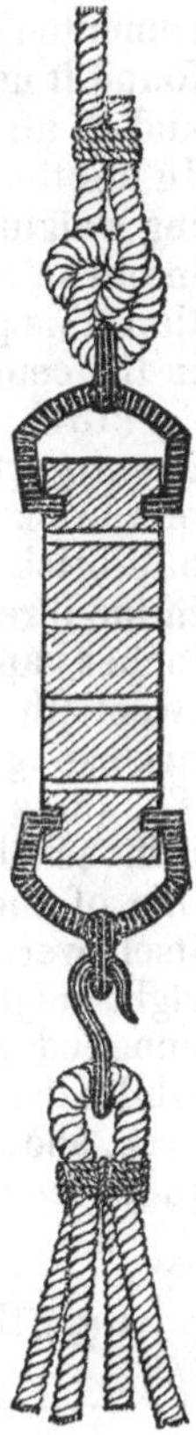

(133) Having proved its strength in the manner alluded to, we afterwards tried what weight was required to separate the joints of five bricks cemented together by our artificial cement, which we suspended vertically under an artillery gyn, or triangle used for mounting guns, by means of a pair of iron nippers clasping the sides of the upper brick, and having its claws let into mortises cut about three quarters of an inch deep in the sides of the brick by a chisel. A similar pair of nippers reversed as to position clasped the lower brick in the same manner, to which was hooked on a large scale board for containing the weights, which were gradually increased until fracture took place. The annexed figure represents this apparatus with the exception of the gyn and scale board, which would have occupied too much room, without making the description clearer.

In our first experiment we were obliged to put 1195 lbs. into the scale, in order to tear the brickwork asunder; but as one brick only out of the five was separated in this first trial, we continued repeating the same experiment a second and a third time with the remainder of the mass, and it required successive weights of 955 and of 1099 lbs to produce fracture, in the last of which experiments one of the bricks was torn asunder instead of the cement joint giving way.

Trying the same experiment with another mass of four bricks connected by the same artificial cement, it required successive weights of 451, and of 1875 lbs. to produce fracture; but on examining the first joint of cement that gave way, it was found imperfect, the two bricks that had been connected by it being united only at the edges, without having any cement in the middle.

As the bricks used in these last experiments were the same that had been attached to the wall of the casemate as before described (131), we concluded that if all the joints had been perfect, more than 12 might have been made to adhere to the wall. It did not occur to us to examine the joints at the time of our first experiments, which took place about four months before the last were tried.

We next tried similar masses of bricks that had been

cemented by Harwich cement in the same manner, and we found it necessary in one trial to apply a weight of 1003 lbs, and in another a weight of 1219 lbs to produce fracture. In another trial, one of the joints separated, without applying weights at all, during the process of cutting a mortise in one of the bricks to receive the nippers, but on examination, this proved to be an imperfect joint without any cement in the center, like that before described.

(134) Our next experiment was intended to ascertain the strength of our artificial cement, as applied to the joints of a brick arch. For this purpose we built a flat segment half brick arch, that is an arch of bricks placed horizontally in the direction of their length, having a span of 15 feet 4 inches, and a rise of only 9 inches, being less than one twentieth part of the span, and the thickness of the arch according to this arrangement being not quite 4½ inches. The piers from which it sprang, were one foot high and only 9 inches thick, but they abutted against the strong walls of one of the casemates; and these piers as well as the arch itself were each 18 inches or two bricks wide, measuring at right angles to the span of the arch. The first of the annexed figures is an elevation or longitudinal section, whilst the second is a plan of the top of this experimental arch, showing the manner in which we thought it expedient to toothe the bricks of the arch, in order to break joint.

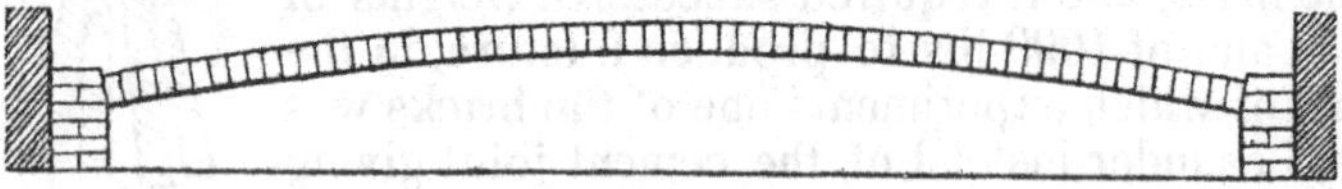

Fig. 1.

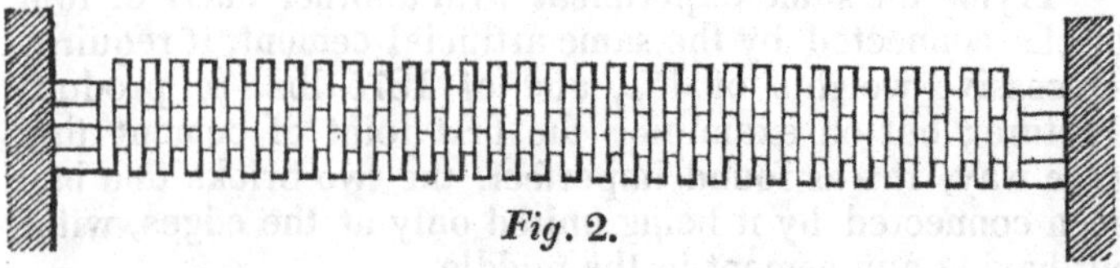

Fig. 2.

To try the strength of this arch, we loaded it four months after it had been built, with dry bricks laid over it in pairs, in the same manner in which the bricks of the arch itself had been placed, that is horizontally in the direction of their length, and in successive courses of half a brick in depth or thickness, each course consisting of 138 bricks and weighing probably about 640 lbs, which were all laid very loosely with wide vertical joints, and not in a compact mass, lest they should become jammed between the two walls, which would have defeated our object. No effect whatever was produced upon our arch, until ten courses of loose bricks weighing 6400 lbs were laid on, which caused a deflection of about 2 sixteenth parts of an inch. The 11th course caused a deflection of 3 sixteenths, the 12th of about 7 sixteenths, and the 13th of 10 sixteenths, but not long after the 14th course had been applied, at which time the total weight pressing on the arch amounted to about 8960 lbs, or to four tons,* it broke down with a tremendous crash, whilst we were about to measure the deflection. The third figure annexed represents the arch thus loaded with the 14 courses of loose bricks, which overwhelmed it in such a manner, that we could not ascertain in what point or points it gave way; but we observed, that the extreme brick had been broken at one end of the arch, for a fragment of this brick was found adhering to the skewback or oblique brick forming part of the upper course of the pier

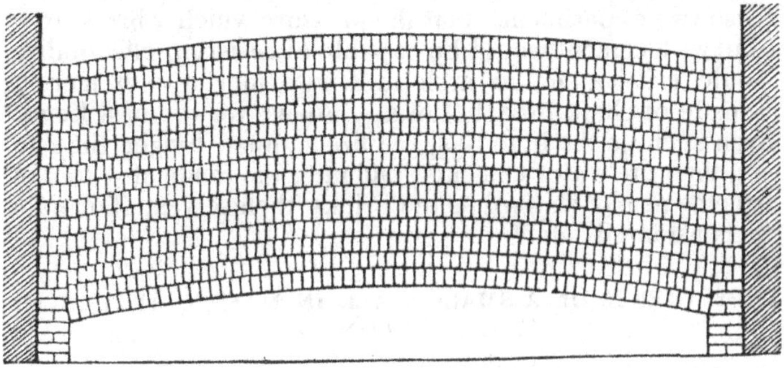

Fig. 3.

on that side. In fact such was the cohesive power of the cement, that the whole arch might be considered as a solid beam of brick, and it was broken just as a beam of iron might have been fractured by excess of weight, gradually applied in the same manner.

(135) We also built a similar brick arch with the best chalk lime mortar, immediately before we commenced the cement arch that has been described, using

* We did not weigh the bricks used in this experiment, and therefore in naming 4 tons as the breaking weight, I have assumed an average of $4\frac{5}{8}$ lbs, bricks, per brick, which is a good deal less than the average weight of bricks, which may be estimated at nearly 5 lbs.

as a centering for both a mass of rammed earth fashioned to the proper curved form, and resting on the floor of the casemate. We cut away the earth from underneath the cement arch the day after it was finished, but we allowed that of the mortar arch to remain during the whole period of four months before mentioned, after which we began to remove it, proposing afterwards to load this arch also in a similar manner, in order to compare the strength of our mortar joints and of our cement joints. But when we had cut away only a part of the earthen centering from under the mortar arch, it fell to pieces without giving us time to finish the removal of the whole of the earth, so that the mortar arch, though equally well built, proved incapable even of supporting its own weight.

(136) These results afford the most conclusive proofs of the expediency or rather necessity of using cement, not only in the brickwork of tunnels exposed to the action of water, such as the Thames Tunnel, but also in tunnels cut through much drier ground, such as those of the London and Birmingham Railway; for it must be evident from these two experiments, that the pressure which a brick arch built with cement would be capable of resisting effectually, would crush a similar arch built with mortar, not merely if made of chalk lime, but of any other; for the best limes, even such as the blue lias, require a much longer time to set, than the water cements; and until the joints of a tunnel become nearly as hard as the bricks themselves, it cannot be considered safe.*

EXPERIMENTS ON A SMALL SCALE IN MAKING ARTIFICIAL STONE, OR CONCRETE.

(137) As the experiments tried on a small scale, both with the natural and artificial cements, proved so very satisfactory, and were such as enabled us with little trouble and expense to form an accurate judgment of the qualities of those useful substances, it occurred to me, that it might be possible to judge in the same simple manner, how far any given sort of lime was fit for making good concrete (28), or good artificial stone on Mr. Ranger's principle (30).

* Mr. Stephenson the Engineer of the London and Birmingham railway, now in progress, who has also had the advantage of having been employed previously, with his father, in executing the justly celebrated Liverpool and Manchester railway, informs me that he makes it a rule, deduced from experience, never to use lime mortar in the arches of tunnels, but to build them with cement exclusively.

With this view early in the present year 1836, I caused four wooden moulds to be made 4 inches long, 2 inches wide and 2 inches deep in the clear, each consisting of two side pieces s s, having grooves cut in them to receive two end pieces, which were fitted in very loosely to allow for the expansion of the wood, which was always to be used wet. The parts of this mould, which were about an inch thick, were held together by two clamps c c, clasping the ends of the side pieces and tightened by wedges, the heads of which are seen in plan, in the annexed figure, which represents the mould put together, and supposed to be filled with concrete.

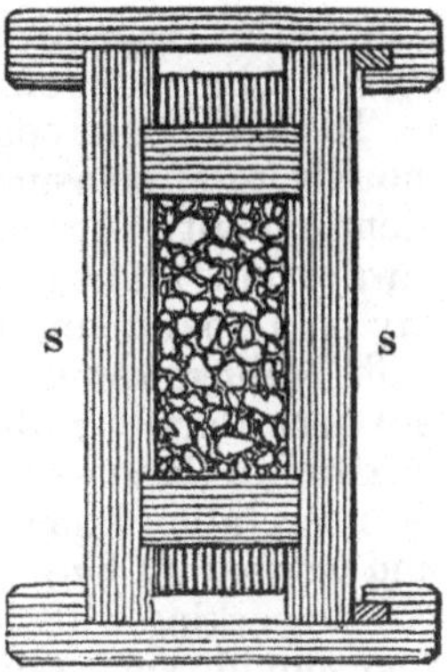

We provided also a number of thin pieces of board, about 8 inches long and 4½ inches broad for temporary bottoms to these moulds, the ends of one of which are shown in the figure, upon which the small artificial stones made in them were left to dry for a day or two, and then placed on skids.

(138) Having prepared a mixture of sharp pit sand and small gravel, having no part larger than a horse bean, we washed it well; and then sifted it, first through a screen with openings of one eighth of an inch in the clear, all that would not pass through which was set aside as *Gravel.* The remainder being sifted a second time through a very fine wire sieve, having 30 meshes to the inch, was thereby divided into *Coarse sand,* and *Fine sand.* The various sorts of limes experimented upon, all well burned and fresh from the kiln, were pounded in a mortar, and preserved in bottles filled quite full and corked up, in order to exclude the air and prevent spontaneous slaking.

In these experiments we sometimes used only the gravel, sometimes only the coarse sand, and sometimes only the fine sand; at other times we used combinations of two of these, or of all three, in various proportions by measure, which we mixed up with each sort of quicklime powder, also in various proportions, between 6 measures of siliceous matter to 1 measure of lime, and 9 or 10 measures of the former to one of the latter. The limes used were chalk lime and Halling lime both of the best quality, and three artificial hydraulic limes made by us, by mixing 5, 6, and 7 measures respectively of chalk lime powder with 1 measure of blue clay, on the principle alluded to in Article 83.

In making our concrete or artificial stones in these little moulds, after measuring the several ingredients, in a sufficient quantity of each mixture, to fill two moulds, for we always made them in duplicates, we first mixed them up dry on a board, and then worked them up well, generally with boiling water, according to Mr. Ranger's principle, but occasionally with cold water, after which we threw the mixture in three or four portions into the mould, where it was carefully rammed, and finally the surface was stricken and worked up by a small wooden float.

In working with boiling water, after having used our four moulds successively, we found that by the time the fourth was filled, the concrete in the first had set sufficiently to admit of this mould being taken to pieces, which we did by taking out the small wedges, disengaging the clamps, and then the side pieces, and lastly the end pieces, one at a time. Having thus made our small artificial stones with the several natural and artificial limes above-mentioned, we put them under water in a long tin trough, with cells, each containing one piece, in periods of not less than 16 nor more than 33 days afterwards; and we have kept them immersed ever since, that is for about seven months, never changing the water entirely, but merely replacing the portion lost by evaporation.

(139) The results of these experiments seemed at first rather favourable to chalk lime, for all the concretes formed with it, in such very small moulds as have been described, had time to set before they were immersed, and as the water was never changed entirely, it had only a very moderate and slow action upon them (12), so that for several months after immersion, they seemed scarcely inferior to the concretes made of Halling lime; but recently towards the close of the above-mentioned period of seven months, the inferiority of the chalk lime became apparent, for a mixture of chalk lime with rather more than three parts of fine sand, which we made in the same moulds, to represent chalk lime mortar, and immersed after it had set to a certain degree, was friable and easily rubbed to pieces by the hands, and the concrete formed of one measure of chalk lime and six measures of gravel proved to be in a state of pulp in all the joints, on being broken, and a considerable portion of the lime had quitted the joints, especially the external ones,

from which it had almost entirely disappeared. The intermediate concretes between these two extremes of chalk lime mixed with gravel alone, and with fine sand alone, were much better, but as the extremes both gave way, these experiments confirmed us in our former opinion, that chalk lime should never be used for concrete; because in such large masses, as are required for the practical purposes of building, it would never have time to set in a damp situation, and wet would destroy it, of which I have known many instances. (See from Article 10 to 15 inclusive).

(140) The concretes formed with the Halling lime seemed on the contrary to be rather improved by immersion, excepting one mixture only, which was formed of 1 measure of lime to 6 of gravel, which was the worst of all the concretes made with this lime, in like manner as a similar mixture was by far the worst of all those made with the chalk lime. In this mixture the Halling lime had partially quitted the outside joints of the artificial stone formed of it, but inside it was not soft or in a state of decomposition like the other. Of the other concretes formed with the Halling lime, and containing sand as well as gravel, those in which the fine sand rather predominated over the coarse were the best; and Halling lime mortar formed in the same mould, with 1 measure of lime to 3½ measures of fine sand was excellent, it having improved very much after immersion. Hence these experiments confirmed the opinion before stated that sand strengthens lime,* for the only concrete of Halling lime, which deteriorated under water was that before mentioned, in which the large joints which separated the particles of gravel, consisted of lime paste or putty alone, unmixed with and unaided by sand, and were therefore weak.

These experiments confirmed another opinion also before stated (34), that good concrete cannot he made with any of the hydraulic limes, mixed with gravel alone. A proportion of sand must necessarily be added, and contrary to the opinion, which I have heard advanced by some practical men, it appears, that coarse sand entirely unmixed with fine, is not to be recommended, either as an ingredient for concrete, or for making mortar.

(141) Having tried a similar course of experiments with our three sorts of artificial hydraulic lime made as before described (138) at the same time and treated precisely in the same way, we could not decide which of them was the

* Article 42, under the head, *Thirdly*.

best, for they proved nearly alike in quality, notwithstanding the different proportions of chalk in their composition, and they all seemed fully equal if not superior to the Halling lime, for the purpose of making concrete or artificial stone.

(142) On afterwards comparing together the concretes made with boiling water, and those made with cold water, we could not discover that the former were in any respect better than the latter, although they had always begun to set sooner, so that as our number of moulds was limited we had been able to work more quickly with boiling water, than with cold.*

(143) At the same time we had attempted to form similar concretes, both with our artificial cement and with Sheppy cement of good quality, mixing one measure of each in the state of calcined powder, with the same various proportions of gravel or of coarse or fine sand, which we had before used. Some of these in which the gravel and sand were in excess, would not adhere together at all, but crumbled into rubbish, as soon as the mould in which they had been made, was taken to pieces. Others did not absolutely fail in this manner, but none of them were good for any thing, and all fell to pieces when put under water in fifteen hours after they were made. In fact, the proportion of gravel and sand used, not being less than 6 to 1 of cement powder in any of these mixtures, was enough in all of them to kill the cement, which circumstance, together with other observations and experiments, led me to the conclusion before stated, that cement is always weakened by sand.† The goodness of the artificial and natural cements used in these fruitless attempts to make concrete with gravel and sand in the proportions stated, could not be doubted, since both of them when made up into moist balls of pure cement in the usual manner, set under water in ten minutes after they had been mixed, and on being taken out next day, we were obliged to use a hammer for the purpose of breaking them.

(144) We afterwards tried another course of experiments on concrete with the same small moulds, in which instead of varying the proportions of gravel and of the two sorts of sand used, we only employed one kind of sandy gravel, namely Thames ballast, of the same description

* I cannot discover any other advantage than this in the use of boiling water, and I doubt still more the benefit supposed to arise from heating the gravel and sand for artificial stones, which is another expedient sometimes resorted to by Mr. Ranger.

† See Article 42, under the head, *Thirdly*.

actually used by Mr. Ranger in making his artificial stones, for the new works in Chatham Dock-yard, excepting that considering the smallness of our own moulds, we excluded all the pebbles that would have been disproportionately large, by screening the ballast through clear openings of half an inch in width, and using only the sand and finer parts of the gravel which had passed through the screen. Thus the sandy gravel in this second course of experiments was all exactly alike, which we considered necessary in order to ascertain, what proportions of the same sort of siliceous matter were most suitable to the various kinds of cement or lime experimented upon; for in consequence of former experiments, in which we found that cement and the more powerful hydraulic limes would not bear so much of the same kind of sand, in making them into mortar, as the less powerful hydraulic limes, and still less than pure lime; it seemed reasonable to suppose, that the same property would hold good in concrete.*

To ascertain this point, in the second course of experiments with our little moulds now alluded to, we mixed one measure of Sheppy cement with from 2 to 9 measures of the screened ballast, and one measure of each of the various sorts of lime, with from 3 to 10 measures of ballast; and on examining all the concretes thus formed at the end of about six weeks, without putting them into water at all, we found that 2 measures of ballast to 1 measure of Sheppy cement, 5 or 6 measures of ballast to 1 of blue lias lime, 6 or 7 measures of ballast to one of Halling lime, and 7 or 8 measures of ballast to one of chalk lime, appeared to be the best proportions for making concrete with these several sorts of cement or of lime, and we were of opinion, that they would all be nearly equally efficient in those several proportions, if used in dry soil, but not in damp or wet soil, where the chalk lime concrete would be sure to fail. The very small quantity of gravel or sand which the cement in those experiments proved capable of bearing, when compared with the limes, confirms the opinion before advanced and supported by other observations and experiments, that cement should never be used for concrete at all (61), but if so used, that

* That is provided that the gravel and sand should not vary in their quality or proportions, as was the case in the first course of experiments just recorded, for this variation in the siliceous matter used in concrete, renders the true proportion of lime brought into efficient action as a cementing substance very uncertain, from the circumstances before taken into consideration, in Articles 34 and 36.

no fine sand should be allowed to form a part of the mixture, for as I have repeatedly mentioned, as an important distinction, the water cements are weakened, whilst all limes are strengthened by fine sand (42).

(145) I shall here take the opportunity of introducing a remark on stucco, which perhaps may apply also to the external surface of Ranger's artificial stones. It appears to me, that good water cement mixed with very coarse sand is far more suitable for stucco, even than the best and strongest limes, and that to make it perfect nothing more is required than to improve upon the present careless mode of applying it, before commented upon in Articles 121 and 122, and to work it up into so smooth a surface as to make it water-tight, which would prevent vegetation or moss forming on it. Next to cement, but as I repeat far inferior to it, in their power of forming a good stucco, are those limes which approach the nearest to cement in their properties, such as the blue lias lime, which sets into a much harder substance, and resists the vicissitudes of air, rain and frost, better than the weaker hydraulic limes, such as the Dorking and Halling limes. In fact the weak hydraulic limes, though good for concrete in the dampest situations, and also good for the mortar of walls and foundations, are comparatively unfit for stucco, since they approach too nearly in their quality to the pure carbonates of lime, such as chalk, &c, which are altogether inapplicable to that purpose. I think it probable, that there are many excellent hydraulic limes in various parts of England, which from the expense of carriage cannot be brought into the London market; but the only lime of this kind that I am personally acquainted with, and of which I sent for some, after hearing it praised by an experienced builder of London, was from Rosehill in Sussex. I never saw this lime in its natural state, as the only specimen I could obtain, which was forwarded to me in a bag, had been previously burned and probably ground, for it was in fine powder and so very stale, that it threw out scarcely any heat on being mixed with water.* I therefore made it up into balls with water, which when dry I burned in crucibles, knowing that reburning restores stale or slaked lime to its original state of quicklime. On trying the Rosehill lime in this improved state, in the usual manner, by making it up

* From the description of it in Mantell's Geology of Sussex, I infer that it is a hard lime stone. The specimen of it sent to me was from the estate of the late well-known John Fuller, Esq. long one of the Members of Parliament for that county.

into moist balls after pounding it when fresh from the fire; it appeared to approach very nearly to a water cement, for on allowing these balls, which now threw out great heat, about a quarter of an hour to cool and then immersing them in a basin of water, they did not go to pieces for several weeks, so that this lime seemed rather superior in its hydraulic powers, even to the blue lias lime of Lyme Regis, with which I had an opportunity of comparing it.

These two limes possess such strong hydraulic properties, that if formed into artificial stones on Mr. Ranger's principle, and allowed some months to set, I do not think that the exposure to running water as in wharf walls would gradually wash away the lime from the surface, which effect I anticipate if such walls be faced with artificial stones made of the weaker hydraulic limes.*

FINAL EXPERIMENTS ON ARTIFICIAL CEMENT TRIED IN 1836.

(146) All our former experiments having been performed by manual labour, I resolved before I published this little Treatise, to ascertain practically the best mode of mixing artificial cement on a great scale, and it appeared to me, that the washmill, such as is used by whitening makers, would be the best machine for grinding the chalk, and that the pugmill would be the best mode of mixing it with the blue clay; for I did not approve of the system that had been adopted by Mr. Frost, in his artificial cement works near Northfleet, of passing not only his chalk but his clay through the washmill, and afterwards drying the mixture by long exposure to air, both of which processes I believed to be injudicious; for to the former I ascribed the failure in 1829, and to the latter the failure in 1830, of my first attempts to make artificial cement on a great scale, which were before recorded. (See from Article 108 to 111 inclusive).

(147) *Mode of grinding Chalk into Whitening.* The process of making whitening is this. After the chalk pre-

* Since the two first sheets of this treatise were sent to press, I have inspected the wharf wall in Woolwich Dock-yard, faced with Mr. Ranger's artificial stone, made I believe of Halling or Dorking lime, alluded to in Article 31, and part of the outer surface of the wall has detached itself for about an inch in depth, which took place as I was informed after the severe frosts of the winter of 1835-6, which acted on the surface previously wet at high water, and exposed to air by the ebbing of the tide. This fact sufficiently proves, that such artificial stone, though a very useful invention for many practical purposes, ought not to be allowed to supersede good bricks, or good natural stone altogether, neither of which would have given way under the like circumstances.

viously broken into small pieces has been ground with excess of water, in a washmill, where it is subjected to the action of iron rollers and harrows in an annular groove, the lighter parts of this fluid mixture are allowed to run off by a small channel into a *Back* or reservoir, from which they are afterwards taken out, and made up into lumps or cakes of *Whitening,* which from the experiments of former years, I knew to be a good ingredient for artificial cement, but far too expensive. The heavier portion of the same fluid mixture of crushed chalk, which settles at the bottom of the groove, and which by a curious misnomer is called *Sand,* is rejected by the whitening makers, and considered of no value. Having examined some of this substance which seemed fine enough for the purpose, I caused it to be dried and reduced to powder, in order to try whether it might not, as an ingredient for artificial cement, be made of some use, instead of being carted away for rubbish.

(148) Having made a little pugmill, and put into it alternately small portions of this whitening maker's sand and of blue clay fresh from the river, until it was quite full, the aggregate of these ingredients having been previously measured in the proportion of 5 parts of the former to 2 of the latter, and having passed the whole through the pugmill two or three times, after which it seemed exceedingly well mixed, we burned it in the usual manner, expecting that this experiment would prove whether the proposed system of using the washmill for grinding chalk, and the pugmill for mixing it afterwards with clay, was likely to prove satisfactory, in making artificial cement on a great scale.

As the calcined cement obtained by this process would not set under water at all, it seemed probable that whitening maker's sand might be an adulterated form of chalk, and therefore we next tried the whitening itself and afterwards pure chalk from the quarry, reducing them both to fine dry powder, and mixing them with blue clay, in the same approved proportion of 5 parts of the former to 2 of the latter, by means of the pugmill; but as these also formed imperfect cements, that would not set under water, the fault seemed to rest with the pugmill, and therefore we tried the same ingredients mixed by hand with the greatest care, in the same proportion, and to our great surprise the cements formed with them, which had never failed in former years, now invariably went to pieces on immersion in a basin of water. These contradictory results at first appeared to me perfectly unaccountable, for I had no suspicion that the blue clay of the Medway could have changed its chemical character, and I could not doubt the accuracy of our former experiments of 1828 and the following years, in all of which the proportion of 5 measures of chalk to 2 measures of blue clay had yielded a cement, which appeared by the experiments on its strength before recorded in Articles 131, 133, and 134, to be fully equal if not superior in tenacity to the average of the natural cements of approved quality, as sold by the most respectable manufacturers; and which, as I never omitted to put every batch of it to that test, had always set under water, on being immersed, in about five minutes after it had been made up into moist balls. Finding therefore that this same mixture now invariably failed, I proceeded, but under these perplexing circumstances, almost in despair of arriving at a more satisfactory result, to try a number of other proportions of chalk and of blue clay, always retaining 5 measures of chalk powder, but using no less than six several proportions of blue clay, mixed with the above, and gradually increasing from 2¼ to 5 measures of this clay.

(149) The mixture of 5 to 5 in which the clay was in excess would neither set in air nor in water. The other proportions, between 5 to 2¼, and 5 to 4½, inclusive, formed water cements; but both the former which set quickly and with great heat, generally into a very light brown, but sometimes into a dirty white or light drab colored cement, and the latter which set very slowly and with scarcely any perceptible heat, into a very dark brown coloured cement outside, and sometimes of a dark bluish green colour inside, occasionally failed. All the intermediate proportions between these two extremes were better, but 5 to 3 appeared to be one of the best. The balls of about one inch in diameter, made of this mixture set into an excellent water cement, which when broken by a chisel in about 24 hours proved equally hard inside and outside, and when allowed to remain two or three days before breaking them, were of a stony hardness throughout.

(150) In this new course of experiments for obtaining proportion, we observed that the best mixtures were those which when immersed were of a nut brown colour, and which in setting became rather warm than hot, attaining their highest temperature in from 7 to 10 or 12 minutes after being made up into balls with water. Those which set more quickly and heated violently from having an excess of chalk, as well as those which set much more slowly and without perceptible warmth from having an excess of blue

clay in their composition, and which were respectively of a much lighter and of a much darker colour than the above, were both of inferior quality.

(151) On now trying the whitening-maker's sand two or three times, mixed with a greater proportion of clay than before, we found that it made a tolerably good water cement, but apparently not equal to that formed with purer chalk. It is possible, however, that it might have succeeded better, if tried oftener, but on consideration, I did not think it worth while to give it a further trial, as it appeared probable, that the manufacturers of artificial cement would prefer grinding their own chalk, instead of purchasing the refuse of the whitening-maker's washmills. Having compared the specific gravity of the whitening-maker's sand, with that of pure chalk, after reducing both to a fine dry powder, we found that the former exceeded the latter in the proportion of 5 to 3. Hence this sand must contain a good deal of foreign matter, probably of ground flints mixed with the chalk of which it is chiefly composed.

(152) It is to be remarked, that in the experiments now alluded to, all those mixtures, which formed a good artificial water cement succeeded equally well, whether the ingredients were mixed by hand or in our little pugmill, but of these two modes, the latter was decidedly the best.

(153) At this period, I remained for some time of opinion, that I must have been too hasty in adopting the rule laid down by me in 1828, and followed in the succeeding years, as to the best proportion of pulverized chalk and of blue clay for an artificial water cement, for although I had always taken the greatest pains to be accurate, I thought it much more probable that my own first experiments in 1828, which so far as regarded the determination of the said proportion were very few indeed, compared with other objects of inquiry, should have been in error, than that the alluvial clay of the Medway should have changed its chemical character within the last six or seven years. I conceive however, on further reflection, that a partial change in the nature of the blue clay must necessarily have taken place, without which, the general success of our original mixture in former years, is perfectly incompatible with the constant failure of the same proportion of chalk and blue clay in the present year. The impression upon my mind until the failures alluded to, always was, that the blue alluvial clay of the Medway contained very little calcareous matter, if any: but on referring to my original analytical experiments

of 1828 and 1829, I cannot find any record of my ever having tested the blue clay at that time, in such a manner as to ascertain this point with proper precision. But after the more recent course of experiments now alluded to, having subjected the blue clay of the Medway obtained from the usual spot, just below Chatham Dockyard, to the action of the diluted muriatic acid, a brisk effervescence took place, which continued so long, as to prove that this clay is now impregnated with a considerable proportion of calcareous matter, which most probably has always been the case, in a river flowing through a considerable tract of country having a subsoil of chalk, and abounding in lime works, which recently have been multiplied on both banks of the river from Rochester towards Aylesford. Hence the difference between the results of the first and last experiments tried by me may be accounted for, by supposing a small increase to have taken place in the proportion of calcareous matter contained in the blue clay, within the last six or seven years.* The blue clay of 1828-9 being rather purer alluvial clay, formed a good water cement, when mixed with chalk in the proportion of 2 measures of the former to 5 of the latter, that is as one of clay to $2\frac{1}{2}$ of chalk; whilst the alluvial clay of 1836 having more calcareous matter mixed with it, requires a smaller proportion of the latter substance to produce a like result, the present proportion, which produces a water-cement nearly of the same quality as our original artificial cement, being nearly as $2\frac{1}{4}$ to 5, or as 1 of clay to $2\frac{2}{9}$ of pulverized chalk. Thus the blue clay of the Medway of 1828-9 formed a good water cement, when mixed with about 11 per cent more of pure carbonate of lime, than is required to be mixed with an equal quantity of the more calcareous blue clay of the same part of the river in 1836, in order to produce a similar cement.

(154) Hence, also though the best proportion of pure chalk and of pure alluvial clay for making artificial cement might be laid down with great accuracy; one cannot determine before hand the proportion necessary, for the same

* If so, it appears to me most probable, that the additional proportion of calcareous matter has been chiefly obtained from the new works in progress, and from the wharf walls in Chatham Dock-yard, whilst being pulled down and rebuilt, the lime of both having been subject to the action of land springs in the foundations and of the tide above that level. The great quantity of chalk rubbish collected near the mouth of the Gravesend Canal Tunnel, opposite to Rochester, and the lime works at Cuxston, Halling, and Burham, &c. on the banks of the river above that city, may also have had their effect.

purpose, in working with chalk mixed with siliceous or argillaceous matter, such as whitening maker's sand or marl, or in working with clay mixed with calcareous matter, either of which, although if fine enough, they may be excellent ingredients for an artificial water cement, will require very different proportions from those which are suitable for pure chalk mixed with pure clay.

During the period that I was doubtful of the accuracy of my first experiments with chalk and blue clay, in 1828 and the years immediately following, it appeared necessary not only to fix a new and more judicious proportion of those ingredients, which was accomplished as related in Article 140, but also to repeat other experiments of those years, in order to ascertain, whether the failures of our first attempts to make artificial cement on a great scale, in 1829 and 1830, were really owing to the circumstances to which we then ascribed them, or whether they might not have arisen partly or chiefly from the then defective proportion of our ingredients, namely 5 of chalk to 2 of clay, which was apparently proved to be erroneous by our subsequent experiments of 1836. With this view we proceeded as follows.

(155) First, we mixed 5 measures of chalk with 3 measures of the brown alluvial clay or surface clay of the Medway, this being the new approved proportion for our cement mixtures of the present year, which on being burned and treated in the usual manner, promised remarkably well at first, but always failed in the course of a few days, in precisely the same manner as the mixture of the like ingredients, but somewhat differently proportioned, had always failed in our original experiments (84).

(156) Secondly, we mixed 5 measures of chalk powder with 3 measures of blue clay, which when burned without loss of time, had always formed an excellent cement, and having put part of this mixture into a tub with about four times its bulk of water, we stirred up together two or three times a day for a fortnight, after which the raw cement was allowed to settle to the bottom of the tub, and the clear water from above was drawn off by a siphon. As this mixture, which had been protected from air by the water, had not lost its blue colour, except at the surface, I thought it probable that the fortnight's washing might not have injured it, but such was the case, for on burning the blue part of the mixture, the cement balls formed of it either fell to pieces under water, or held together in an unsatisfactory manner, without setting into a hard substance.

(157) Thirdly, the same cement mixture, when properly prepared for use, and allowed to dry in the sun, retained its blue colour under the surface with so much obstinacy, that we were obliged to bruise it in a mortar, and cut it into very small pieces and spread it out in this state, to admit air freely, before an entire change of colour could be produced. We then mixed it with water into a stiff paste, and burned it in the usual manner, into an artificial cement, which was rather of inferior quality, for although it set well under water, and sounded well when taken up and thrown against the side of a basin, yet it remained soft at the surface, and stained the hand when rubbed against it, much longer, than any of our good artificial cements had done; and the experimental balls formed of it, were two months in attaining a satisfactory state of hardness, such as the same mixture, if burned without loss of time, and consequently without allowing the clay to become stale by exposure to air, would have exhibited in the course of a week.

(158) Fourthly, on taking part of the same mixture, which had been washed for a fortnight, as described in Article 156, and which by that process alone appeared to be very much injured, and afterwards spreading it out to dry, until it completely lost its blue colour, and burning it; the experimental balls formed of it, instead of being entirely ruined by the second process, which the experience of former years had taught us to reprobate, never failed in setting under water; although they indurated rather more slowly, than cement balls made of the same mixture had done, when burned immediately without washing and drying. As the experiments described in this and the two preceding articles appeared rather contradictory; we thought it desirable to repeat them afterwards with a different mixture, which we ascertained to be a still more powerful artificial cement than the former.

(159) This was composed of 5 measures of chalk powder, mixed with $2\frac{7}{9}$ measures of blue clay only, being a little less than the proportion of clay last used. On subjecting this mixture successively to the processes, first of drying until it lost its blue colour, secondly of washing and stirring for a fortnight, and thirdly of washing and stirring and afterwards drying to entire loss of colour; the experimental balls formed of it under all those circumstances, invariably set well under water, without staining the hand for any considerable length of time. Those subjected to the first process only did not appear at all injured, and the others also attained by degrees a satisfactory state of induration, but

rather more slowly, than balls of the same mixture burned immediately, and without washing or drying, had generally done.

(160) Thus of our two new mixtures of 1836, one was occasionally but not generally spoiled, and when not spoiled, it was only deteriorated by the three processes now under discussion; whilst the second was never spoiled, and apparently very little injured by any of those processes. It seems therefore reasonable, to ascribe the difference between the effects produced by washing, and by drying with change of colour, upon our cement mixture of 1829 and 1830, and our second mixture of 1836, to the circumstance, that the former was calculated to produce a very quick setting cement, and the latter to produce a cement of medium quality, equally removed from the two extremes, of those that set too quickly with violent heat, and of those that set too slowly without heat; the former of which are liable to fail from having an excess of calcareous, and the latter from having an excess of argillaceous matter, in their composition. Now as the cement mixture of 1829 and 1830 approached nearly to the former of these extremes, whilst the last described mixture of 1836 was equally removed from both, it may be allowed, that any circumstance having an injurious tendency, might entirely ruin the former, whilst it might only slightly deteriorate the latter.

(161) Without now having the means of procuring any blue clay, in precisely the same state as that of the Medway from 1828 to 1830, I know that the original approved proportion, then adopted by us, formed a very quick setting water cement, bordering upon one extreme, liable to failure; because we endeavoured to make it as similar as possible to the specimens of the natural Sheppy cement stone which we had procured from Sheerness, and which set so quickly, when burned by us and made into experimental balls, that they did not always allow us time for marking them with a few printer's types,* and with so much heat, that they were sometimes painful to the hand. Having little or no experience at that time, we naturally concluded from the high reputation of the Sheppy cement, that these symptoms were a proof of excellence,† and that the incon-

* As examples of the mode of marking our experimental balls, C 5 B 2, F 27, 29, denotes 5 parts of chalk mixed with 2 of blue clay on the 27th February 1829, whilst S, N 7, 30 denotes Sheppy cement, November the 7th, 1830. These symbols were marked in three lines to avoid confusion.

† This opinion is distinctly stated in Article 69, for it is to be observed, that the first part of this Treatise, from the commencement to Article 146, inclusive, was printed, before the final experiments of the present year, which proved the fallacy of that opinion as to the superiority of very quick setting water cements, had been tried.

venience of setting too quickly was only to be surmounted by mixing the calcined cement with excess of water (47), or by allowing it to become rather stale by temporary exposure to air. I do not think now, that those specimens could have been in the best state, in which that excellent cement may be found, for they must have had rather more of the carbonate of lime in their composition, than was desirable. As a proof that they could not bear much more, I may observe that in the experiments tried by me several years ago, in order to ascertain by what additional substances of various kinds, the natural cements might be spoiled, we found that the Sheppy cement stone then in our possession was so much deteriorated by being mixed with one fourth of its own bulk of chalk, both being previously reduced to and measured in the state of dry powder; that after this addition it entirely lost the property of setting under water; which experiment ought to have been inserted together with the others mentioned in Article 101, but was unintentionally omitted.

(162) To return to our more recent experiments, having reason to believe from these, that our artificial cement of 1829 and 1830, was not the best, that might have been made with the same ingredients, it now appeared desirable to ascertain, which was the strongest of four different proportions of chalk and blue clay, all capable of forming a water setting cement, by attaching bricks to a wall with each of them, until they broke down, in the mode represented in the lower projecting brick pier of the figure at the end of Article 131, marked *Row of bricks standing out on edge.* This mode of trying the comparative strength of these mixtures was adopted, not only because experiments on a smaller scale would have been unsatisfactory, but also with a view to obtain a direct comparison of the tenacity of our own artificial cement with that of the best natural water cements of England, the former of which had never been tested in this manner, but the latter frequently (132).

(163) But as it appeared probable, that the measurement of chalk in the state of powder might not be convenient, in working on a great scale, and as it is well known that weight is generally a more accurate test of quantity than measure,* I determined to fix the most suitable proportions

* As has been proved in several parts of a Treatise published by me in 1834, entitled " Observations on the expediency and practicability of sim- " plifying and improving the Measures, Weights and Money, used in this " country, without materially altering the present standards."

for forming the best artificial cement with chalk and blue clay by weight, on a sufficiently large scale, and with so much precision, as to leave nothing doubtful for the future; and in adopting this system, it appeared best always to weigh the blue clay fresh from the river, where it was obtained at low water at the depth of a foot or 18 inches below the brown surface, but not to weigh the chalk, until previously well dried and pounded: and as the smallness of our apparatus required the raw cement, necessary for each of the four experimental brick piers, to be burned in several batches, making more than 20 in all; I directed a small but correct cube measure of each successive portion of blue clay and of chalk powder, used in those batches of raw cement, to be carefully weighed in troy grains, before the whole quantities required were weighed in pounds, in order to ascertain the average specific gravity of these two ingredients, which now proved to be 90 lbs, per cubic foot for the blue clay, being a little more than our estimate of former years before stated (85), and 40 lbs per cubic foot for the dry chalk powder, measured lightly, according to our invariable practice in determining proportions.

(164) The necessity of weighing the chalk in a perfectly dry state, arises from its extraordinary retentiveness of moisture, from which I doubt whether it is ever obtained entirely free, even in the driest weather, and which cannot well be expelled from it, without breaking it into very small pieces, and drying them in the sun or before a fire. For example, a lump of pure chalk from any quarry near Chatham seldom weighs less than 100 pounds per cubic foot at any time, and after immersion in water it will weigh 113 lbs, of which 23 consist of moisture, for when perfectly dry its average weight is only 90 lbs. Thus the block chalk of this neighbourhood is capable of absorbing rather more than one fourth of its own weight of water. More compact chalk equally pure, for in this respect it varies, would weigh more when dry, and gain less after immersion, than the above proportion.

(165) We have found by repeated experiments, that 2½ cubic feet of pure chalk powder perfectly dry and measured lightly, when afterwards made up with water into a stiffish paste, of the consistency of mortar, will compress into the space of one cubic foot only, so that chalk paste, thoroughly moist, but not quite so wet as the alluvial clay of rivers and lakes, contains as nearly as possible the same quantity of chalk which if dry would weigh 100 lbs; and

this mode of ascertaining the quantity of chalk previously pulverized, by wet measure, was considered by us after due trial, to be rather more accurate than dry measure.

(166) Having thus explained, how far these three modes of ascertaining the true quantity of pulverized chalk by weight and measure agree, I shall describe the four new mixtures all determined by weight, which we now proceeded to try, the first denoted by the emblem C B, in which the letter C denotes dry chalk powder and B fresh blue clay, implying a mixture of equal weights of those two ingredients, whilst the next C 4 B 5 implies a mixture of 4 parts of chalk, with 5 of blue clay by weight, and so on, as stated in the following table, in which the several symbols of each mixture are inserted and explained by weight; and their equivalents by measure, with the chalk both dry and wet, are also entered opposite; in reference to the data in the three preceding articles.

TABLE OF CEMENT MIXTURES TRIED IN 1836.

PROPORTIONS BY WEIGHT.		EQUIVALENTS BY MEASURE.	
Mixture of C, chalk, and B, blue clay.	Or of dry chalk powder to fresh blue clay as	Dry chalk powder, measured lightly, to blue clay as	Stiff chalk paste to blue clay as
C B	1 to 1	5 to $2\frac{1}{4}$	1 to $1\frac{1}{8}$
C4 B5	1 to $1\frac{1}{4}$	5 to $2\frac{7}{9}$	1 to $1\frac{7}{18}$
C3 B4	1 to $1\frac{1}{3}$	5 to $2\frac{26}{27}$	1 to $1\frac{13}{27}$
C2 B3	1 to $1\frac{1}{2}$	5 to $3\frac{1}{3}$	1 to $1\frac{2}{3}$

As the experiments, which I proposed to try with these mixtures, demanded too great a quantity of cement, to admit of its being prepared in a common fire place, but not enough to require the use of the small lime kiln at Gillingham before described (112), which was also at an inconvenient distance from the Barracks, I had a little iron furnace made as a substitute for a kiln, under the superintendence of Mr. Robert Howe,* whom I requested to keep the journal of the proposed experiments, as I expected to be absent part of the time that they were likely to occupy; he having also previously assisted me in the most interesting of the experiments tried some years before, for ascertaining the strength of our artificial cement, (133 and 134), and in all the recent experiments on a small scale in respect to artificial stone or concrete (from 137 to 144 inclusive), in which he was himself the manipulator.

* Who has been Clerk of Works at the Royal Engineer Establishment, Chatham, since its first formation, and also subsequently Professor of Practical Architecture to the junior Officers, to which duty Mr. Howe was appointed, in consequence of his superior qualifications and merit.

As the little furnace alluded to, and indeed the whole of the apparatus now employed by us, and also made under Mr. Howe's superintendence, might be extremely useful hereafter, not only to any gentleman, who might be desirous of making similar experiments upon artificial cement, but also to the future manufacturers of that important article, who may prefer testing the quality of any new or doubtful mixture upon a small scale, instead of burning a large kilnful for this purpose, it may be useful to describe our apparatus.

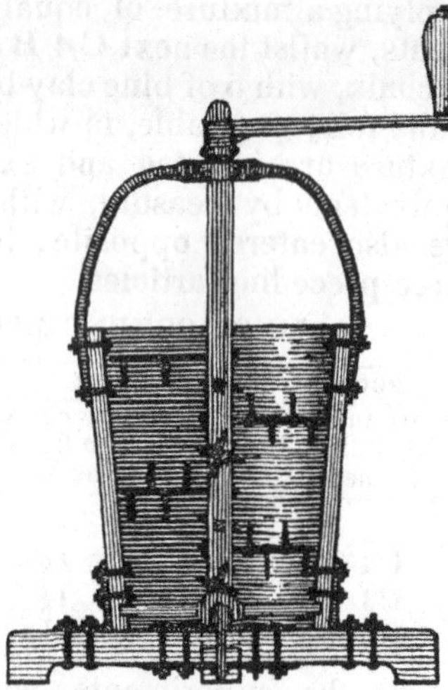

(167) The pugmill in which our chalk powder and clay were mixed, which is represented in section in the annexed figure, was a correct model of the most approved form of that useful machine, excepting the little iron lever at top, which was cut short for a man to work it by hand, instead of representing the usual wooden lever, of much greater proportional length, worked by a horse. The essential parts of the pugmill, are first, a vessel for holding the ingredients to be mixed, which in our model was like a small bucket 8 inches deep, and measuring 5 inches at bottom, and 6½ at top, in clear diameter; and secondly, a vertical iron spindle, with knives and teeth attached to it, which is secured to an iron bow at the top of the vessel, from which it can be disengaged by drawing out a forelock, and has a socket, working upon a pivot at the bottom of it. There were eight iron double-edged knives, each about an inch broad at the end, placed at equidistant heights, within the vessel, and radiating from the four sides of the iron spindle, to which they are fixed, towards the sides of the vessel, which they nearly meet. Four of these appear in full length in the figure, whilst the others are foreshortened, being seen in transverse section or in end elevation. The uppermost knife had two much smaller knives or teeth fixed nearly at right angles to it, and pointing downwards, as shown in the figure. All the other knives had four similar teeth, two pointing down-

wards and two upwards, excepting the undermost which had only two pointing upwards. It will be observed, that the two edges of each knife are not on the same level, but have a moderate slope, though they are all placed horizontally in the general direction of their length. Hence as the axis revolves, the mixture that is to be incorporated, is cut through and through in a vertical direction by the teeth, and in a horizontal direction by the knives, the obliquity of which serves at the same time to force it lower and lower down the vessel by continual spirals, until it is squeezed out at a small hole on one side of the bottom.*

(168) In preparing the ingredients for the pugmill, the dry chalk powder and the proper proportion of blue clay were weighed separately, after which the dry chalk was mixed with water, until it became a stiff paste,† and the whole quantity of chalk paste and of blue clay were then each divided into 16 equal portions or lumps, with as much accuracy as this could be done by the eye. These small lumps of chalk paste and of blue clay were then put into the pugmill, alternately and in regular succession, until the whole space was filled with these ingredients, which were mixed by working the lever; and as fast as the raw cement was forced out from the bottom of the mill, it was taken up by a spatula, and put in at the top again, until the whole of it had passed about four times through the mill, which was done in order to keep the vessel always full, without which the cutting apparatus does not act, it being understood, that in using the real pugmill, new materials are always added, instead of passing the same through it more than once.

The pugmill contained conveniently about one twelfth part of a cubic foot, of raw cement, which after being properly mixed, was made up into nearly equal balls of about two inches and a half in diameter, after which they were burned in the small furnace, which shall next be described.

(169) This was 2 feet high and 11¾ inches square in the

* From 3 to 3½ feet in clear diameter at top, and also in height, are the usual dimensions of pugmills. When filled from wheelbarrows, they are sometimes raised 9 inches or a foot higher than the above on one side. They seldom have a continued bottom, but merely a strong cross of wood work resting on the ground, as a base for the vessel, in the center of which cross the spindle moves. Recently iron pugmills have come into use.

† The quantity of water was not usually weighed, but we have ascertained that our chalk powder perfectly dry required on an average about 1-4th of its own weight, or about 1-6th of its own bulk, of water, to bring it into this state. This water was necessary, because the clay, though wet itself, had not spare water enough to bring the compound formed by adding the chalk to it, into a sufficiently moist state for use.

clear, being made of one entire plate of 3-eighth-inch boiler-plate-iron of the standard dimensions of 2 feet by 4, full measure, cut into four pieces for the sides, which were held together by two square iron hoops outside, and prevented from collapsing by iron studs inside. A small square grate, having half inch bars and half inch openings, was placed 8 inches from the bottom of this furnace, with an ash hole below it, and a small movable iron door above it, fixed upon a hook towards the top of the furnace, and kept close shut by the weight of a handle, long enough to allow a person to take it off when hot, without inconvenience. The space below the level of the grate was contracted by a mixture of siliceous road scrapings, with a small proportion of clay, well rammed round an upright wooden cylinder or core, in the centre of the furnace, to which an opening was cut horizontally by a knife from the mouth of the ash hole, and in the same form. The wooden core was then taken out, and the grate placed and bedded upon the mass of rammed earth, the form of which was altered or improved a little, by paring away with the knife, or adding by the hand. A moderate fire was then lighted, and kept up for a few hours to dry and partially burn the mixture of road scrapings, &c, below the grate, after which the kiln was ready for use. The following figures will sufficiently explain these particulars.

SMALL IRON FURNACE USED AS A CEMENT KILN.

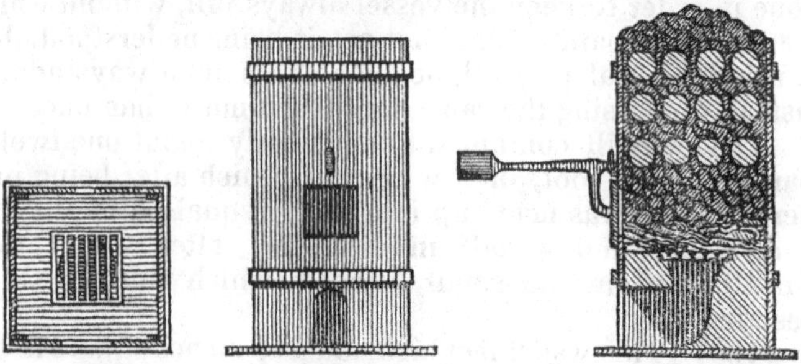

1 *Plan.* 2 *Elevation.* 3 *Section.*

Fig. 1 is a plan, or horizontal section of the furnace, taken above the level of the grate. Fig. 2 is an elevation, showing the ash hole and opening of the door, with the hook above it, but the door itself is supposed to be off. Fig. 3 is a vertical section of the kiln, having the door fixed, and supposed to be filled with fewel and raw cement in alternate layers, ready for burning, which were arranged as follows.

(170) Shavings and wood were laid first upon the grate, with coals or coke over them, to the height of five or six inches, which being lighted, the fire was blown by a pair of bellows through a hole left in one side of the furnace, a little above the level of the grate, until it burned well. Then a layer of cement balls was laid over the whole area, upon which a second layer of fewel was thrown, and then a second layer of cement balls laid, and so on; it being understood that the uppermost as well as the undermost of these alternate layers or courses always consisted of fewel. The little furnace could just hold 3 courses of cement balls, thus alternating with 4 courses of fewel, each course of those balls being nearly equivalent to the twelfth part of a cubic foot, or to one pugmillful of raw cement, whilst each course of fewel measured only about 2 inches in depth, except the undermost and uppermost, the latter of which was heaped.

(171) After trying both, we abandoned the use of coals altogether, because as our little furnace, though placed on the hearth of a Smith's forge,* could not conveniently be brought under the chimney, the coals caused an intolerable smoke, and required the bellows to be kept blowing almost incessantly, and as they generally burned unequally, they either did not calcine the whole of the cement sufficiently, or spoiled a part of it by overburning. Moreover the coals were liable to cake, and required poking, which could not be done without breaking the cement balls. By burning coke on the contrary, the nuisance of smoke was avoided, and after a clear fire with a strong red heat had been obtained, by using the bellows for about a quarter of an hour, it continued in the same satisfactory state, without any necessity either for blowing or poking it, until the cement was properly burned, which was usually accomplished in about three hours; and as part of the calcined cement could be drawn or taken out from the bottom, whilst more raw cement and fewel were added at top, without disturbing the fire, in the same manner as is done in a common lime kiln, this little furnace might have been kept constantly burning night and day; but we seldom found it necessary to renew its contents, or to fill it more than once in the same day, although it might have been conveniently filled three times during the usual working hours. In *drawing* this little kiln, that is in raking out the undermost of the calcined cement

* For this reason we made our bellows hole on the far side of the furnace, opposite to the door, which would not have been convenient any where except in a Smithy.

balls through the door, a small fragment of each was always tested by the diluted muriatic acid, and if effervescence took place, the balls were put in again at the top, along with more fewel to be reburned (45). But if non-effervescence in the acid showed that they were burned enough, any clinkered or overburned portions, which occasionally occurred, were broken off and thrown away; and the remainder of the calcined cement was then ground and carefully bottled up for use, and every bottle labelled.

(172) The raw cement from the pugmill ought to yield an equal quantity or even a little more of calcined cement powder measured lightly; but owing to waste from overburning, &c, the average produce of 10 measures of our raw mixture was only 9½ measures of the cement powder.

(173) To grind our calcined cement, we used a very simple and ingenious little mill, represented in the following figure, which did the same work in much less time and more effectually, than the common pestle and mortar.

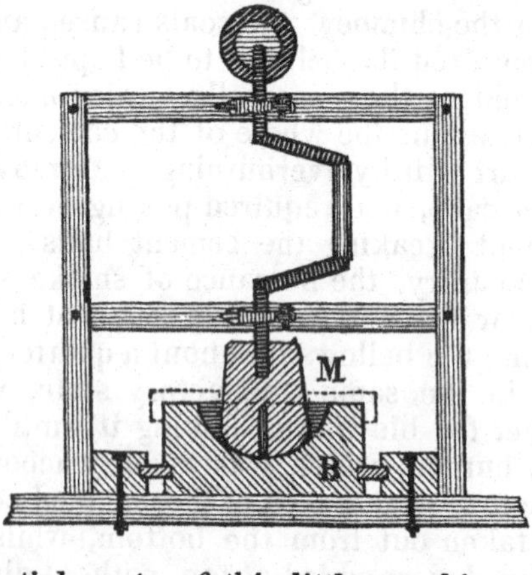

The essential parts of this little machine, said to have been invented for grinding indigo dry, by Mr. Rawlinson of Derby,* are first the stone bed B, having a cavity cut in the upper part of it, in the form of a portion of a hollow sphere, serving as the mortar, and secondly, the upright stone or muller M, nearly of a cylindrical form, but having its bottom convex to suit the former, and acting as the pestle, in grinding the calcined cement, which it does by being made

* See the Article Drugmills, in Sir David Brewster's Edinburgh Encyclopædia. This little machine is said to have been honoured by a medal of the Society for encouraging Arts, Commerce, and Manufactures.

to revolve by turning a crank on its own vertical axis or spindle, which is loaded with a weight at top to give it greater power, for which purpose we used a 5½ inch shell filled with melted lead.

The Muller, which was 10 inches high, and about 6 inches in diameter at bottom, had a rectangular hole cut entirely through the lower part of it, which was essential for the proper action of the machine; and to prevent the fragments and dust, which were whirled violently round, from flying out, the lower stone B, which was square in plan, was covered with a movable wooden lid in two pieces, shown by the dotted lines, and embracing the muller.

The moving spindle was fixed to a wooden frame, and in order to prevent shaking, the whole apparatus shown in the figure was firmly bolted down to the top of a strong well framed wooden tressel, the other parts of which have been omitted for the sake of clearness.

(174) We tried the same little machine for grinding our chalk, but its performance was not satisfactory. In working dry, the chalk became caked between the two stones, and on trying to grind it wet as in the washmill, we found that solid particles remained unbroken, corresponding with the whitening maker's sand (147), which the form of the machine did not afford us the means of separating. We therefore after having previously broken our chalk into small pieces, which we dried by a fire, or by placing them round our small furnace when in use, pounded it with an iron pestle, in a broken ten inch shell, which we used in preference to a common mortar, sifting the chalk powder thus produced through the very fine sieve before described (108); for except in working on a very small scale, the use of a sieve cannot be dispensed with.

(175) It was before stated, that our object now was to compare the strength of four different mixtures, by setting out piers or rows of bricks on edge from a wall, with each sort of them; and as our recent experiments had made us of opinion, that very quick setting was not desirable in a water cement, we resolved that we would only add one brick every day, sundays excepted, in the following manner, to each of our four experimental piers, which were attached to one of the interior walls of what was originally the Farrier's Shoeing shop in the North Stable-yard, Brompton Barracks.

In commencing each of our piers, a small rectangular portion of the wall sufficient for receiving the first brick had the whitewash scraped off, and the original joints of the brick-work raked out about half an inch in depth, which were filled with pure cement, and then the first brick was attached to the face of the wall by more cement, applied before that in the joints had begun to set. Whenever a new brick was about to be added, it was immersed for half a minute in a bucket of water, and the face of the wall or brick already placed, to which it was to be attached, was wetted at the same time by the brush, after which cement was applied to both surfaces by a bricklayer, first in a thin coat to the wall or fixed brick, and then in a thicker coat to the new brick, pressing it down at first into the hollow in the center of this second brick, in order to prevent the formation of those cavities or air cells, which sometimes occur in the joints of bricks, when put together with cement in a hasty manner, as we had observed in former experiments of the same nature (133). The new brick was then applied to the former, with cement against cement, by another man, who adjusted and pressed them together, first with his hands, and then by leaning forward and holding the point of a trowel against the center of the back of the new brick, and as this pressure squeezed out a part of the cement on all sides, the bricklayer finished the joint by working up these projecting edges of moist cement all round with his trowel, pressing them in, and at the same time removing superfluous parts, so as to produce a smooth surface; during which process, and for some time afterwards, the other man still remained steadying the new brick, which he did not quit, until from 5 to 10 minutes after the cement for the joint had been mixed, allowing the longest time to those mixtures, which had the largest proportion of clay. The average thickness of the cement joints thus formed was only a quarter of an inch, but owing to the hollow on one side, with which English bricks are purposely moulded, they were about half an inch thick in the center. The bricks used in all our piers were handsome *Malms* or *Marle bricks,* selected from a brickfield near Chatham, and were previously rubbed together to take off the glaze from the surface, which was done without any perceptible diminution of their thickness. Each cement joint required about 18 cubic inches of cement powder, excepting the first joint connected with the face of the wall, which took double that quantity, in consequence of the original joints of chalk lime mortar, which would have been too weak, having been raked out and made good, as before stated.

(176) In this manner four brick piers were built simultaneously, adding one brick to each daily, by the same men, who commenced at four o'clock every afternoon, sundays excepted as before said, with mixture C B, being the quickest setting mixture, and finishing with C 2 B 3, the slowest of the four, and continued this operation, until the whole of the piers fell down, which occupied not less than 35 days from the placing of the first brick on the 28th of September, to the downfal of the last. During the whole of this period, the little kiln was burning every day, also with the exception of sundays, sometimes containing only one pugmillful of raw cement, sometimes three, which was the utmost that it was capable of holding, and which was equivalent to the fourth part of a cubic foot, but more usually containing two, or the sixth part of a cubic foot. It required a good deal of care and arrangement to avoid error, in preparing and serving out at the same time, so many different sorts of cement for our experiments, which was done quite to my satisfaction by an acting Corporal,* who took charge of every process necessary in our little temporary cement manufactory, from the procuring of the raw materials, to the mixing of the calcined cement powder for the use of the bricklayer, whom he also assisted in pointing the joints. It rained almost incessantly during this period, which included three days of frost and snow, but very few of clear dry weather.†

(177) At the time that every new brick was added, we made one or two experimental balls of each mixture, and put them under water which we changed daily, in order to judge also of their hydraulic properties, and with the exception of C B, in which chalk predominated, and which occasionally fell to pieces in the course of five or six days after immersion, all the others set remarkably well, and so did this mixture itself generally, for those balls which did not fail became very hard. Hence if we had only tried these mixtures on a small scale, we would have considered

* Lance Corporal John Down, who executed most of the other experiments of this year also, assisted by three military boys, of whom he had charge, he being the acting Bugle-Major. Private John Carter was the bricklayer employed. Private Edward Lowe, a mason, held every new brick whilst being placed.

† This damp state of the weather must have been unfavourable to our experiments, because in reference to the circumstance of Messrs. Francis, White and Co. having set out 29 bricks from a wall by their best natural cement, as stated in Article 132, one of the partners informed me, that this could be done in one month in summer, but that it required two months to accomplish it in winter.

C B rather inferior from this circumstance, but we might have thought all the others excellent, and supposed them all to be equally good. This however was far from being the case, as we discovered from the results of our four experimental piers, which proved that there was a considerable difference in the strength of all these mixtures, and consequently convinced us of the great utility of this mode of comparison. The figures afterwards subjoined show the state of each pier before it fell, with its length at the average rate of only 2⅔ inches per brick, which is much less than the average thickness of the courses of common brickwork, and its weight at the average of about 6 lbs, per brick.*

Mixture C B, which was generally of a light brown but sometimes of a dirty white colour, which last we had before ascertained to be an unfavourable symptom of a water cement (149), supported only 21 bricks, and in about half an hour after the last was placed, the pier gave way at the second joint from the wall, leaving two bricks standing.

Mixture C 2 B 3 supported 26 bricks, and mixture C 3 B 4 supported 28 bricks, after which each of their respective piers fell down some time in the night, separating at the wall joint, but tearing out small portions of the original red bricks belonging to the wall, which did not seem so good as those with which we operated. The above three piers, after detaching themselves from the wall by one joint only, were fractured in several other parts by falling, from the height of nearly 5 feet, down upon a floor paved with small hard stones; and in several of these last fractures, two adjacent bricks were both broken obliquely, instead of the separation taking place uninterruptedly, in the continued direction of the cement joint, by which they had been connected.

* The average weight of each of these bricks when quite dry was 5 lbs 5 oz. The surplus weight was caused chiefly by the cement joint, but partly also by their being each saturated with water when placed, so that the last built portion of the pier was always in a damp state.

We also tried to set out a line of bricks from a wall, placing them with their greatest width downwards but end to end, or with the whole length of every brick at right angles to the wall. On allowing only a quarter of an hour for each, it was impossible for any of our cements to set out more than one brick, as the pier always broke down on the addition of a second. We therefore allowed one hour for each brick, and then mixture C B set out three bricks end to end, but on the addition of the fourth, when the pier measured three feet in length, it broke down. C 4 B 5 also set ont four bricks, after which the pier broke down. Upon the whole we did not consider this sort of experimental pier to be at all satisfactory. The smallness of the cement joints renders them much more liable to be deranged, during the placing of each new brick than in the former method, and the leverage abruptly added, by one whole brick's length at once, is too great.

(178) Mixture C 4 B 5 supported 31 bricks, which projected 6 feet 11½ inches from the face of the wall, and weighed 189 lbs, and in 16 hours after the last brick had been placed, the pier gave way at the first joint from the wall, leaving one brick standing. The remainder of the pier, composed of a mass of 30 bricks cemented together, was prevented from breaking after detaching itself from the wall, by a plank covered with bags full of shavings, placed a little below it for this purpose.

SECTIONS *of the* EXPERIMENTAL PIERS *before they fell.*

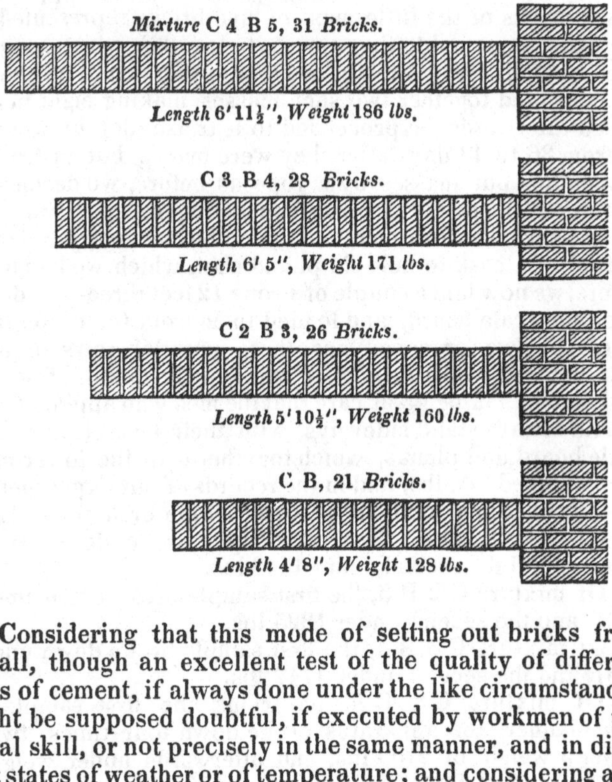

Considering that this mode of setting out bricks from a wall, though an excellent test of the quality of different sorts of cement, if always done under the like circumstances, might be supposed doubtful, if executed by workmen of unequal skill, or not precisely in the same manner, and in different states of weather or of temperature; and considering also that a pier of this sort exhibited at the premises of any cement manufacturer might not be considered satisfactory, by persons who had not been actually present at the time the bricks were placed; since there can be no doubt, that by permanently supporting such piers for several months, a much greater effect would be produced, than by the mode I have described of only holding each successive brick a few minutes by a trowel; it therefore appeared desirable, when we first prepared for commencing those experimental piers, to make arrangements also for trying the comparative strength of the same four artificial mixtures, by tearing them gradually to pieces by dead weight, in the manner before described in Article 133.

(179) For this purpose we cemented together little masses of three bricks each, having mortises cut in the upper and lower bricks for receiving the iron nippers before described, exactly similar to the mortises represented in the upper and lower bricks of the little mass of five bricks represented in the figure annexed to the same Article (133), which has just been quoted. With each of our four experimental mixtures, we cemented together two such masses, making eight in all, which afterwards we proceeded to tear asunder by weights in from 36 to 43 days after they were made; but instead of suspending our masses by a rope as before, we deemed it more convenient to use iron connecting rods, and as the scale board was too small to hold the necessary weights, instead of a cask to hold 24 pound shot, which we had used before, we now laid a couple of strong 12 feet three-inch deals upon the scale board, and loaded them from the center outwards by two or sometimes three parallel rows of iron half-hundred weights, laid on one at a time, until fracture took place; taking great care that the men who applied them should always stand sideways, with their feet clear of the scale board and planks, which together with the lower nippers weighed 205 lbs, and in the records of our experiments, this was of course included as a part of the breaking weight.

Of mixture C B, the first sample broke down under 989 lbs, and the second under 1549 lbs.

Of mixture C 2 B 3, the first sample broke down under 1437, and the second under 1325 lbs.

Of mixture C 3 B 4, the first sample broke down under 1661, and the second under 1717 lbs.

Of mixture C 4 B 5, on trying the first sample in this manner, the apparatus broke down four times, twice under a weight of 2109 lbs, and afterwards under weights of 2389 and of 2333 lbs, but on examination we found in all those cases that the iron work only had given way, whilst the bricks and cement remained perfect. On trying this same sample two days afterwards with stronger iron work, it gave way soon after 44 iron half-hundred weights were

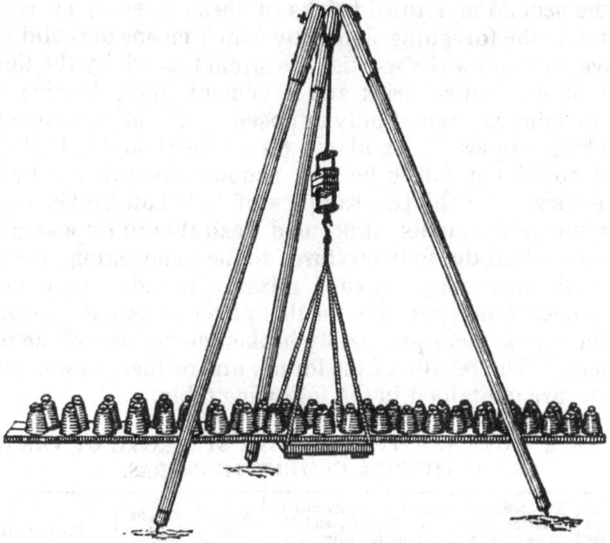

placed upon the planks over the scale beam, as represented in the above figure, the total weight at this period being 2669 lbs, but on examination, the upper brick was found to have been torn entirely across from one mortise to the other, whilst the two cement joints were both perfect. The second sample of the same mixture gave way under 2109 lbs, but in this case also, the uppermost brick only was torn to pieces.

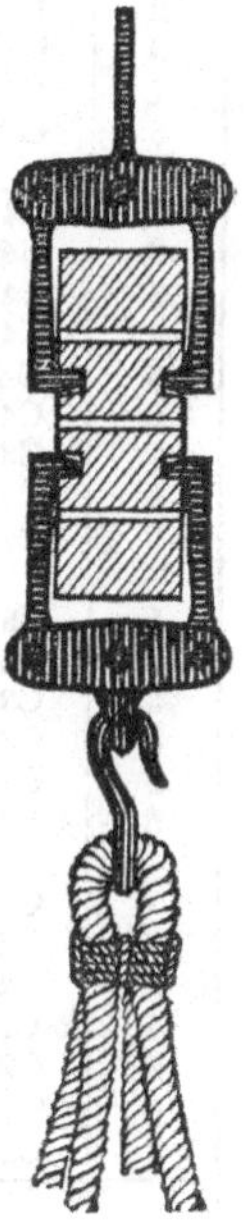

(180) Thus the experiments for tearing asunder the cement joints of these little masses appeared to confirm the general results of the former, inasmuch as the average weights for rupturing them were smaller or greater, in proportion as each mixture had previously stuck out a smaller or greater number of bricks from the wall. These last experiments also seemed to prove that C 4 B 5, the best of our four experimental mixtures, was much stronger at the age of 38 or 40 days, but that the three others were weaker, than the select bricks used by us. In order to fracture the joints of this superior mixture, it therefore appeared necessary to adopt a different arrangement, by forming masses of four bricks, and making the mortises for the nippers in the middle

of the second and third bricks of these masses, as represented in the foregoing figure, by which means the solid part above and below the mortises is strengthened by the thickness of one entire brick and a cement joint, leaving the center joint of cement only exposed to the full action of the breaking apparatus; and we therefore thought, that this joint could not fail to be torn asunder, before the bricks gave way. As the two samples of C B had broken under very unequal weights, it seemed desirable to expose more samples of all the four mixtures to the same action, in order to obtain an average for each mixture founded upon more experiments than two, and for this purpose we used portions of the brick piers previously broken down, as well as new cement. The results of the former and of these new experiments, are contained in the following table.

TABLE I. COMPARATIVE COHESIVE STRENGTH OF VARIOUS ARTIFICIAL CEMENT MIXTURES.

No. of the Experiment with each Mixture	Propoitions by weight of the Artificial Cement Mixture, of C, Chalk, and B, Blue Clay, and whether the Brickwork was an original Mass, or part of a Mass, or of a Pier, previously experimented upon.	Age of the Cement in days.	Weight that tore asunder the Mass, in lbs.	The fracture took place in the
1	C B. No. 1, Mass	35	989	*Joint.*
2	C B. No. 2, Mass	35	1549	*Joint.*
3	C B. Part of Pier	58	2105	*Joint.*
4	C B. Remainder of No. 1 ..	68	1549	BRICK.
5	C B. Remainder of No. 2 ..	68	1885	*Joint.*
1	C4 B5. No. 1, Mass........	38	2109	APPARATUS.
2	C4 B5. The same.	38	2109	APPARATUS.
3	C4 B5. The same.	38	2389	APPARATUS.
4	C4 B5. The same.	38	2333	APPARATUS.
5	C4 B5. The same.	40	2669	BRICK.
6	C4 B5. No. 2, Mass	40	2109	BRICK.
7	C4 B5. Part of a Pier.	65	3257	BRICK.
8	C4 B5. Remainder of No. 1	74	2725	BRICK.
9	C4 B5. Remainder of No. 2	74	3117	BRICK.
1	C3 B4. No. 1, Mass	39	1661	*Joint.*
2	C3 B4. Remainder of No. 1	39	1101	BRICK.
3	C3 B4. No. 2, Mass	39	1717	*Joint.*
4	C3 B4. Remainder of No. 2	39	1325	*Joint.*
5	C3 B4. Part of a Pier	69	2525	BRICK.
6	C3 B4. Part of a Pier	74	4455	BRICK.
1	C2 B3. No. 1, Mass	39	1437	*Joint.*
2	C2 B3. No. 2, Mass	42	1325	*Joint.*
3	C2 B3. Part of a Pier ...	65	1965	*Joint.*
4	C2 B3. Part of a Pier	74	4009	BRICK.
5	C2 B3. Remainder of No. 1	75	2277	*Joint.*
6	C2 B3. Remainder of No. 2	75	1605	*Joint.*
7	C2 B3. Part of a Pier	91	3981	*Joint.*

(181) On examining the contents of the table with attention, it will be evident that the induration of cement is more gradual than is generally supposed, since all those mixtures increased greatly in cohesive strength, by allowing some weeks longer for the cement to set, before the joints were subjected to the breaking apparatus. Thus for example, whilst all were improved by age, mixture C 2 B 3, which had not borne more than 1437 lbs. when about 40 days old, now bore 4009 lbs. in one experiment, in which the bricks gave way and not the joint. Also mixture C 3 B 4 which had only borne 1717 lbs. when 39 days old, bore 4455 lbs. when 74 days old, in one experiment, in which also the bricks broke down, for the new arrangement of the brick masses did not render them stronger than the cement, as we had expected; and unfortunately C 4 B 5, which proved to be the strongest of all our mixtures at 38 or 40 days old, was associated, after the whole had attained the age of about 70 days or more, with much weaker bricks, than either of the two other mixtures that have been mentioned. But supposing this cement still to retain its superiority over those two at the above increased age, which at this period was probable though not absolutely proved, it appeared to me, that the average weight for tearing a joint of this mixture to pieces at 70 days old, could not reasonably be estimated at less than 5000 lbs.

It may be observed, in respect to the general increase of average cohesive strength by age, that the quickest setting cements appear soonest to come to a stand; for C B increased by far the least of all in proportion to its age. We do not know what the increase of C 4 B 5 the next in order in this quality may have been; but the increase of C 3 B 4, and of C 2 B 3 the two next in order, and the slowest setting cements of the whole was very remarkable. And unless experiments of the same kind were tried with hard stone instead of bricks, it is impossible to say, whether these three last mixtures may not, by trying them at still greater ages than 70 days, require much greater weights than 4000 or 5000 lbs. to break them down. It is also doubtful, whether as they all grow still older, the slowest setting cement C 2 B 3 may not gain upon C 3 B 4, and both upon C 4 B 5. In the mean time, it may be observed, that this last cement acquired in the space of 40 days much greater cohesive strength, than that of the best bricks that we were able to procure in this neighbourhood, which I believe to be as good as any made in England.

ADDITIONAL EXPERIMENTS ON THE COMPARATIVE COHESIVE STRENGTH OF QUICK SETTING AND SLOWER SETTING CEMENTS, BY STICKING OUT BRICKS FROM A WALL.

(182) Having some doubts, whether the common mode of setting out bricks from a wall in shorter periods than one day, was a fair trial of the comparative cohesive strength of cements, I tried C B the quickest setting, against C 4 B 5, then considered the strongest, of our cement mixtures, by sticking out bricks from a wall with each, in the same manner that was before described in Article 178, but in a much shorter time, allowing only a quarter of an hour, instead of one whole day, for the setting of each brick.

Adopting this arrangement, C B supported 12 bricks, after which the pier fell down, but on examining one of the joints, it had evidently been injured by moving the last brick in placing it, whilst the cement was setting; and therefore we gave this proportion a second trial, taking care to mix the cement very thin, and indeed almost in a fluid state, in order to prevent its setting before each new brick was definitely placed. Using this precaution, mixture C B set out 18 bricks from the wall.

C 4 B 5, the strongest of our mixtures, on the contrary, only supported 12 bricks when applied at intervals of a quarter of an hour. On trying the same mixture by setting out bricks at intervals of half an hour, it supported 18 bricks, after which it fell down. But on trying it a third time, with intervals of one hour per brick, it supported 24 bricks before the pier fell down, which was a greater number than C B had supported at intervals of one day per brick.

THAT THE MODE OF TRYING THE COMPARATIVE COHESIVE STRENGTH OF TWO DIFFERENT SORTS OF CEMENT, BY STICKING OUT BRICK PIERS FROM A WALL WITH EACH, IS UNCERTAIN, UNLESS MORE PIERS BE MADE, AND MORE TIME AND TROUBLE GIVEN TO THIS SORT OF TRIAL, THAN WOULD GENERALLY BE CONVENIENT.

(183) From these last experiments, it may be inferred, that in trying the comparative cohesive strength of two sorts of cement by setting out brick piers from a wall, the intervals allowed for the placing of each brick should be in proportion to the time required by the slowest setting cement of the two; otherwise an erroneous result may be obtained; for at shorter intervals, the quickest setting cement, though from that very property less fit for the purposes of Practical Archi-

tecture, and in reality the weakest of the two, may stick out the greatest number of bricks.

But as two experimental piers only, one for each sort of cement, might be liable to error, from the great risk of the last cement joint of each being disturbed, whilst every new brick is being placed, which risk increases in proportion to the length of the pier; I consider that to insure any thing like certainty, six such piers, three for each sort of cement, would be necessary, in order to afford a just average of comparative strength; and as the intervals for placing successive bricks in both piers ought not to be less than one hour, or rather two hours for each brick, which would require the persons employed to work night and day; this arrangement would be attended with much trouble and inconvenience. If the workmen left off placing bricks during the night, which might be considered fair enough, as both cements would be treated alike, it would require two days to complete the trial at intervals of one hour, and three days to complete it at intervals of two hours, per brick: and in so doing, if the cements were both tolerably good, one cannot reckon upon using fewer than an average of 20 bricks for each pier, connected to the wall and to each other by the same number of cement joints, and amounting to 120 bricks and 120 joints in all. Besides which, no person who took a great interest in ascertaining the comparative goodness of the two cements accurately, would be satisfied with the result, unless he saw every brick placed himself. Hence this sort of trial, if executed fairly, would give more trouble and take up more time, than any man of business could conveniently devote to it.

(184) After the last experiments which have been recorded, though there seemed good reason for believing that C 4 B 5 was the strongest of all the cement mixtures, that we had yet tried; yet as it was not impossible, but that some intermediate mixture between this and C B, which differed from it in containing less, or between this and C 3 B 4, which differed from it in containing more blue clay, might be still stronger; it seemed desirable to ascertain this point by further experiment, by trying the same mixture in competition with some others, containing rather smaller or rather larger proportions of clay. But for the reasons stated in the preceding article, we determined to relinquish the embarrassing process of setting out bricks from a wall, and adopted a much simpler, less troublesome, and altogether more satisfactory method, in the new series of experiments, which shall next be described.

FINAL EXPERIMENTS TRIED IN 1837, FOR ASCERTAINING THE STRONGEST ARTIFICIAL CEMENT, THAT CAN BE MADE OF CHALK AND BLUE ALLUVIAL CLAY.

(185) The new mixtures were formed by combining the same invariable quantity of dry chalk powder, namely 4 lbs, with nine different proportions of fresh blue clay, differing from each other by one tenth of a pound, and varying from 4 pounds 8 tenths, to 5 pounds 6 tenths, inclusive. It was not my intention to have gone farther than 5 lbs 3 tenths, but the results of the three higher proportions, though prepared by mistake, shall not be omitted; especially as they tended to set one of our former conclusions aside, which proves the great advantage of multiplied experiments, in trials of this kind. By each of these mixtures, 4 or 6 small masses of bricks, or sometimes more, were cemented together in pairs, thus having only one joint in each mass, which we considered the safest mode of proceeding, because there is no risk of its being disturbed, which might be the case, if a third brick requiring a second joint were added. Mortises for the nippers were cut in each pair of bricks in the usual manner, which in order to give more strength to the bricks, were cut nearer to the common joint, than to the top or bottom of the little mass, as shown in the annexed figure.* We fixed upon 10 days, as the proper period for experimenting upon these little masses, by the breaking down apparatus before described, which time we considered more than enough to afford a just estimate of the comparative strength, even of the slowest setting cements, being in fact five times longer, than I had ever known to be allowed for any one joint, in the common mode of trying the strength of cements.† At the same time,

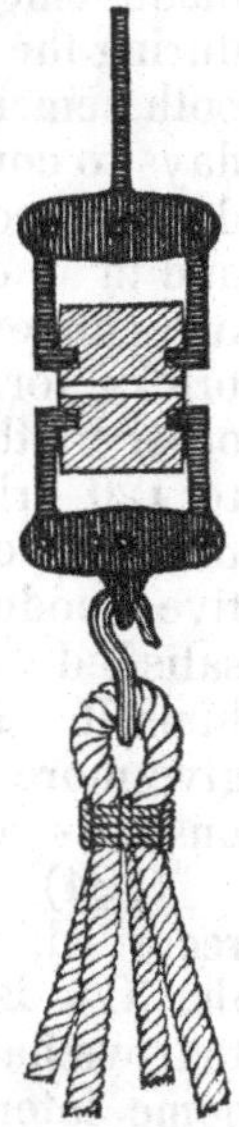

* A piece of cord, not shown in the figure, was always used to connect the lower nippers with the gyn, and prevent them from falling to the ground also, after the scaleboard and weights fell down.

† I allude to the usual mode of setting out bricks from a wall, for which purpose I have never known more than two days per joint to be allowed, which was only done in winter, when cement sets most slowly. See the second note to Article 176. It was not my intention to have allowed a longer period than 10 days for trying any of these mixtures, but the men employed understood, that I meant what they termed 10 clear days for the cement to set, and therefore did not prepare the breaking down apparatus, &c, for my inspection until the 11th day. Thirty days make a great difference in the cohesive strength of cement; but 2 or 3 additional days, or even 4 or 5, appear to make very little difference after the 10th day.

from our former experiments, we concluded that the strongest cement, when only from 10 to 12 days old, would be so much weaker than well burned bricks, that we anticipated no trouble from any of the bricks giving way. The process now adopted was also very convenient, because after the mortises in the sides of the bricks are cut, it only requires one bricklayer and his assistant to be employed for a couple of hours, in putting 10 masses together, consisting of 5 pairs of bricks each, which are quite sufficient for trying any two cements properly; and at the end of the ten days after the apparatus is prepared, the whole of them may be broken down also in a couple of hours, by the assistance of three or four labourers; both of which operations any man may attend personally without inconvenience.

In preparing for this experiment, we had selected the very best bricks that could be procured, partly clamp and partly kiln-burned, not only from the former brickworks, but from two other brickfields, all of which were well burned, ringing well when struck together, and most of them being paving bricks, weighing no less than 6½ lbs, on an average, when perfectly dry. We were therefore disappointed to find, that an unusual number of these bricks gave way, instead of the cement, under weights much less than what bricks, apparently not superior to these in outward appearance, had born without injury, in our former experiments. In fact a sufficient number of cement joints did not break down, to render this new set of experiments conclusive, the results of which, such as they are, have been entered in the following Table.

TABLE II. COMPARATIVE COHESIVE STRENGTH OF VARIOUS ARTIFICIAL CEMENT MIXTURES.

No. of Experiment.	Proportions of the Mixture by weight of C Chalk, and B Blue Clay.	Age of the Cement in days.	Weight that tore asunder the Mass, in lbs.	The fracture took place in the	Average breaking weight in lbs.
1	C4 B4·8	11	936	*Joint.*	997
2	C4 B4·8	11	1304	*Joint.*	
3	C4 B4·8	13	926	*Joint.*	
4	C4 B4·8	13	821	*Joint.*	
1	C4 B4·9	11	1409	BRICK.	1220
2	C4 B4·9	11	1374	*Joint.*	
3	C4 B4·9	13	1108	BRICK.	
4	C4 B4·9	13	989	BRICK.	

TABLE II. CONTINUED.

No. of Experiment.	Proportions of the Mixture by weight of C Chalk, and B Blue Clay.	Age of the Cement in days.	Weight that tore asunder the Mass, in lbs.	The fracture took place in the	Average breaking weight, in lbs.
1	C4 B5	11	1227	*Joint.*	1359
2	C4 B5	11	1101	*Joint.*	
3	C4 B5	11	1493	*Joint.*	
4	C4 B5	11	1073	*Joint.*	
5	C4 B5	11	1409	*Joint.*	
6	C4 B5	11	1437	BRICK.	
7	C4 B5	11	1577	BRICK.	
8	C4 B5	11	1416	BRICK.	
9	C4 B5	13	1402	*Joint.*	
10	C4 B5	13	1155	*Joint.*	
11	C4 B5	12	1465	BRICK.	
12	C4 B5	12	1549	*Joint.*	
1	C4 B5·1	11	709	*Joint.*	751
2	C4 B5·1	11	541	*Joint.*	
3	C4 B5·1	13	989	*Joint.*	
4	C4 B5·1	13	765	*Joint.*	
1	C4 B5·2	11	933	*Joint.*	1412
2	C4 B5·2	11	1521	*Joint.*	
3	C4 B5·2	11	1465	*Joint.*	
4	C4 B5·2	11	1353	BRICK.	
5	C4 B5·2	11	1297	*Joint.*	
6	C4 B5·2	11	1941	*Joint.*	
7	C4 B5·2	11	2193	BRICK.	
8	C4 B5·2	11	1605	BRICK.	
9	C5 B5·2	13	427	*Joint.*	
10	C4 B5·2	13	1388	*Joint.*	
1	C4 B5·3	12	1199	BRICK.	1277
2	C4 B5·3	12	1388	BRICK.	
3	C4 B5·3	12	1332	BRICK.	
4	C4 B5·3	12	1113	BRICK.	
5	C4 B5·3	12	1113	BRICK.	
6	C4 B5·3	12	1514	BRICK.	
1	C4 B5 4	12	1584	BRICK.	1413
2	C4 B5·4	12	1328	BRICK.	
3	C4 B5·4	12	1493	*Joint.*	
4	C4 B5·4	12	1325	BRICK.	
5	C4 B5·4	12	1325	BRICK.	
6	C4 B5·4	12	1423	BRICK.	
1	C4 B5·5	12	1493	BRICK.	1453
2	C4 B5·5	12	1409	BRICK.	
3	C4 B5·5	12	1325	BRICK.	
4	C4 B5·5	12	1633	*Joint.*	
5	C4 B5·5	12	1528	BRICK.	
6	C4 B5·5	12	1325	*Joint.*	
1	C4 B5·6	11	1115	*Joint.*	987
2	C4 B5·6	11	641	*Joint.*	
3	C4 B5·6	11	1115	BRICK.	
4	C4 B5·6	11	1409	*Joint.*	
5	C4 B5·6	11	821	*Joint.*	
6	C4 B5·6	11	821	*Joint.*	

(186) On examing the contents of the foregoing Table with attention, and comparing it with the results of our former experiments, some anomalies appear. For example, mixture C4 B5·1 being an intermediate proportion between C4 B5, and C4 B5·2, if it had been equally well burned and equally well managed with these two, ought undoubtedly to have been stronger than the former of them, and amongst the nine mixtures experimented upon, it ought to have ranked the fifth in point of strength, instead of being the weakest of the whole. Moreover, in our last experiments, C4 B5 proved to be weaker than C4 B5·2, C4 B5·4 and C4 B5·5, whereas it ought to have been stronger than any of them, according to the conclusion drawn from our former experiments recorded in Articles 178 and 180, for it had then proved superior to C3 B4, which agrees nearly with mixture C4 B5·2 of our last experiments.

In these it was a great subject of regret to us, that we should have been so unfortunate in respect to the quality of our bricks, which now broke down under one half, and sometimes even under one third part only, of the average weights that were required to tear asunder the excellent bricks used in our first experiments.* Hence our last experiments afforded no criterion at all of the comparative strength of mixture C4 B5·3, and a very imperfect one of the two mixtures next in order in the Table.

More numerous experiments tried with better bricks, if we could procure them, would undoubtedly enable us to fix the comparative strength of every mixture with sufficient precision, and at the same time would reconcile the anomalies alluded to; for upon the whole, the results of the experiments already tried are in sufficient accordance with each other, to satisfy us, that these anomalies do not arise from any irregularity, in the comparative strength of mixtures prepared and managed exactly in the same manner, but from the imperfections of our own manipulation. In fact it is so difficult to mix, burn and use, the same artificial cement mixture always equally well, that nothing but the average of a very great number of experiments can yield satisfactory

* The weakest bricks now alluded to were marle bricks, of the very same earth as the strongest of the bricks, that we had experimented upon before, made in the same field and under the superintendence of the same Master Brickmaker. On examining the fractures of these weak bricks, it appeared that the ingredients, namely brick earth with a small proportion of chalk to colour it, had not been so well mixed as in the others; for small white lumps of imperfectly ground chalk were visible in many parts. This defect, which was scarcely perceptible in the fractured parts of the strong bricks, cannot be discovered by merely examining the outside.

results; but I was deterred from trying more numerous experiments for this purpose, partly from having no hope at the time of obtaining bricks stronger than the last, and partly from the consideration, that the experiments already tried having proved that our artificial cement after a short time becomes stronger than the strongest bricks, it was unnecessary, as far as brickwork was concerned, to attempt to prove more than had thus already been proved, namely that it was stronger than strong enough.

FINAL DEDUCTION FROM THOSE EXPERIMENTS ON THE COMPARATIVE COHESIVE STRENGTH OF OUR VARIOUS ARTIFICIAL CEMENT MIXTURES OF CHALK AND BLUE ALLUVIAL CLAY.

(187) Upon the whole, on comparing the results of all our experiments attentively, and weighing probabilities where their evidence was imperfect or contradictory, the superiority seems to rest with C4 B5·5, and by reducing this proportion to a more convenient form, it therefore appears that a mixture of 10 parts by weight of pure chalk perfectly dry, with 13¾ parts, also by weight, of blue alluvial clay fresh from the Medway, will produce the strongest artificial cement that can be made by any combination of these two ingredients, and it has the advantage of not setting so quickly either as the artificial cement prepared by us in 1830, or the natural cement produced from the pebbles of the Isle of Sheppy.

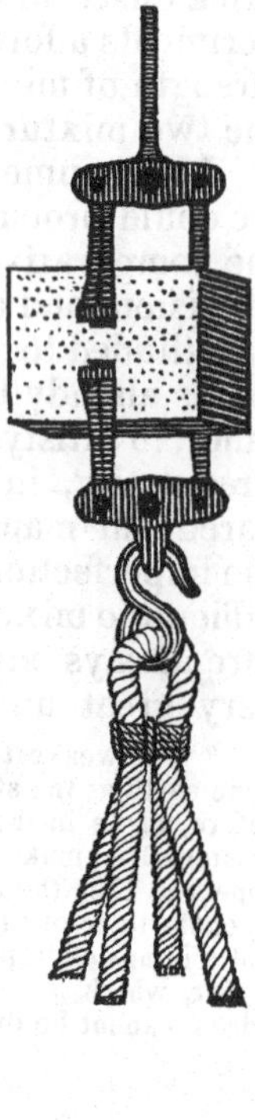

(188) After having discovered what very little dependence could be placed on the cohesive strength of bricks, it appeared desirable, in the event of further experiments on the same subject, to prepare what may be termed stone-bricks, of some kind of stone that should be easily worked, and yet stronger in the cohesiveness of its particles, as opposed to a tearing force, than our best artificial cement. In order to ascertain what sort of stone would answer best, I therefore previously caused pieces of stone of various descriptions, two of each sort, to be cut to the same length and width as common bricks, but to more than double their depth, and to be prepared with four mortises, two on each side, one

over the other, allowing 1¾ inch of solid matter above and below these holes, being a quarter of an inch more than in our brick experiments. On suspending each of these blocks to the triangle by the upper nippers, and attaching the scaleboard to them by the lower nippers, as shown in the preceding figure, and applying weights in the usual manner until they all successively broke down, the results inserted in the following Table took place.

TABLE III. COMPARATIVE COHESIVENESS OF VARIOUS SORTS OF STONE.

Description of Stones experimented upon.	Weight in lbs, that tore them asunder, in Experiment No. 1.	No. 2.	Average fracturing weight, in lbs.
1. Portland Stone	2964	5045	4004
2. Cornish Granite	3741	3941	3841
3. Kentish Rag Stone	3549	3997	3773
4. Yorkshire Landing	3597	3688	3642
5. Craig Leith Stone	2103	2775	2439
6. Bath Stone	1549	1268	1408
7. Pure Chalk	323	623	473

Hence Portland appeared to possess the strongest cohesiveness of all those stones, and being easily worked, and also apparently of a favourable texture for uniting with cement, it seemed the most convenient sort of stone that we could use for further experiments of the nature before described, in which our bricks had failed.*

At this period, our experiments had been sufficiently numerous to enable us to lay down rules that may be very useful to the purchasers of cement.

FIRST. RULE FOR JUDGING WHETHER THE CEMENT SUPPLIED BY A MANUFACTURER IS IN A PROPER STATE FOR USE.

(189) The natural cements obtained from the same spot are not all equally good even if burned alike, and in Go-

* In both trials with the Craig Leith stone, it was torn asunder horizontally according to the grain, opposite to the bottom of the lower mortise, as if it had been split by wedges. In respect to all the other stones, if the solid part above one of the mortises gave way, the fracture diverged upwards, if below downwards, at rather an obtuse angle. Hence when two opposite mortises gave way, the original level surface of the stone above or below them was changed into a ridge in the center, and reduced much narrower, so as to approximate to that form, towards the ends. The same sort of fracture was also exhibited by the bricks previously experimented upon by us, whenever they gave way. It may be remarked that the best of our bricks seemed not to be inferior to Portland stone in cohesive strength, since some of them bore more than 4000 lbs, although they had less solid matter beyond the mortises cut in them. But one cannot depend on the average strength of bricks, with so much confidence as on that of good stone.

vernment Departments especially, where the lowest tender is generally accepted, cements much inferior to the terms of contract are sometimes offered. The mode of experimenting adopted by us at Chatham, affords a sufficiently correct criterion, for judging whether the cement is in a proper state for use, that is whether it is good, well burned and well preserved.

For this purpose, mix up the cement powder with water, and make four or five experimental balls of it, not exceeding an inch in diameter (47). Allow them to set with warmth and cool again, which in good but rather slow setting cement, will require about half an hour, after which put two or three of them into a small basin of water. If they continue to set in this state, and become very hard in the course of a day or two, the cement is good. If on the contrary, they do not become very hard in this time, both inside and out, whether previously kept in air or in water, the cement is not in a state fit for use, and should be rejected.

(190) *Secondly. Rule for ascertaining whether Cement of improper quality, has originally been good, but injured by becoming stale, or whether it may not have been either the produce of bad cement stone, or of good cement stone adulterated after calcination, which two last cases cannot be discriminated, in cement prepared for sale.*

If the experimental balls, made as described in the last article, will not set properly either in air or in water, the next point is to ascertain the cause of their failure. For this purpose burn the same balls in a crucible for two or three hours in a common fire place, exposing them to a full red heat, until they cease to effervesce with acids (45). Then pound the calcined cement in a mortar till you reduce it to an impalpable powder, and mix it up once more into experimental balls with water. If the cement, after being thus reburned, should set well both in air and under water, it is a proof that it was originally good, but had been spoiled by exposure to air or damp.

(191) If on the contrary the cement, supplied by a manufacturer in a state unfit for the purposes of practical architecture, should not be improved by the process of reburning it, which completely restores the virtue not only of stale cement, if originally good, but also of stale lime, as was before explained (146), it is a proof that it has either been made of cement stone of such very bad quality, as would not repay the expense of burning it, or if the pro-

duce of good stone, that it must have been adulterated, by mixing the calcined cement powder with earth or sand of the same colour, before it was casked up.*

RULE FOR JUDGING OF THE COMPARATIVE COHESIVE STRENGTH OF DIFFERENT SORTS OF CEMENT.

(192) Having ascertained that the cements to be compared together are in a serviceable state, by the simple process described in the preceding articles, without which it is of no use to take any further trouble, two modes which have already been described offer themselves for judging of their comparative strength; first the usual mode of setting out bricks from a wall, which I consider objectionable for the reasons stated in Article 183.

Setting aside the above, I know no method so good and attended with so little trouble, as to prepare a breaking down apparatus, such as has before been described, consisting of a scale board, planks and weights, and a couple of pairs of nippers, to be used with a gyn or triangle, as shown in the figure to Article 179, which gyn might be dispensed with, if the cap of a large tressel, or some beam sufficiently strong for suspending the apparatus, were available for the same purpose.

But instead of using bricks, which we found to be so very precarious, I would recommend a number of pieces of Portland or Kentish rag stone, or of any other sound hard stone, each exactly 10 inches long, 4 inches broad, and about 4 inches deep, more or less, for this last dimension is of minor importance, to be prepared with mortises in the sides, each one inch wide and one inch deep, and from three quarters to half an inch in height, to receive the nippers, after being cemented together in pairs, as shown in the following figure, which mortises should have

* Having recently been requested to examine some cement supplied for the works in Chatham Dock-yard, which the receiving Officers had rejected, I not only confirmed their opinion that it was unfit for use, because the experimental balls made of it would neither set in air or water; but as the same failure took place after reburning the cement, I assured them that in all probability it must have been adulterated, by mixing it with some dark coloured earth, for as one must give the person who supplied it credit for common sense, though not for the faithful execution of his contract, it is scarcely to be supposed, that he could have been foolish enough to burn a quantity of bad cement stone. I did not inquire the name of this person, as I rather wished not to know it. I also examined some other cement supplied for the use of the same Dock-yard at the same time, by other manufacturers whose names were made known to me, and found it to be very good. This last was the same cement, with which the Master Bricklayer of the yard had once set out 14 bricks from a wall (132).

at least 2½ inches of solid stone above and below them. The rectangular surface of each stone, where the cement is to be applied, should be roughened every where to the depth of about an eighth of an inch, by furrowing it all round with a half inch chisel, and dotting it in the center by cutting a number of small holes into it with a mason's pick or point, and all the surfaces should be carefully compared, to see that they are as nearly alike as possible. Ten such stones should be provided for each sort of cement that is to be experimented upon, that is 20 for 2 sorts and 30 for 3 sorts of cement, which must be cemented together in pairs, and left 10 days for the cement to set. At the end of this period, let them all be broken down by successive weights, which will allow 5 trials for each sort of cement.

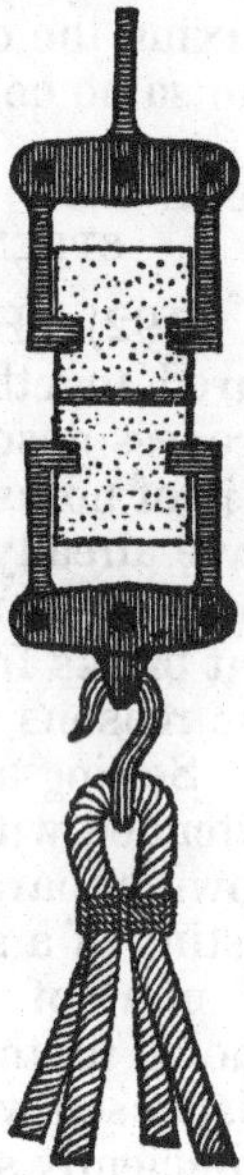

Probably less than 10 days might suffice for a fair trial of comparative strength, and possibly also if the period of trial were increased to 2, 3, 4, or more months, the slower setting cements might gain upon that which was the victor of the ten days. But however that may be, there can be no doubt that when cements, all of good quality and each ten days old, are tried against each other, that which proves to be the strongest of the whole, whether it should be able to retain its superiority at a much greater age or not, cannot fail to be excellent. The small stone-bricks here described may be used for years, if cleaned after each trial, by making the nippers and mortises fit rather tightly, and by tying small strings round the nippers to keep them and the stones together, and prevent them from falling when the cement joint is fractured.

(193) To ascertain more precisely the effect, which sand produces upon the cohesion of cement, which we had long known to be unfavourable, I afterwards caused three little brick masses to be connected by pure cement of mixture C 4 B 5, and three others to be connected by a mixture of the same cement with clean sharp sand in equal parts, by measure; and in order to compare their cohesiveness with that of old chalk lime mortar also, I caused some little masses to be cut out of the best brick walls built with mortar of this description, that I could find within

Chatham Lines, and having prepared these small specimens of brickwork also, with proper mortises to receive the nippers, the whole were torn asunder in the usual manner, by successive weights, as stated below.

TABLE IV. COMPARATIVE COHESIVE STRENGTH OF PURE CEMENT, OF CEMENT MIXED WITH SAND, AND OF COMMON CHALK LIME MORTAR.

No. of the Experiment.	Whether with Cement, or with Chalk Lime Mortar.	Age in Days, or Years.	Weight that tore the joint asunder, in lbs.	Average fracturing weight, in lbs.
1	Pure Cement......	11 days	1241	
2	Pure Cement......	17 days	1003	1092
3	Pure Cement......	17 days	1031	
1	Cement and Sand..	11 days	205	
2	Cement and Sand..	11 days	257	225
3	Cement and Sand..	17 days	313	
1	Chalk Lime Mortar	30 years	334	
2	Chalk Lime Mortar	30 years	64	
3	Chalk Lime Mortar	30 years	75	155
4	Chalk Lime Mortar	30 years	47	
5	Chalk Lime Mortar	30 years	205	
6	Chalk Lime Mortar	30 years	204	

(194) Hence it appears that pure cement is more than four times as strong at the same age, as the customary mixture of cement and of sand in equal parts by measure, which is in common use in and near the metropolis. Mr. Brunel was therefore quite right, in employing only pure cement in the arches of his Tunnel, and I think that the same ought to be done in all arduous works, where the sudden downfal of an arch is to be guarded against, as in tunneling in bad soil, even in dry situations, where an overwhelming inrush of water from above is not to follow. At the same time, in all other situations, where no extraordinary risk is to be apprehended, as in building wharf walls, locks of canals, &c, with cement, the addition of an equal volume of sand is not to be reprobated; because this proportion, whilst it renders the cement mortar cheaper, is not sufficient to take away its hydraulic properties, and even when only 17 days old, it appears from those experiments to equal if not to exceed the strength of the best chalk lime mortar of 30 years of age. I conceive it probable, that if tried at a greater age, the cement mortar would exceed the strength of the chalk lime mortar in a five-fold ratio; for I know that blasting has been required to separate old brickwork built with the former, whilst the latter has no strength

at all in water (15), and so little in air, that several of the chalk lime mortar joints of brickwork of the age before mentioned, separated merely on being handled, after we had cut them out of the walls alluded to for the sake of experiment, without any weights being applied to them, which circumstance I do not consider as a proof that the strength of those joints was absolutely null; because they may have been shaken and weakened in the process of cutting out the bricks round them, which would not have injured old cement at all. For we had before observed, that cement was much less brittle than bricks (177).

(195) Having prepared a number of stone-bricks, as they may be termed, of various sorts of stone, of the dimensions and description explained in Article 192, we cemented them together in pairs by mixture C 4 B 5, and on breaking them down, generally at the end of 11 days afterwards, we obtained the results contained in the following Table.

TABLE V. COMPARATIVE ADHESIVENESS OF CEMENT TO BRICKS, AND TO VARIOUS SORTS OF STONE.

Bricks or sort of Stone used.	When tried.	No. of the Experiment.	Age of the Cement, in days.	Fracturing weight, in lbs.	Average fracturing weight, in lbs.
Bricks	From Dec. 1836, to Feb. 1837, incl.	Twelve	11 to 13	See Table II.	1359
Bath Stone........	April 17th, 1837	1	11	961	1103
	" "	2	11	1031	
	" "	3	11	1010	
	" 28th	4	11	1086	
	" "	5	11	1429	
Cornish Granite	March 28th ..	1	11	657	900
	" "	2	11	520	
	" "	3	11	895	
	April 29th	4	11	1310	
	" "	5	11	1116	
Portland Stone	April 17th	1	11	779	856
	" "	2	11	954	
	" "	3	11	975	
	" 28th	4	11	757	
	" "	5	11	813	
Yorkshire Landing Stone	Dec. 31st, 1836	1	12	1101	823
	April 17th, 1837	2	11	485	
	" "	3	11	604	
	" "	4	11	1010	
	" 28th	5	11	596	
	" "	6	11	1142	
Kentish Rag Stone	April 28th	1	11	1373	1349
	" "	2	11	1240	
	" " ...	3	11	2157	
	" "	4	11	764	
	" "	5	11	1212	

TABLE V. CONTINUED.

Sort of Stone used.	When tried.	No. of the Experiment.	Age of the Cement, in days.	Fracturing weight, in lbs.	Average fracturing weight, in lbs.
Crag Leith Stone ..	May 13th, 1837.	1	11	876	855
	" "	2	11	764	
	" "	3	11	596	
	" " ...	4	11	1093	
	" "	5	11	946	
Cornish Granite polished	May 19th, ...	1	11	1100	928
	" "	2	11	757	

(196) In respect to the above experiments with the small stone bricks, part of the cement was burned in March or in the beginning of April, part towards the end of that month, and part in the beginning of May, and the stones, though all accurately squared to the same gage, were prepared by different masons, and some were much smoother than others, which we ought to have guarded against beforehand, but did not remark the difference until after the first 26 pairs of stones had been experimented upon. Considering therefore that these two causes combined, namely the possible difference in the cohesive strength of the cement, and the known inequality in the state of the surfaces of the stones connected by it, but especially the latter, might have given rise to the principal irregularities in the above results, we resolved, in order to determine this point with more precision, to polish the surfaces of the same four pieces of Cornish Granite used in experiments 4 and 5, and to cement them together again by the same mixture, and to break them down once more in the usual manner, after the cement should have attained the proper age. To our great surprise, the results of these new experiments, which appear as the last in the same Table, prove that the cement adhered to the polished surfaces, with nearly as much energy, as it had done to the same pieces of granite when they were rough, for its average adhesiveness to the polished granite in these two experiments is measured by 928 lbs, whilst its average adhesiveness to the same pieces of granite when rough, as determined by the two experiments Nos. 4 and 5, tried on the 29th of April was 1213 lbs. Moreover the average adhesiveness of the cement to the polished granite exceeded by one half its average adhesiveness to the rough granite, in three out of the first five experiments, tried with the granite in a rough state. These discrepancies are very curious, but the difference in the state of the cement may have caused them.

It is also to be remarked that the adhesiveness of cement to Kentish Rag Stone, and to Bath Stone in four experiments out of five, to Yorkshire Landing Stone in three experiments out of five, and to Cornish Granite in three experiments out of six, is not much inferior to its average adhesiveness to bricks. In the remaining experiments with these stones, the comparative adhesiveness of cement is considerably less; but so far as regarded the Kentish Rag Stone in particular, we ascertained that the great inferiority of adhesiveness in Experiment No. 4 must undoubtedly have proceeded from the cement having been mixed by the workmen in that experiment, in a much more fluid state, than in most of the others, which happening to observe at the time, I caused the circumstance to be noted in the journal of our experiments, and the effect to be watched and recorded afterwards, when all the little masses of that stone were broken down. Never having particularly attended to this circumstance before, I conceive it very probable, that in our former experiments recorded in Table II, Article 185, some of the most glaring inequalities in the cohesive strength of our various cement mixtures, as for example, Experiment No. 9 with mixture C5 B5·2, in which the joint broke down under 427 lbs, though the average strength of 9 other experiments tried with the same mixture exceeded 1400 lbs, may have proceeded from the same cause, namely the too great fluidity of the cement; for we always measured 18 cubic inches of cement powder for each joint, and if the men happened to pour a little too much water over any such portion, we did not correct it properly, by immediately adding more cement powder as we might have done, because we could not afford to waste our cement, and therefore some few of our joints were much thinner than the generality of others.

(197) Upon the whole, I do not consider the experiments recorded in the preceding Table, to be absolutely conclusive, as to the comparative adhesiveness of the same species of cement to various sorts of stone, to determine which with sufficient precision, much more numerous experiments would be desirable. It may however be safely inferred from them, that the adhesiveness of cement to the least congenial sort of stone is more than one half and from thence to two thirds of its adhesiveness to bricks, but that its adhesiveness to some sorts of stone, such as the Kentish Rag Stone is very nearly equal to its adhesiveness to bricks, and it appears to adhere to some of the hardest

stones with greater energy than to much softer ones; and so far as regards the former, our few experiments with the Cornish Granite seem to prove, that the state of the surface whether more or less smooth is of very little importance, so that in building with stone and cement, every mason may be allowed to dress the beds of the stones as he pleases, provided that he adhere to some general rule, laid down in the Architect's or Engineer's Specification.

(198) On comparing these last experiments with those previously recorded in Table IV, Article 193, a fact of the greatest importance is fully proved, namely that *pure Cement attaches itself to the most refractory sort of Stone, with five times as much adhesive force, in eleven days, as the best Chalk Lime used in Brickwork is capable of attaining in* 30 *years.*

(199) At this period, as the small artificial stones in the form of square prisms, made by us for experiment, in in the manner described in from Article 137 to 144 inclusive, which were all kept in a dry apartment, had attained a sufficient age, generally more than 12 months, to enable us to judge of their comparative strength with some degree of confidence, we determined to break them down by weights applied to the middle of each when supported at both ends, by a couple of iron stirrups, at the clear distance of 3 inches apart. Before the prism was placed upon these bearers, another iron stirrup inverted was passed over one end as far as the middle of the prism, upon the upper surface of which the top of this stirrup rested, whilst the bottom of the same stirrup being open we hooked on a very light scaleboard to it, for containing the weights, which were gradually increased till fracture took place. The upper iron work of this apparatus was suspended to the gyn, as in our former experiments. The surface of the two stirrups that supported each prism was flat, but that of the inverted stirrup, which pressed upon the top of it, was in the form of a blunted knife edge.

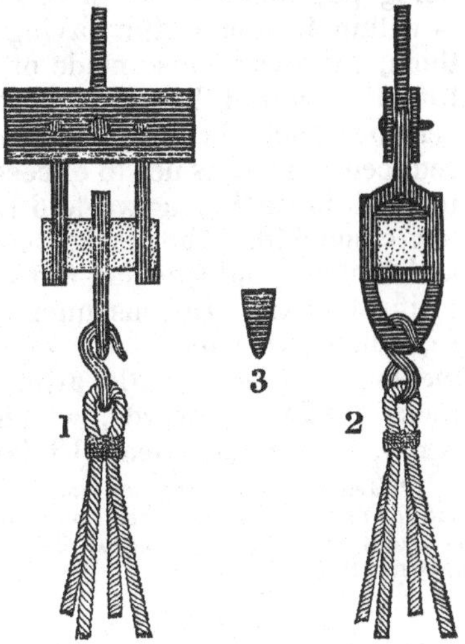

Figs. 1 and 2 in the preceding page represent one of these small artificial stones, or square prisms, about to be experimented upon by this apparatus, the former in side, the latter in end-elevation, whilst Fig. 3, shows the section of the knife edge of the inverted stirrup before mentioned, on a larger scale.

This mode of proceeding is on the principle that had previously been adopted by General Treussart, one of the French writers before alluded to (4), in his elaborate course of experiments on common and hydraulic mortars, and is perhaps the best that could be used for determining their resistance, or comparative strength, but not their comparative adhesiveness to bricks and different sorts of stone.*

(200) Of the two Tables hereafter annexed, relating to these experiments, the first, which is numbered Table VI, refers chiefly to the artificial stones made of Gravel, marked by the letter G, coarse Sand marked by the letter S, and fine Sand marked by a small s, which were described in from Article 138 to 143 inclusive. Some few experiments only in the same table, in which the sand is not specially divided into fine and coarse, imply that the sort of sand used was of the quality proper for making mortar.

The whole of the artificial stones made by us were not forthcoming, some of them having split, or having been broken by accident, or having crumbled into dust from excess of sand, on being merely handled: others, though well adapted for setting in air, having gone to pieces from being put under water in small tin troughs too soon, that is within 15 hours after having been made, which none of them, not even those made of that very powerful lime the blue lias, nor of the still more powerful natural or artificial cements could bear, unless the proportion of sand were reduced so low, as not to exceed the lime or cement powder in much more than a two-fold ratio by measure, which does not come within the usual definition of a concrete at all, and which would be far too expensive to be used as an artificial stone. In consequence of the failures now alluded to, which occurred at the commencement of our experiments upon those little artificial stones, we allowed all those made afterwards, which we wished to try under water, to remain at least 16 days, and some of them even

* General Treussart's experimental prisms were 5 centimetres square, and the distance between their points of support 10 centimetres, which dimensions are very nearly equal to 2 English inches square, and 4 inches between the bearings.

32 days or more in air, previously to immersion, which enabled them to set to such a degree, that most of them seemed proof against the further action of the water, which however was any thing but violent, as they were kept in small tin cases not much larger than the prisms themselves, and the water was not changed, but only made good when lessened by evaporation. The precise number of days previously to immersion are noted opposite to each, in one of the columns of the following Table, it being understood that where no number is entered opposite to any artificial stone, it was not put under water at all.

(201) From the experiments recorded in Table VI, I should be inclined to suppose, that even chalk lime concrete, if kept dry by means of cofferdams or otherwise, until it had set to a certain degree, and if protected afterwards from the action of water in mass, would not lose a certain moderate degree of hardness, whether used for the foundation or the backing of the wall of a wharf or wet dock; but as chalk lime mortar made and used with equal care, is well known to be much inferior in strength to Halling lime mortar: and as I myself never saw any wharf wall laid open, in which the backing, if originally built with chalk lime mortar, was not in a state of pulp (15), I should be very sorry to try the experiment of using chalk lime concrete on a great scale in damp situations. In respect to our experimental artificial stones made with Halling lime, it will be observed that the same mixtures which had been kept under water, were generally stronger than those which had been kept in air. With regard to blue lias lime, considering the many well authenticated reports of its very strong hydraulic properties, we did not think it worth while to expose any of the concretes formed with it to the action of water.

(202) Table VII. refers to those artificial stones or concretes mentioned in Article 144, in which the same sort of screened Thames ballast only was used, in combination with various sorts of limes and cements, and which from this circumstance it seemed best to keep separate from the former, although the ballast may have been very nearly similar to some of the mixtures of gravel and of coarse and fine sand, which are described in Table VI. None of our small artificial stones, made with the screened Thames ballast, were exposed to the action of water.

TABLE VI. COMPARATIVE RESISTANCE OF EXPERIMENTAL ARTIFICIAL STONES, COMPOSED OF DIFFERENT SORTS OF LIME OR CEMENT, MIXED WITH VARIOUS PROPORTIONS OF GRAVEL OR SAND, OR BOTH.

When made.	No. of the Experiment.	Quick Lime Powder or Cement Powder by measure, mixed with G Gravel, S coarse Sand, and s fine Sand.	Whether made with hot or cold water	If kept under water, age when immersed in days.	Age when broken down, in days.	Proportion of Lime or Cement to the Sand only.	Proportion of Lime or Cement to the Gravel and Sand.	Fracturing Weights in lbs.
1836.								
Feb. 2	1	Chalk Lime 1, Sand $3\frac{1}{4}$..	cold	19	432	1 to $3\frac{1}{4}$	......	48
Mar. 10	2	,, 1, Sand $3\frac{1}{4}$..	,,	..	396	....$3\frac{1}{4}$	......	109
,,	3	,, 1, Sand $3\frac{1}{4}$..	,,	..	396	....$3\frac{1}{4}$	......	98
Feb. 6	4	,, 1, G3 S1 s$\frac{1}{2}$.	,,	..	428	...$1\frac{1}{2}$	1 to $4\frac{1}{2}$	87
Jan. 15	5	,, 1, G6	hot	18	457	......	6	crumbled.
,,	6	,, 1, G4 s2....	,,	18	457	2	6	67
18	7	,, 1, G4 s2....	cold	33	454	2	6	69
,,	8	,, 1, G4 s2....	,,	..	463	2	6	60
15	9	,, 1, G5 s2....	hot	18	457	...2	...7	48
19	10	,, 1, G3 S2 s2.	,,	..	446	...4	7	20
15	11	,, 1, G6 s2....	,,	18	457	2	8	56
18	12	,, 1, G6 s2....	cold	..	463	2	8	94
Feb. 6	13	,, 1, G4 S2 s2 .	,,	..	428	..4	8	27
Jan. 19	14	,, 1, G4 S2 s2 .	hot	..	446	4	8	67
,,	15	,, 1, G6 S2 s1 .	,,	32	446	3	9	50
,,	16	,, 1, G5 S3 s1 .	,,	..	446	4	.. 10	39
Mar. 18	17	,, 1, G6 S2 s2 .	,,	..	388	4	.. 10	76
,,	18	,, 1, G6 S2 s2 .	,,	..	388	4	.. 10	71
21	19	,, 1, G6 S2 s2 .	,,	..	385	.. .4	.. 10	crumbled.
Jan. 18	20	,, 1, G7 s3....	,,	..	463	3	.. 10	39
May 9	1	Halling Lime 1, Sand 3 .	cold	..	342	1 to 3	......	179
,,	2	,, 1, Sand 3 .	,,	..	342	3	......	83
Feb. 1	3	,, 1, Sand $3\frac{1}{2}$	,,	20	433	...$3\frac{1}{2}$	......	211
Jan. 21	4	,, 1, G3 S1 s$\frac{1}{2}$	,,	..	444	....$1\frac{1}{2}$	1 to $4\frac{1}{2}$	57
19	5	,, 1, G3 s2 ..	hot	..	462	2	5	87
15	6	,, 1, Sand 6 .	,,	36	457	6	......	197
,,	7	,, 1, G 6....	,,	18	457	......	6	67
Feb. 6	8	,, 1, G 6....	cold	..	444	......	6	56
Jan. 15	9	,, 1, G4 s2 ..	hot	18	457	2	6	209
18	10	,, 1, G4 s2 ..	cold	33	454	2	6	142
15	11	,, 1, G5 s2 ..	hot	18	457	...2	7	151
,,	12	,, 1, G5 s2..	,,	36	457	2	7	69
18	13	,, 1, G5 s2 ..	cold	..	463	2	7	120
19	14	,, 1, G4 S2 s2	hot	..	462	4	8	70
21	15	,, 1, G4 S2 s2	cold	..	444	4	8	86
15	16	,, 1, G6 s2 ..	hot	18	457	2	8	132
18	17	,, 1, G6 s2 ..	cold	32	454	2	8	76
,,	18	,, 1, G6 s2 ..	,,	..	463	2	8	70
19	19	,, 1, G5 S3 s1	hot	32	453	4	9	165
,,	20	,, 1, G6 S2 s1	,,	32	453	3	9	183
15	21	,, 1, G7 s2 ..	,,	18	457	2	9	116
Feb. 22	22	,, 1, G6 S2 s2	,,	..	412	4	.. 10	76
Jan. 15	23	,, 1, G7 s3 ..	,,	18	457	3	.. 10	132
18	24	,, 1, G7 s3 ..	cold	33	454	3	.. 10	64
15	25	,, 1, G8 s3 ..	hot	18	457	3	.. 11	crumbled.
,,	26	,, 1, G8 s3 ..	,,	18	457	3	.. 11	84
19	27	,, 1, G8 s3 ..	cold	32	453	3	.. 11	48

TABLE VI. CONTINUED.

When made.	No. of the Experiment.	Quick Lime Powder or Cement Powder by measure, mixed with G Gravel, S coarse Sand, and s fine Sand.	Whether made with hot or cold water.	If kept under water, age when immersed in days.	Age when broken down, in days.	Proportion of Lime or Cement to the Sand only.	Proportion of Lime or Cement to the Gravel and Sand.	Fracturing Weights in lbs.
April 30	1	Blue Lias Lime 1, G4⅔ s1⅓	hot	..	352	1 to 1⅓	1 to 6	188
„	2	„ 1, G4⅔ s1⅓	„	..	361	....1⅓	6	39
„	3	„ 1, G5 s2 ..	„	..	345	2	7	118
„	4	„ 1, G5 s2 ..	„	..	361	2	7	136
„	5	„ 1, G6 S2 ..	„	..	345	2	8	111
„	6	„ 1, G6 S2 ..	„	..	361	2	8	131
„	7	„ 1, G7 s2 ..	„	..	345	2	9	102
May 9	1	Rosehill Lime 1, Sand 2..	hot	..	342	1 to 2	......	188
„	2	„ 1, Sand 2..	cold	..	342	2	......	210
		Sheppy & Harwich Cements, mixed.						
Mar. 21	1	Francis's.... 1, G1½ S1 s1	hot	..	385	1 to 2	1 to 3½	118
„	2	„ 1, G2 S1 s1	„	..	385	2	...4	48
„	3	„ 1, G4 S1 s1	„	..	385	2	6	102
„	4	„ 1, G3 S2 s2	„	..	385	4	7	24
		Artificial Cement.						
Mar. 21	1	(C5 B2)......1, G1½ S1 s1	hot	..	385	1 to 2	1 to 3½	20
„	2	„ 1, G4 S1 s1	„	..	385	...2	6	27
Jan 26	3	„ 1, G4 S2 s2	„	..	439	4	8	55
„	4	„ 1, G4 S2 s2	„	..	439	4	8	48
„	5	„ 1, G5 S2 s2	„	..	439	4	9	52
		Artificial Hydraulic Lime, 1st sort.						
Feb. 3	1	(C5 B1)......1, G3 S2 s2	hot	18	431	1 to 4	1 to 7	97
„	2	„ 1, G3 S2 s2	„	..	431	4	7	69
„	3	„ 1, G4 S2 s2	„	18	431	4	8	83
„	4	„ 1, G4 S2 s2	„	..	431	4	8	62
„	5	„ 1, G6 s2 ..	„	18	431	2	8	132
„	6	„ 1, G6 s2 ..	„	..	431	2	8	104
„	7	„ 1, G5 S3 s1	„	18	431	4	9	90
„	8	„ 1, G5 S3 s1	„	..	431	4	9	86
		Artificial Hydraulic Lime, 2nd sort.						
Feb. 4	1	(C6B1)......1, G3 S2 s2	hot	17	430	1 to 4	1 to 7	86
„	2	„ 1, G3 S2 s2	„	..	430	4	7	125
„	3	„ 1, G4 S2 s2	„	17	430	4	8	64
„	4	„ 1, G4 S2 s2	„	..	430	4	8	97
„	5	„ 1, G6 s2 ..	„	17	430	2	8	83
„	6	„ 1, G6 s2 ..	„	..	430	2	8	101
„	7	„ 1, G5 S3 s1	„	17	430	4	9	58
„	8	„ 1, G5 S3 s1	„	..	430	4	9	65
		Artificial Hydraulic Lime, 3rd sort.						
Feb. 5	1	(C7 B1)......1, G3 S2 s2	hot	16	429	1 to 4	1 to 7	109
„	2	„ 1, G3 S2 s2	„	..	429	4	7	97
„	3	„ 1, G4 S2 s2	„	16	429	4	8	76
„	4	„ 1, G4 S2 s2	„	..	429	4	8	90
„	5	„ 1, G6 s2 ..	„	16	429	2	8	90
„	6	„ 1, G6 s2 ..	„	..	429	2	...8	125
„	7	„ 1, G5 S3 s1	„	16	429	4	9	55
„	8	„ 1, G5 S3 s1	„	..	429	4	9	98

TABLE VII. COMPARATIVE RESISTANCE OF EXPERIMENTAL ARTIFICIAL STONES, COMPOSED OF DIFFERENT SORTS OF LIME OR CEMENT, MIXED WITH VARIOUS PROPORTIONS OF SCREENED THAMES BALLAST.

When made.	No. of the Experiment.	Proportion of Quick Lime Powder, or Cement Powder, and Ballast, by measure.	Experiment, No. 1. Age in days.	Experiment, No. 1. Breaking weight in lbs.	Experiment, No. 2. Age in days.	Experiment, No. 2. Breaking weight in lbs.	Average breaking weight in lbs.
1836. July 30th	1	Chalk Lime 1, Ballast 3	240	47	270	60	53
..	2	.. 1, .. 4	240	106	270	95	100
..	3	.. 1, .. 5	240	142	270	126	134
July 29th	4	.. 1, .. 6	241	89	271	109	99
..	5	.. 1, .. 7	241	93	271	94	93
..	6	.. 1, .. 8	241	61	271	49	55
..	7	.. 1, .. 9	241	47	271	39	43
..	8	.. 1, .. 10	241	35	271	36	35
July 30th	1	Halling Lime 1, Ballast 3	240	169	270	175	172
..	2	.. 1, .. 4	240	178	270	133	155
..	3	.. 1, .. 5	240	159	270	112	135
..	4	.. 1, .. 6	240	75	270	..	75
..	5	.. 1, .. 7	240	61	270	84	72
..	6	.. 1, .. 8	240	69	270	70	69
..	7	.. 1, .. 9	240	61	270	77	69
..	8	.. 1, .. 10	240	79	270	54	66
Aug. 6th	1	Blue Lias Lime 1, Ballast 3	234	75	263	87	81
1st	2	.. 1, .. 4	239	40	268	..	40
..	3	.. 1, .. 5	239	72	268	68	70
..	4	.. 1, .. 6	239	138	268	100	119
..	5	.. 1, .. 7	239	108	268	53	80
..	6	.. 1, .. 8	239	47	268	44	46
..	7	.. 1, .. 9	239	56	268	52	54
..	8	.. 1, .. 10	239	47	268	40	43
Aug. 1st	1	*Sheppy & Harwich Cements, mixed.* Francis's .. 1, Ballast 2	239	124	268	143	133
..	2	.. 1, .. 3	239	110	268	108	109
6th	3	.. 1, .. 4	234	80	263	54	67
..	4	.. 1, .. 5	234	47	263	20	33
..	5	.. 1, .. 6	234	32	..	..	32
..	6	.. 1, .. 7	234	16	..	..	..
..	7	.. 1, .. 8	234	Accidentally broken			
..	8	.. 1, .. 9	234	Crumbled into dust			
..	9	.. 1, .. 10	234				

(203) It will be observed that the experiments in Table VI, and those in Table VII, are rather inconsistent with each other. In the former, all our concretes made with chalk lime were very much inferior to those made with Halling lime, not only under water, as might be expected, but in air also; whereas in the latter, the chalk lime concretes were generally very little inferior to those of Halling

lime. Besides which, it might be inferred from both of these Tables, that the blue Lias or Aberthaw lime, was rather inferior in strength to the Halling lime. And it might also be supposed from Table VII. in particular, that the blue Lias lime forms a stronger concrete with six times its volume of gravel and sand, than with any smaller proportion of the same ingredients. The first of these conclusions, as to the superiority of the Halling lime over the blue Lias, is in direct contradiction to the experience of the most celebrated Civil Engineers of this Country; and the second, as to the quantity of sand that this lime will bear, is in contradiction to my own observation, as well as to the experience of Captain Savage, who used blue Lias lime not only at the Cobb or Pier of Lime Regis which was built by tide work (26), but at Hobbs' Point near Pembroke, in a much more difficult work executed by the diving bell, which will be noticed in the Appendix. We therefore could not help doubting the accuracy of these conclusions; and afterwards we discovered that the usual mode of making concrete in England, adopted in our first experiments, was not calculated to yield an accurate estimate of the comparative resistance, of artificial stones formed with different sorts of lime.

(204) The mode of proceeding now alluded to, as being objectionable, is the same that has been approved by all our British Architects for making concrete for foundations, and which has more recently been adopted by Mr. Ranger in making his artificial stones, in which he has altered nothing in the former practice, except by substituting boiling water instead of cold. For this last purpose, I now consider the system of working with quick lime fresh from the kiln to be erroneous. Though reduced to a fine powder, and notwithstanding the use of hot water, both of which expedite the slaking, some small portions or particles of lime, being always more sluggish than others, will not slake until the greater part of the mass of the moistened quick lime powder shall have undergone that process. But all calcined calcareous stones swell in slaking, except cement. Hence after the artificial stone moulded in the usual manner, has begun to set, owing to the slaking of the greater portion of the lime with which it has been formed, the more refractory portions of the same lime, in slaking afterwards, have a tendency to split the artificial stone, or to produce cracks, at the surface. But although some of our small artificial stones consisting of blue Lias lime mortar

and concrete had gone to pieces from this cause, namely the mortar alluded to in Article 26, and some concretes also made in 1836, which are recorded in the journal of our operations for that year as having been made but destroyed afterwards, from proving imperfect; and although in respect to the mortar in particular, a remark had been entered in the same journal at the time, that the lime of which it was made, appeared to be imperfectly slaked; yet such is the power of habit, that it never occurred to me, that the customary mode of making concrete in England was objectionable, until I had tried the following experiments.

(205) From the circumstance, so often remarked by me, that the strength of cement is much impaired by adding sand to it, and that the blue Lias lime, which approaches nearer to cement in its water-setting properties than any other of our English limes, in like manner cannot bear so much sand as any of the weaker hydraulic limes or common lime, it occurred, that as pure cement is much stronger than cement mortar, so the blue Lias lime used pure or nett, and mixed with water alone, might in like manner be stronger than the mortars made with the same lime and sand. I had also remarked in reading the experiments of General Treussart, and of M. Vicat in France, that the former had found, in some cases though not in all, that the hydrates of several hydraulic limes, experimented upon by him, were stronger than the mortars of the same limes mixed with sand, whilst the latter stated that small square prisms or cubes of common lime of the same quality as our English chalk lime, mixed with water alone, were much stronger than the mortars of the same sort of lime. Before I commenced my own researches, I should not only have considered M. Vicat's opinion to be quite erroneous, but I should have doubted the correctness of those statements of General Treussart, which were at variance with the general opinion of practical men in this country, who all agree in believing the mortar of every lime to be much stronger than the hydrate or putty of the same lime. But after having tried the experiments with cement and small artificial stones, that have been recorded in the foregoing parts of this Treatise, I became disposed to place confidence in the results of those experiments of General Treussart, which I should otherwise have considered paradoxical, and though not equally inclined to assent to M. Vicat's opinion, yet I felt that I was neither justified in adopting or rejecting the one or the other, without subjecting both to the test of further experiment.

I therefore proceeded to make some small artificial stones of the hydrates of blue Lias as well as of chalk lime, in the same little moulds and by the same process before described (137), using quick lime powder mixed with water whilst fresh from the kiln, and putting it into the mould immediately, and allowing it to heat there. This process, proper for cement only, but which as applied to concretes made of any sort of lime, mixed with a certain excess of gravel and sand, does not always develop its injurious tendency, proved to be quite inapplicable to nett lime mixed with water alone; for no sooner had the lime putty or paste, introduced moist into the mould, begun to slake with heat, than it violently burst to pieces. I therefore afterwards allowed the lime paste to go through the process of heating, swelling, and bursting or cracking first, and mixed it up again with more water before putting it into the mould. Even on taking this precaution with the blue Lias lime, it always split some time afterwards, unless we allowed several hours to intervene between the slaking and the moulding. For example, the first artificial stone or small square prism of blue Lias lime, which I attempted to form one Saturday afternoon, split into three pieces in the course of an hour, though it was formed of lime paste, which had been allowed to become perfectly cool, after having been slaked with great heat in a basin just before. I therefore broke it down again and remixed it with more water, and remoulded it that same evening. On inspecting it on the Monday morning following, I found that this second attempt had also failed, for a large fragment of this new prism had split off during the interval. Though the lime at this time had set into a hardish substance, I caused it to be scraped or pared down into very thin slices by a knife, and then pounded in a mortar, and remixed with an additional quantity of water, and remoulded a third time, after which it set without splitting, and at the end of 14 days from its being first mixed, but only 12 days from its being finally remoulded, we subjected it to the same breaking apparatus, by which we had acted upon the small artificial stones which form the subject of Tables VI. and VII.; and to our great surprise, it required a weight of 328 lbs. to break it down, whereas the same lime mixed as a concrete with six times its volume of gravel and sand, as recorded in Table VI, bore only 188 lbs, though nearly a year old, and when mixed with 6 times its volume of screened Thames Ballast as in Table VII, it bore only 119 lbs, though nearly eight months old; but as other artificial stones afterwards made by us of the same hydrate or lime putty, did not evince the same extraordinary degree of strength, we cannot draw any inference from this remarkable but solitary experiment.

(206) On trying similar experiments with small artificial stones, some made of chalk lime putty, and others of chalk lime mortar, I found contrary to M. Vicat's opinion, that the resistance of the hydrates, to a breaking or crushing weight, was inferior to that of the mortars.

(207) It may however perhaps be admitted as an axiom in Practical Architecture, that whether the strength of any lime be diminished by the addition of sand, which applies to cement, and from General Treussart's experiments, probably to some of the hydraulic limes also; or whether it be increased by the addition of sand, which is believed to be the case with common limes; that the peculiar mode of slaking and using each lime, which produces the strongest hydrate when mixed with water alone, will also produce the strongest mortar or concrete of the same lime, when mixed with the most suitable proportion of sand and gravel also, in addition to water. Hence I am now of opinion, from considering the results of our more recent experiments, that the blue Lias lime used in making our small artificial stones in 1836, should have been allowed not merely to slake with water and cool again, but to slake thoroughly, and for several hours, previously to mixing it with the gravel and sand; instead of mixing the ingredients immediately, and allowing the lime to heat in the mould and cool again, without disturbing it afterwards. There must have been a tendency in the lime of all these little artificial stones to crack, after the greater part of it was set, which process is extremely strong in the hydrates, but as I said before is checked by the interposition of gravel and sand, in proportion to the quantity of those ingredients. But although when thus checked, it may not be powerful enough to split the small artificial stones to pieces, or even to crack them at the surface, I have no doubt that the same action, however diminished, must tend to produce internal defects, and consequently a certain degree of weakness.

(208) Although common limes, which slake much more quickly and readily than the blue Lias, are less liable to be injured by the same mode of proceeding, yet since my attention has been drawn to this subject, I have recently ascertained, that 1 out of 3 artificial stones made with chalk

lime and water, split in the course of 24 hours, and 2 out of 3 artificial stones, made of 1 measure of Halling lime and 3 measures of sand, cracked in several places four days after they were moulded, though all made of lime that was supposed to have been thoroughly slaked. In fact the number of artificial stones experimented upon by us in 1837, was less by about one third than the number actually made the year before; and as we did not always record the circumstances, that caused the defective ones to be thrown away, excepting such as failed from too early immersion, which was so remarkable a fact that it could not fail to draw pointed attention; and as we did not at that time suspect, that the common mode of making artificial stone in this country was objectionable; I can only now state as a matter of opinion, that in all probability a considerable proportion of them must have gone to pieces, from the cause now under consideration. The circumstance of even chalk lime, being liable to injury from the tardy slaking of refractory particles, is so well known to the plasterers, whose inside work is often cracked from this cause, that although this is the quickest slaking lime known, they always drown their lime by *running it*, as it is technically termed, with great excess of water, and leave it a long time before they use it, in order to obtain it perfectly stale, and thus to guard against this unfavourable contingency.

That the usual mode of forming Concrete in England is good for Foundations, but not for Artificial Stones for the walls of Buildings.

(209) Though for the reasons that have been stated, I consider the usual mode of forming concrete with quick lime powder, to be inferior to the system of working with lime thoroughly slaked, as adopted in the construction of the sea wall at Brighton, yet as applied to the foundations of buildings, or to the backing of wharf walls, the use of quick lime powder fresh from the kiln does not appear to be by any means objectionable; for in such very large masses, if a temporary weakness should take place in any part of the concrete, from partial slaking, it may be remedied by the weight of the superincumbent mass afterwards added in a moist state, which will cause the whole to consolidate; whereas in building dwelling houses, as Mr. Ranger proposes with single artificial stones of the same substance, which set at the surface before they are used, a defect in any one stone, arising from the same cause, cannot be remedied in this manner. If it should be split in two, which

I have sometimes observed to be the case, this stone must either be entirely rejected, or the fracture patched with mortar in an unsatisfactory manner.

(210) The impropriety of using lime in the same manner as cement was further proved, when we attempted afterwards to unite some stone bricks, by means of blue Lias lime in its nett state, mixed with water alone, with a view to ascertain the comparative adhesiveness of this lime. By way of experiment, in our three first trials we reduced the blue Lias lime fresh from the kiln to the state of quick lime powder, by pounding it in a mortar; and then making it up into a stiffish paste, we applied it immediately to the proper surfaces of both stones, which we cemented together, and pointed the joint all round, in the same manner precisely, as we had done in all our cement experiments; so that the union of the two surfaces was completely effected, before the lime paste had thrown out any perceptible warmth. In the course of less than an hour, the blue Lias lime heated, and swelled in all directions whilst slaking, so that the edges of the lime joint projected perceptibly beyond the sides and ends of the stone, with which they had originally coincided, which alteration of form, combined with its upward expansion in the direction of least resistance, caused it to crack horizontally throughout nearly its whole upper surface, and to detach itself from the upper stone entirely or nearly so; at the same time, that it also separated but more partially from the lower stone. On removing the upper stone, the furrows and holes cut in the roughish surface of it by the masons, left their impression on the cement in relief, in as perfect a manner as could have been obtained by means of a plaster cast: and on separating the lime from the lower stone also, which it quitted with little difficulty and without breaking, the same appearance was exhibited on the under side of the thin cake of hydrate of lime, which had for a short period been the common joint connecting these two stones. This specimen, on being preserved, became exceedingly hard, and appeared to set into as firm a substance nearly as the joints of similar experiments, in which pure cement had been used. And I have no doubt, but that the blue Lias lime would unite stones together nearly as powerfully as the best cements are capable of doing, if it did not swell in all directions, when the attempt is made to combine the two processes of slaking and setting, which take place simultaneously in cements, without increase of bulk, and consequently without forcing them to split or to detach themselves from stones or bricks.

(211) Upon the whole, I now attach little value to the results of the experiments recorded in Tables VI. and VII. but whilst I regret that our investigation of the comparative resistance of artificial stones or concretes, as attempted in those experiments, should be so unsatisfactory; yet considering that it would require at least six months from the present period, to try them again in a more accurate manner, I am unwilling to defer the publication of this Treatise any longer, my chief object in first undertaking it having been rather to ascertain the properties of calcareous cements, than of calcareous mortars or concretes, although they are too intimately connected, to allow either to be discussed without some reference to the other.

(212) As the value and importance of artificial stone used for the walls of buildings, or for those of wharfs or docks must depend upon its strength in opposition to a breaking weight, it now appeared desirable to ascertain its resistance in competition with that of the common building stones of this country, as well as with that of bricks and of pure chalk from the quarry, for which purpose, I caused a number of similar small prisms each 4 inches long and 2 inches square to be cut out of all those substances, which being subjected to the proper breaking apparatus, yielded the results contained in another Table, No. VIII, and in order to render this more complete, the cohesiveness of the same stones has been repeated from Table III, whilst that of well-burned bricks and of inferior bricks have been estimated, from the average of the strongest and some of the weakest results, recorded in Tables I and II.

TABLE VIII. COMPARATIVE RESISTANCE OF VARIOUS NATURAL STONES, BRICKS, AND CHALK, REDUCED TO SQUARE PRISMS OF THE SAME DIMENSIONS AS THE SMALL ARTIFICIAL STONES BEFORE EXPERIMENTED UPON, WITH THEIR COMPARATIVE COHESIVENESS ALSO.

Description of Stones, &c.	Weight of Prism, in Troy Grains.	Weight per cubic foot, in lbs.	Breaking weight in lbs, in several successive Experiments.			Average resistance, in lbs.	Average cohesiveness, in lbs.
1. Kentish Rag	10739	165·69	4286	3817	5099	4581	3773
2. Yorkshire Landing.	9571	147·67	2976	2500	3185	2887	3642
3. Cornish Granite....	11164	172·24	3179	2801	2445	2808	3841
4. Portland	9598	148 08	2195	2892	2958	2682	4004
5. Craig Leith	9383	144·77	1940	1786	1961	1896	2439
6. Bath	7945	122·58	708	694	596	666	1408
7. Well burned Bricks.	5944	91·71	704 955 722	795 622 706	717 640 823	752	3007
8. Inferior Bricks	....	....	204	262	522	329	1105
9. Pure Chalk (dry) ..	6157	94·99	414	265	314	334	473

(213) From the above experiments it appears, first, *that the* RESISTANCE *of various stones in opposition to a breaking force, is not in proportion to their specific gravity, nor in any direct proportion to their* COHESIVENESS *in opposition to a tearing force.*

Secondly. An examination of the results recorded in this last Table, compared with those of Tables VI. and VII, will perhaps convince the reader as it has done me, *that Artificial Stone made according to Mr. Ranger's system is very much inferior, not only to all the natural Building Stones in common use in this country, but even to sound well-burned Bricks.* For in the walls of buildings, which are exposed to weather, and may be subject to external violence; that sort of resistance in which this substance seems to be deficient, is one of the most important qualities, though not in foundations and backing, which are exposed to dead weight only.* In fact those experiments prove, that artificial stone after allowing a year for the lime to set, is inferior in strength even to common chalk in its dry state, which has been rejected for many centuries as being unfit for the purposes of building.† Supposing that my objections to the usual method of forming artificial stone with quick lime powder as developed in Articles 204, 205, and 207, be well founded; it follows, that a stronger artificial stone may be formed by working with thoroughly slaked lime. But admitting this much, it must also be allowed, that no conglomerate substance composed of small ingredients, can be stronger than its weakest part, and therefore the resistance of an artificial stone is not to be measured by that of the flinty pebbles or gravel which are its strongest part, but by that of the comparatively weak lime or mortar, by which those pebbles are connected together, and which I believe can never rival the hardness of any sort of stone, except in a very thin coating or skin just at the surface. Upon the whole therefore, I would recommend that even Bath stone, the weakest build-

* Mr. Ranger having at all times, with great civility, afforded me every information I requested relative to his proceedings, and he being the only person in England, who makes artificial stone, from these circumstances combined, I really feel considerable reluctance in giving my opinion of it, which is much more unfavourable, than it was when the first sheets of this Treatise were sent to press. But as he is not merely a maker of artificial stone, but a builder of skill and reputation, I sincerely hope, that if the former branch of his business should decline, the latter may prosper, so as to make him ample compensation.

† Chalk was used probably before the Norman Conquest, for the filling in of thick walls, and also occasionally for inside vaulting, of which specimens may be seen at Canterbury and Rochester.

ing stone in England, or where stone is not to be had, that good brick-work, stuccod with cement, should be used in preference to the best artificial stone that has hitherto been invented, not only for public edifices, but even for the mansions of opulent private individuals.* At the same time, I shall repeat my former observation, that *I consider Concrete, if kept within proper bounds, to be a most excellent and useful Expedient.* Rejecting it for the walls of buildings, and even for every description of arch or vault, whether above or below ground, it should be confined to the foundations of buildings chiefly, in addition to which it may however also be used occasionally, but with judgment, for the backing of wharf walls, and for the formation of retaining walls, such as that of the Eastern Cliff at Brighton. It may also as I said before be used for Military Revetments, in peculiar situations, as for example in fortifying sandy islands (32).

EXPERIMENTS TO ASCERTAIN WHETHER CEMENT MAY BE USED TO ADVANTAGE IN MASONRY, ESPECIALLY THAT COMPOSED OF VERY LARGE STONES.

(214) The circumstance of cement having hitherto been used almost exclusively for brickwork, in and near the Metropolis, has produced an impression on the minds of many practical men, that it is not applicable to stone walls, especially if to be built of very large heavy stones, for two reasons; first from a supposition that it does not attach itself well to stones, and secondly because it sets too quickly to admit of a very large heavy stone being properly adjusted, without disturbing the bed of cement upon which it is to be laid. The first supposition has been proved to be erroneous, by the experiments recorded and commented upon in the preceding articles. The second, which I considered no less unfounded, induced me to try the following experiments.

Having provided two large Bramley-fall stones of the same size and weight nearly, I caused the top of one and the bottom of the other to be dressed in the same manner, as would be proper for the common joint of the stones of

* I do not consider that the resistance of the concrete casemate at Woolwich to the fire of heavy Artillery, as recorded in Articles 32 and 33, invalidates this opinion, because we found by our experiments at Chatham, that the penetration of a 24-pound shot into a parapet of loose fine sandy soil was only double its penetration into the wall of that casemate. In the absence of positive experiments, I conceive that the resistance of the weakest sort of stone now used for the purpose of building, would exceed that of loose sand in much more than a twofold ratio.

two successive courses of large ashlar masonry, suitable for the wharf wall of an important Pier or Harbour, or for the wall of a Light-house exposed to the action of the sea. This was done by cutting them all round by a mallet and a bosting tool or chisel 1½ inch broad, and afterwards cutting three intermediate spaces, parallel to and equidistant from the long sides, and by then picking down the remaining rectangular portions of the surface by a pointed tool, as shown in the annexed figure, in which the small furrows cut by the chisel are represented by short parallel lines, whilst the pick holes are dotted, neither of them penetrating more than one eighth of an inch below the general surface of the two stones, each of which after being thus dressed, measured nearly 39 inches long, 29 inches broad, and 26 inches deep, and weighed about 2662 lbs.

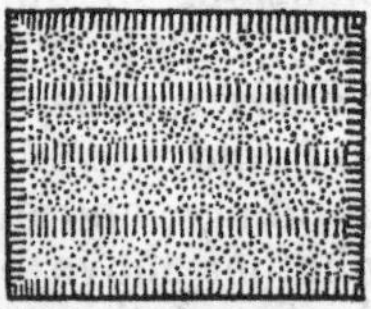

That stone which was to represent the lower course of masonry in our proposed experiment, had four iron bearers, each 14 inches long, 4 inches deep, and three quarters of an inch thick, let into it, so as to project 10 inches from opposite sides, and towards the ends of the stone, in a horizontal direction. These bearers were fixed with the same artificial cement about to be experimented upon. The upper stone had an eyebolt sunk into it to the depth of 16 inches, for raising it when necessary, which was of a dovetailed form at bottom, and wedged into a hole cut in the stone, of much greater diameter than the shank of the bolt, by small fragments of granite.*

After being thus prepared, the first stone was set upon strong wooden skids, to represent as before said one stone of a finished course of masonry, and the other being raised by hooking on the eyebolt to the tackle of an artillery gyn, was let down upon the former, in such a manner, that the two

* This eyebolt was made of 1¼ inch bolt iron, and let into a hole 3 inches in diameter cut in the stone. At the distance of 4 inches from the lower end of the bolt, its thickness was made to increase from thence downwards, in a conical form, till it became about 3 inches in diameter, and thus nearly fitted the bottom of the hole. After inserting the eyebolt into its proper place, small pieces of broken hard stone were poured in, which filled the cavity round the upper part of the hole, and became jammed there as soon as a purchase was applied. This simple arrangement suggested by Mr. Howe prevented the bolt from being drawn out of the upper stone, when afterwards acted upon by a total weight exceeding 39,000 lbs. avoirdupois. He had previously recommended it to Captain Chesney, as a good mode of fixing moorings or warping eyebolts in rock, for the proposed navigation of the Euphrates, which was afterwards undertaken by that Officer, with the local rank of Colonel in the East.

surfaces which were to be afterwards cemented together, were made to coincide properly, which being of equal size and both out of winding, they could not fail to do.

Having thus adjusted the stones in a dry state, we raised the upper one again to the height of about 1 foot by the gyn, after which the cement, which consisted of mixture C4 B5, was made up with water into a stiffish paste, and thrown upon the top of the lower stone, upon which one bricklayer spread it properly, whilst another applied the superfluous parts of it in a thin coat to the bottom of the upper stone, in the same manner as would be done in plastering a cieling, both working with long-handled tools resembling large garden spuds or small spades, instead of the trowel, which could not be used conveniently in the confined space between the two stones. After the surfaces of both were thinly coated with cement, they were united by letting the upper stone gently down into its proper place upon the lower one, where by its great weight, which exceeded a ton, it pressed down the bed of cement, which was still soft, to a regular thickness, squeezing out a part of it on each side out of the joint, which was pointed all round by the bricklayers with their trowels.

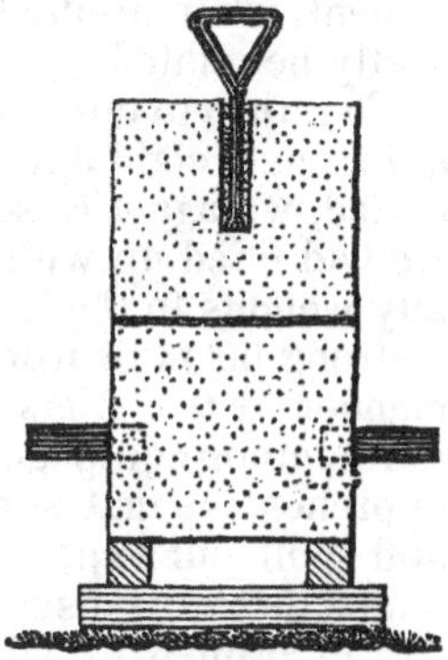

The gyn was then removed, and the stones thus cemented together were allowed to remain upon the skids, in the state represented in section in the annexed figure, which shows two of the four iron bearers that were cemented into the lower, as well as the iron ringbolt that was sunk into the upper, stone.

At the end of ten days, the gyn was again erected, and the upper stone hooked on to it, and raised about an inch, on which the cement joint gave way, being torn asunder by the lower stone alone, which with the iron bearers attached to it, weighed about 2,688 lbs. This first experiment was by no means promising, but as the stones had been cemented together during a frost, it seemed possible that the weather might have affected them. If not, 10 days was evidently too short a period, to enable the cement joint to bear the weight of the lower stone. We therefore determined to try the same experiment again, but allowing more time for the cement to set.

Accordingly having cleaned the proper surface of each stone, we cemented them together again, and at the end of 60 days, we raised them up by the gyn, and having laid two strong deals longitudinally upon the iron bearers on each side, and short sheeting or mining planks transversely over the ends of these, so as to form a sort of double platform, resting upon the lower stone, we laid successive weights upon these deals and planks, until the mass broke down and fell to the ground, through a distance of about 1 foot, for at this period we had removed the skids. The total weight applied was 7083 lbs; but on examination it proved to be the hook of the artillery gyn that had snapped in two, which must have been of bad, probably remanufactured, iron,* whilst the cement joint of the stones, notwithstanding the concussion of the fall, was to all appearance still perfect,† but further experiments afterwards led us to believe that this opinion was erroneous; for if C 4 B 5, the artificial cement now used, had not been seriously injured by the fall, it ought to have proved, as in all other trials, nearly equal to the best of the natural English cements; whereas it did not, after this accident, evince more than one fourth of the adhesiveness of Messrs. Francis's English cement, when applied to the same two stones, as will presently be related.

To return to our first experiments with the large Bramley-fall stones, eight days after the accident that has been mentioned, we again raised the stones by the same gyn, which we had fitted up with a new hook, but before we applied any weights to the lower stone, we suspended the mass to a strong beam of wood, properly supported in a temporary manner by two piers made of the woodwork belonging to our pontoon equipage, and then removed the gyn, as we were apprehensive that some other part of it might break down, and spoil our experiment. We then laid on the longitudinal deals and transverse planks as before, and loaded this sort of framework with paving stones, each weighing from 30 to about 70 lbs, and over them iron 56 lb weights. The annexed figure represents this arrangement, in which the

* Remanufactured iron, which from my experience I consider to have been of very inferior quality, was used for some years in the Ordnance Department, probably from its cheapness; but as nothing that is bad is really cheap, it has since been disused by order of the Board.

† This experiment was tried in presence of Sir James R. Carnac Chairman, and several Directors of the Honourable East India Company, on the 30th of March, 1837, after they had inspected their Depôt under the command of Lieutenant-Colonel Hay, for which purpose they had visited the Garrison of Chatham.

beam that bore the whole weight, and the transverse planks are represented in section, the other parts being distinctly shown in elevation, except the temporary wooden piers, which supported the beam, of one of which the outline only is marked by dotted lines.

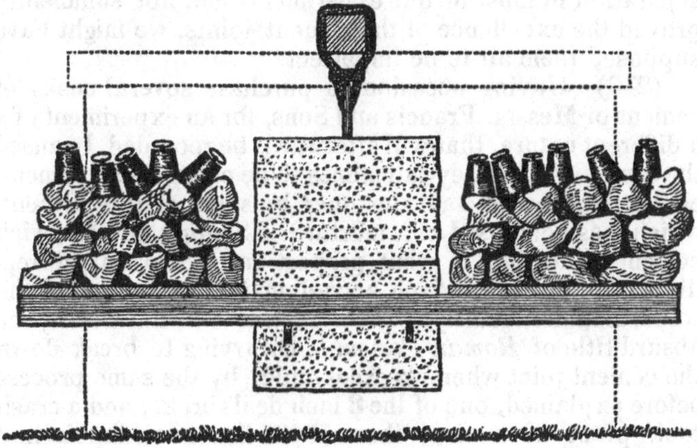

Thus we proceeded as before, continually adding more weights, until the mass broke down under a weight of 9660 lbs, the details being as follows.

Weight of the lower Bramley-fall stone....	2656 lbs.
Four iron bearers	32
Four 3 inch deals laid longitudinally	311
Eight planks laid transversely	224
Sixty-nine pieces of paving stone	3637
Fifty iron half-hundred weights	2800
Total weight that broke down the mass,. .lbs.	9660*

(215) *Usual appearance of cement joints after being fractured.* On examining the cement joint, which was the part that had now given way, it did not average more than one tenth, and no where exceeded one eighth of an inch in thickness, excepting where it entered into the chisel furrows or pick holes; and the greater part of it adhered to the upper stone, leaving the roughish surface of the lower stone generally as clean to all appearance, as if no cement

* This experiment was tried in presence of Lieutenant-Colonel Hay and Major Somerville of the East India Company's Depôt, Captains Jebb and Mackenzie of the Royal Engineers, and other Officers both of the King's and Company's Engineers. Mr. Howe was also present and directed the workmen, in this, as in all the other experiments.

at all had been applied to it; and we had observed the same circumstance in our experiments with bricks and stone-bricks, in which also, whenever a cement joint gave way, one of the two surfaces connected by it, appeared nearly clean. Hence if the great weights required to effect separation in most of our experiments had not sufficiently proved the excellence of the cement joints, we might have supposed them all to be imperfect.

(216) Having occasion to purchase several casks of cement of Messrs. Francis and Sons, for an experiment of a a different nature, that will afterwards be recorded, I caused the same two Bramley-fall stones to be cleaned and cemented together in the same manner by some of this cement, which was composed of a mixture of Sheppy and Harwich cement stone, but in what proportions the manufacturers did not inform me, which according to the custom of the trade, though in reality *English Cement* was falsified by the absurd title of *Roman Cement*. On trying to break down the cement joint when six weeks old, by the same process before explained, one of the 3 inch deals broke, and a crash took place after more than 19000 lbs. avoirdupois had been applied, but the cement joint was uninjured, as part of the weights only had fallen to the ground. Three days afterwards, having suspended the stones by two fir beams, instead of one, which we deemed unequal to support the probable weight, and having laid a couple of oak beams, 14 feet long and 9 inches square, upon the iron bearers, instead of the 3 inch deals, and having laid double rows of short planks transversely across the ends of the beams, to support the necessary weights, we attempted to break down the same cement joint a second time, having previously borrowed a quantity of iron pig ballast from Chatham Dock-yard, which we thought in addition to our former weights, would be quite sufficient to effect this object. Instead of which to our extreme surprise, after exhausting all our iron weights, and ballast, and a number of paving stones, we laid on sand bags filled with earth, until they were heaped so high that no more could be added without oversetting the pile, and therefore we left off for the night, without removing any of the weights, with an intention of borrowing more iron ballast from the Dock-yard next morning. But probably, in consequence of the rain which fell during the night having increased the weight of the sand-bags, a fall accompanied by a loud crash took place about one o'clock in the morning; but on examining the appa-

ratus, and clearing away the confused masses of fallen weights after daylight, the cement joint was found to be still perfect, instead of which the lower stone itself had given way, a fragment of which weighing 203 lbs, of an irregular triangular form had broken off, immediately under one of the four iron bearers, which was bent downwards, but not broken. As soon as this gave way, one of the oak beams being deprived of its support at one end, snapped in two, and the other beam also broke in consequence. The detail of the parts of the load that had been used, which were carefully weighed next morning whilst the sand bags were still damp, was as follows:

Lower stone, and its iron bearers..........	2662 lbs.
2 Oak Beams	679
22 two-inch sheeting planks	611
60 Iron 56 lb. weights.....................	3360
Iron pig ballast of different sorts	15242
Paving stones of Guernsey granite	3097
159 bushel sand bags filled with earth	10893
Total lbs...	36,544

(217) The annexed figure represents in perspective the two stones that had been experimented upon, after the weights that had broken off a part of the lower one were removed. As there was now no possibility of separating the cement joint by weights, in consequence of the mutilated state of the lower stone, we split it by means of chisels and mallets, and were not a little surprised to find, that with the exception of the outer part of the cement that had been

exposed to the air, which was extremely hard, the whole inside of the cement joint was softish, and neither resisted the action of the thumb nail, nor of a sixpenny piece on edge, which being drawn along the surface, scored it to the depth of nearly a sixteenth of an inch. The two fir beams, having a piece of teak over them in contact with the rope, by which the stone and load were suspended, had undergone a deflection of one inch in 6 feet 8 inches of bearing, and the oak beams had undergone the same nearly, but in a contrary direction, being supported near the middle and not at the ends; but the 2 inch tarred rope, which was used in 23 folds was not injured, and not the least extraordinary part of this remarkable experiment was the resistance of the small pieces of broken granite, by which the eyebolt, bearing the whole weight, of more than 29,000 lbs, was jammed in the upper stone. The strength of the iron bearers, each subject to a weight of more than 7000 lbs, was also worthy of notice, being a proof of the efficiency of iron cantilevers set in stone work by cement, for the purpose of supporting balconies, &c.*

(218) The experiments thus tried with the large Bramley-fall stones, having led to no decisive results, except inasmuch as they proved that a large cement joint of good natural English cement, 45 days old, is at least as strong as that excellent sort of stone, if not more so; I next determined to try the strength of two similar cement joints only 11 days old, in competition with each other, one of Francis's best cement, such as had been used in the last experiment, the other of C4 B5·5; for which purpose we prepared four Bramley-fall stones, each measuring 1000 superficial inches on the corresponding surfaces, which were to be cemented together, and each averaging from 11½ to 12 inches in thickness, so that they did not weigh quite half as much as the first large stones, out of which in fact three of them were made, by splitting the former at the middle of their original thickness. This experiment, which we hoped would have been definitive, unfortunately led to no satisfactory result, in consequence of the eyebolts, which were now fixed only 9 inches into each stone, drawing out of the holes, before the cement

* This experiment was tried on the 9th of June 1837, in the presence of numerous spectators, who were perfectly astonished at the extraordinary strength of the cement joint. Besides all the Officers of Engineers, Colonel Warre the Commandant of the Garrison of Chatham, and a great number of Officers of various Regiments attended, as well as Captain Clavell, Superintendant of Chatham Dock-yard, together with several members of the Philosophical and Literary Institution, and other Gentlemen, Tradesmen, and Mechanics.

joints were injured. Hence the two stones united by Francis's cement fell down, together with the platform and weights applied to the lower stone, when the weight acting upon the joint amounted to 4193 lbs, and after falling from the height of about 1 foot upon a couple of wooden skids, the cement joint proved to be broken by the fall. The weight bearing upon the joint of the artificial cement C 4 B5·5 amounted to 5975 lbs, after which the eyebolt was also drawn out of the hole, and the two other stones, with their load, fell through the same space down upon a stone pavement, but the cement joint was to all appearance perfectly uninjured, and we could not separate the two stones without splitting the joints by several smart blows of a mallet and chisels. On examining the two joints, Francis's cement was thinner, being only from $\frac{1}{10}$th to $\frac{1}{8}$th of an inch, and what is termed much shorter or more brittle, than the artificial cement joint, which averaged from $\frac{1}{8}$th to $\frac{3}{16}$ths of an inch in thickness, and was much tougher than the former. The thumb-nail or edge of a sixpenny piece acted on both of them. All the spectators agreed in opinion, after the untoward end of this experiment, that if the mode of fixing the eyebolts had not failed, the artificial cement would have gained the victory in this competition, which they inferred not only from its having sustained the shock of the fall without injury, but from its more perfect state, especially in the center of the joint. The result of this experiment also proved, that the artificial cement joint 10 days old, which was first tried with the Bramley-fall stones must have been spoiled by the frost as we suspected, and that the second more perfect artificial cement joint 60 days old, must undoubtedly have been injured by its previous fall (214); otherwise there seems every reason to believe that no weight much less than 30,000 or 40,000 lbs. could have broken it down at that age. Unfortunately in this last experiment, two of our four stones were broken by their fall, which will require time to replace, and afterwards to retry the same experiment, in which we propose to use the common Mason's Lewis, as Mr. Howe's expedient requires a greater depth to prevent the broken stones, and consequently the eyebolt, from being drawn out of the hole, than the thickness of our last Bramley-fall stones will admit. It is of course impossible from our present data, to estimate the strength of a cement joint of 1000 superficial inches and 11 days old, with accuracy, but I apprehend that such a joint, whether of good natural or artificial cement, would not give way under much less than 12,000 lbs.

The experiments with the Bramley-fall stones connected by large cement joints, which occupied us no less than six months from the setting of the first joint to the drawing of the eyebolts in the last trial, have been related consecutively, in order not to interrupt the thread of our narrative, though other experimental proceedings of prior date to some of them, took place in the mean time.*

REMARKS. ***First, that Cement may be used to the greatest advantage, in Masonry, even of the largest stones.***

(219) This is sufficiently proved by the foregoing experiments with the large Bramley-fall stones, combined with the experiments on the same stone bricks before remarked upon (195), of which the results were recorded in Table V. In working with very large and heavy stones, the only precaution necessary is, to adjust the stones together, first in a dry state, and then to raise the upper one before the bed of cement for the joint is laid on. As the strongest cement, whether natural or artificial, does not set in less than from 15 to 20 minutes, or sometimes more, and as it cannot require more than 5 minutes at the utmost to place the largest stone, after having once adjusted it dry, it is impossible that any difficulty can occur in building with cement, provided that this arrangement be adopted.†

Secondly. That the cohesive strength of calcareous Cements should never be estimated by the superficial inch.

(220) As the surface of the joint, of Francis's best English Cement, connecting our two largest Bramley-fall stones, measured 1131 superficial inches, which resisted 36544 lbs, and then the stone gave way, (216), what farther weight it might have borne, had the stones continued perfect, must be a matter of conjecture; but from the softish state of the cement, which in all our former experiments we had remarked as a proof of weakness, I do not think that it could possibly have resisted more than about 56,000 lbs. at the utmost, which averages only 50 lbs to the cubic inch;

* This experiment was tried on the 15th July 1837, in presence of the Officers of Engineers as usual, and also of Colonels Tremenheere and Lawrence of the Royal Marines, besides Mr. Howe and other Clerks of Works employed under Government, and some Officers of the Dock-yard.

† If the system customary in building large masonry with lime, that is to lay the bed of mortar first, and to get the stones into their places afterwards by a crane or other machine, were adopted in using cement, bad workmanship would necessarily be produced. Even in building with bricks, which are laid so much more quickly than large stones, if the same mode of using cement were adopted, that is usual and proper in building with mortar, the work would be ruined. In short, one mode of manipulation is necessary in using cement, whilst another is better in working with mortar; and if properly understood, the former is as easy as the latter.

whereas one of our artificial cement joints 74 days old, applied to bricks, and measuring about 40 superficial inches, resisted more than 4400 lbs, after which the bricks gave way, thus evincing an actual adhesiveness of 110 lbs. to the superficial inch, which, if the bricks had been stronger, may reasonally be estimated at 125 lbs to the inch, as was before implied (181). Though somewhat of this opinion is conjectural, still I consider it to be a safe and accurate conclusion, that any cement whether natural or artificial, as applied to a very large joint, such as that of the two Bramley-fall stones, may be torn asunder by a much smaller weight in proportion to its surface, than that which would be necessary for separating the joint of two much smaller stones, connected by cement of the same quality and age. But the strength of the same cement used at the same age, and under the same circumstances, must necessarily be equal; so that the apparent difference in the cohesiveness of cement applied to a very large joint, and to a much smaller one, only confirms what I have before so often stated, that cement in large masses, not easily accessible to air, sets very slowly; for this was the case with our large joint, the whole of which except just round the edges being inaccessible to the external air, it was therefore in the same predicament as the central parts of a large mass of cement. Hence the system adopted by some authors, of estimating the cohesive strength of calcareous cements, in some proportion to the superficial inch is only true, when all the joints are equal and similar surfaces, and equally accessible to the external air, or to water if used in hydraulic works; and as this can never be the case in practice, it is best not to estimate the comparative cohesive strength of cements by the superficial inch at all, which is very likely to lead to error. In fact, the cohesive strength of any joint made of the same cement, is directly as the age of the cement, and inversely as the magnitude of the joint, but it is not equal or proportional throughout the whole of very large joints, the outside of which may be at the rate of 125 lbs, whilst I conceive that their inside may not exceed 30 or 40 lbs to the superficial inch, as measured by the weights capable of tearing them asunder. But as there must, after a certain period, be a maximum both of COHESIVENESS, in opposition to forces tending to tear cement asunder in two contrary directions, and of RESISTANCE, or absolute strength as opposed to a breaking or crushing weight, which no cement can exceed: and as both stones and cement are porous, and not absolutely impenetrable either to air or water, I have no doubt, but that in process of time, the central parts of large masonry built with cement, will attain the same perfection as the outside, though much more slowly.

(221) I was partly induced to try the above experiments with those Bramley-fall stones, by reading the account of the construction of the Edystone Lighthouse by Smeaton, and of the Bell Rock Lighthouse by Mr. Stephenson, both of which reflect the highest credit upon those distinguished Civil Engineers, especially the former, whose proceedings the latter professedly adopted as his text book, although many of his own arrangements were perfectly original, and his task upon the whole was by no means less arduous. In considering this subject, I am ready to admit that nothing can be better than these celebrated Lighthouses, as they now stand, but as the powerful water cements of England, absurdly termed Roman cements, were absolutely unknown in Smeaton's time, and the use of them may be said to have been in its infancy when the Bell Rock Lighthouse was built, and even now is by no means general in Scotland; a query occurred to my mind, whether in the event of the erection of some future Lighthouse in such exposed situations, a great part of the difficulties encountered, which arose chiefly from the enormous weight of the stones, might not be avoided, by using large heavy blocks of granite for the outside courses only, and by building the whole inside of the lower or solid part with bricks, or with hard stone if that be preferred, in light and manageable pieces, not requiring the aid of machinery to move them, and all laid in cement. For this internal part of the bottom of a Lighthouse, flat bedded stones, capable of being split to any thickness, and having consequently very regular horizontal joints requiring little cement to make them good, such as were recommended by Smeaton for the backing of walls exposed to the action of water, would be extremely applicable.* The vertical courses of large masonry built with cement might be filled in with cement grout, in a state not more fluid, than just to allow it to run freely into every crevice. In respect to Smeaton's stone joggles, they may be used if necessary, but his oak trenails must be dispensed with as a matter of necessity, because they might disturb the cement in the act of setting. And it appears to me that his dovetails in the stones may also be dispensed with, which will save a great

* The Whinstone of Scotland, recommended for Bridges by Messrs. Smith of Darrick, in an able paper in the first volume of the Transactions of the Institute of British Architects, would perhaps be no less applicable.

deal of expense in cutting and of trouble in placing them, for if all the large outside stones of each course be cut, so as to form a horizontal circular arch, inclosing a solid central mass of smaller materials, and the whole laid in pure cement, the cohesiveness of which, considerable even from the first, will in time equal or exceed that of the stone itself, I see no risk of the heaviest sea being able to displace any of them. In adopting this arrangement, the outside stones of successive courses should be all headers and all stretchers alternately, so as to break bond with each other, but there is no objection to their being all of equal length nearly, in the same course.

(222) In like manner the arches of first rate bridges also might undoubtedly be built with bricks, or with long thin stones of moderate weight, at least in situations not near to the Metropolis or other great cities, and not on the line of much frequented public roads; where I should be sorry to oppose the expensive luxury of fine bridges, such as the Waterloo or new London Bridge, which like other pieces of splendid architecture are not always to be condemned, merely because plainer buildings would cost less money.

DESCRIPTION OF A REMARKABLE BRICK SUMMERHOUSE, THE CONSTRUCTION OF WHICH WOULD HAVE BEEN IMPRACTICABLE WITHOUT CEMENT.

(223) The cement from Messrs. Francis and Sons, of which part was used in the foregoing experiments, had previously been procured for a small summer-house, which I was then desirous of building, in the garden attached to the Field Officer's quarters in Brompton Barracks occupied by me, and which to answer also the purpose of an interesting experiment in Practical Architecture, I determined to build entirely of brickwork and cement, making the walls and roof half brick thick throughout, and constructing the latter as a sort of hollow square pyramid with projecting eaves, and with curved sides, convex internally, but concave externally. In short, the brickwork skeleton was purposely made by me of such a form, as could not possibly have held together but for the cement.

This little building was 7 feet 2 inches square in the clear, and 9 feet 3 inches high from the foundation to the lower side of the eaves. It was built on a bed of concrete 1 foot thick, over which a brick foundation was laid 9 inches thick and 9 inches high, and above this level, the whole of the work was only half brick, or 4½ inches thick, as before mentioned. There were entrances left in two adjacent sides, each 4 feet wide and about 7½ feet high, which were covered by a horizontal line of bricks laid flat, with vertical joints, for I studiously avoided the arch-like form, excepting in the roof, where by inverting that form, as was before stated, not strength but weakness ought to have been produced. The curved sides and the projecting eaves of the roof were formed by four equal quadrants, having their upper radii

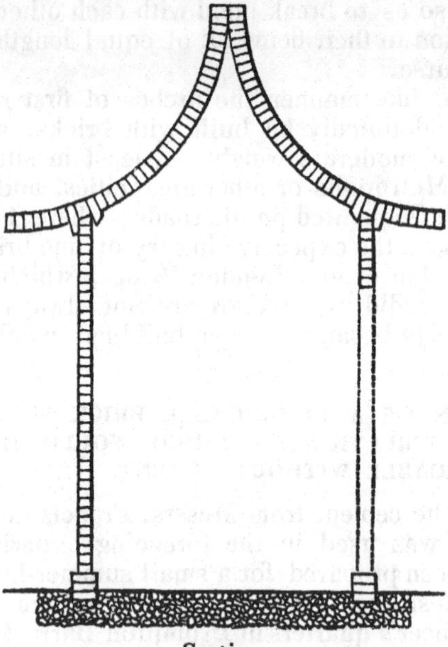

Section.

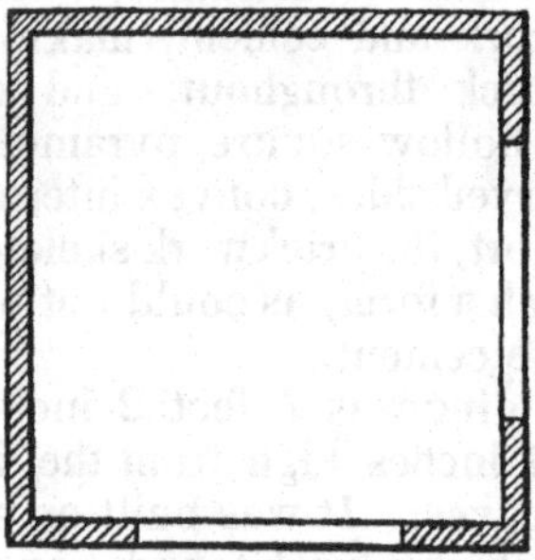

Ground Plan.

all in the same horizontal plane, and their convex extremities meeting at top, each radius being 5½ feet long.

This construction is explained in the two foregoing figures, one of which is the plan of this little edifice. The other is a section of the rough brickwork, in which, notwithstanding the smallness of the scale, every brick is distinctly represented. As the eaves of the rough brickwork of the roof projected 2 feet each way beyond the walls, the roof including those projections, measured 12 feet square at bottom.

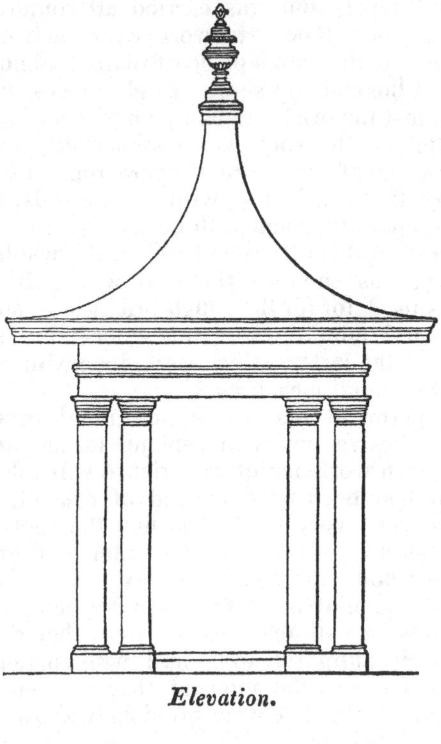

Elevation.

This little building was stuccod inside, but on three sides only outside, the fourth exterior side, both of the walls and roof, showing the rough brickwork, in order that the construction may at any time hereafter be understood by a spectator, to whom it is explained; and as this rough side faces a boundary wall with a very narrow passage between, it is not seen from any of the walks in the garden. The foregoing figure represents the appearance of one of the finished fronts of the building, having ornamental pilasters, and mouldings, and an urn on the summit of the roof.*

We used neither hoop iron, nor stone, in any part of the building. The projecting pilasters, &c, were roughed out by attaching fragments of plain tiles, or of bricks, to the surface of the original brickwork by cement. The eaves were formed entirely of bricks stuck out by cement, in the same manner in which our experimental brick piers had been made to stand out from the face of a wall (175). The roof was built without centering, by a plum line attached to a quadrantal level, and was carried all round by single courses at a time. The brickwork over each of the two entrances was built upon a piece of plank laid horizontally, supported at the ends by short upright pieces, which were pressed against the brick jambs by an intermediate strut at bottom. Before the roof was commenced, four planks, composing a sort of square frame, were applied to brace the walls externally near the top, with iron tie rods, to prevent them from separating, and with horizontal struts of wood inside to prevent them from collapsing, the whole of which were removed as soon as the roof was finished. Pure cement was used for all the rough brickwork, and also for the ornamental mouldings of the stucco, but not in the plain parts of the latter, which were done with cement and sand mixed in equal measures.

This experiment affords an additional proof of the wonderful cohesive power of cement, for any architect or builder, or even workmen of experience will admit, that if lime-mortar had been used instead of cement, the walls would have been very weak, and that the roof could not have been built at all without an interior framework or centering of wood, nor could the eaves have been added without some similar support; and after being finished by this apparatus, it will also be admitted, that the whole of this little fabric, supposed to be built with common mortar, would fall to pieces the moment that the centering was removed, even if the lime were previously allowed 20 years to set.

(224) In a former part of this Treatise, I explained the usual modes of preparing for bold projections in cement fronts, by introducing Yorkshire stones, and by other expedients mentioned in from Article 126 to 128 inclusive, and afterwards in reference to Mr. Brunel's celebrated experiment of the two semi-arches, springing from one central

* The decorations were designed by Mr. Howe.

pier only (59), I remarked, that there would be no difficulty in forming the boldest projections by brickwork and pure cement alone, with the aid of hoop iron in the joints, according to his system. The experiment of the summer-house has since convinced me, that the boldest brick projections necessary in ornamental architecture are practicable, not only with but without hoop iron, and also that the expedient of nails and spunyarn to strengthen a cement moulding (128) is perfectly useless.

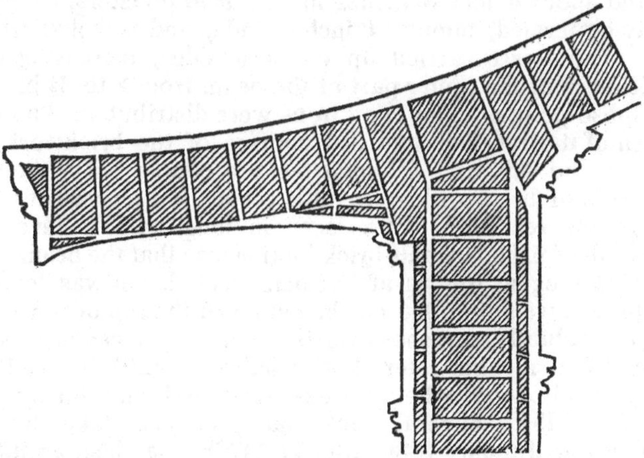

The annexed figure shows in section the details of the projections and mouldings, at the meeting of the walls and roof of the summer-house, on a larger scale. Two bricks laid as headers, and cut to suit the form of the mouldings, and one brick between these, laid on edge, and another brick below this, cut obliquely at the top of the wall, will be observed, which were the only courses of bricks in the whole work, not laid flat and as half bricks. In this figure, the bricks and fragments of tiles or of bricks are shaded dark, whilst the cement joints and stucco are light. The cornices were worked or as it is technically termed *run* by a mould, but some of the smaller mouldings were run separately at a board, and fixed in their places afterwards.

REMARKABLE BRICK BEAMS BUILT WITH CEMENT, AND STRENGTHENED BY HOOP IRON BOND.

(225) Besides his extraordinary semiarches (59), Mr. Brunel tried another experiment upon the strength of pure cement scarcely less remarkable, by building what he

termed a brick beam, which was a substantial mass of brickwork, resting near its extremities upon two points of support, at the clear distance of 22 feet 9 inches apart, and having 12 pieces of hoop iron introduced longitudinally, and distributed amongst 5 of the lower joints. The beam was 3½ bricks thick at bottom for the first 3 courses, then 3 bricks thick for 1 course, 2½ bricks thick for 3 courses more, above which the remainder of the work was 2 bricks thick, there being 17 courses of brickwork in all, so that it measured about 4 feet 3 inches high. Two pilasters, as they may be termed, about 14 inches wide, and 5 feet apart in the clear, were carried up on both sides, increasing the thickness of the upper part of the beam from 2 to 3½ bricks at those points. The hoop irons were distributed, 3 above each of the 1st, 2nd, and 3rd courses of the brickwork, 2 above the 4th, and 1 above the 5th course. The first 7 courses of brickwork were built vertically upwards at both ends, the remaining 10 as seen in longitudinal elevation were diminished by half brick footings, so that the beam was not so long at top as at bottom. This beam was loaded with iron weights piled on the center of the top of it, which were gradually increased from the month of December 1835, about two months after it was finished, until the end of March following, when it broke down under a total weight of 27,025 lbs. Several cracks had previously taken place, the first in January 1836, after 11,147 lbs. had been applied.

(226) A like proof of the great cohesive power of cement may now be seen, at Messrs. Francis and Sons cement works at Nine Elms, Vauxhall, where they have built a similar brick beam resting upon two brick piers, at an interval of 21 feet 4 inches apart in the clear. These piers are 6 feet high, 2 bricks wide as seen in elevation, and 2½ bricks thick as seen in section, and these dimensions are increased at bottom by footings. The beam is 2½ bricks thick for 6 courses of bricks in height, above which it is 2 bricks thick for 13 courses of bricks, making 19 courses of bricks, or about 4 feet 9 inches in all, for its total height.

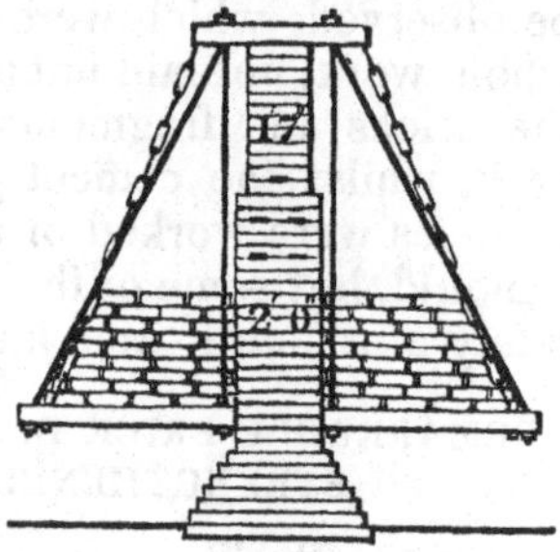

Section.

There are 2 pieces of hoop iron above the second course, 2

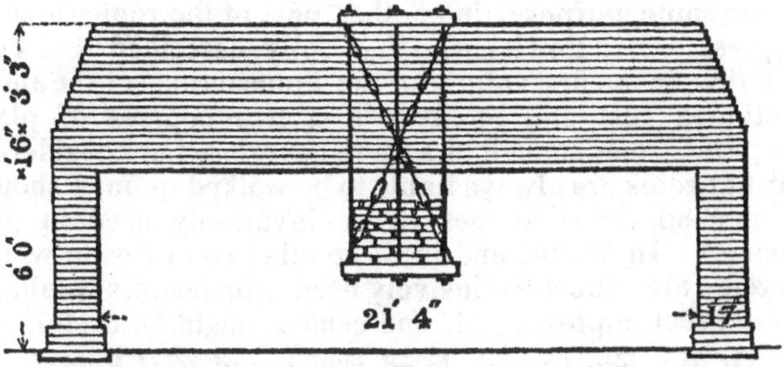

Elevation of Messrs. Francis's Brick Beam.

more above the fifth, and 1 above the eighth course of brickwork of the beam, making 5 in all.

The first of the annexed figures represents this beam in Section, the second in Elevation, by comparing which, the details of the construction, and the mode of loading it afterwards with iron ballast to test its resistance will be clearly understood. When I inspected it in June, 1837, the weight suspended below, but acting on the top of the beam, was 24,000 lbs, which had not produced the smallest crack or deflection. I was informed, that the planking used in building the beam had been removed in 10 days after it was finished.

OF THE APPLICATION OF CEMENT TO THE CONSTRUCTION OF FLAT ROOFS, FIRE-PROOF CEILINGS AND PARTITIONS.

(227) Another proof of the extraordinary strength of pure cement, is the system which has lately been introduced of using it to unite plain tiles together, for the covering of flat roofs, although the beams or horizontal rafters which support them, may be about 4 feet apart. In this sort of roof 3 courses of plain tiles are used, all breaking joint with each other, which even at central intervals of 4½ feet between the bearers, are said to be strong enough to support as many persons as can find standing room. Examples of roofs of this description supported entirely by iron bearers, and consequently fire-proof, constructed by direction of Mr. Fowler, Honorary Secretary of the Institute of British Architects, may be seen over the two Taverns facing the Thames, as well as on the Terraces over the Porticos north and south of the Hall, at Hungerford Market. Laths were used to support the first course of tiles on being placed, which were removed in about half an hour, and used

for the same purpose, in another part of the roof whilst in progress. Mr. Fowler who has fully described this construction, in the first volume of the Transactions of the above mentioned Institute, observes that two courses of plain tiles, are strong enough as a mere covering, but considering that flat roofs are always liable to be walked upon, I should recommend three courses to be invariably used in preference.* In Malta, and in those other countries in which flat roofs are almost exclusively used, iron bearers would be a very great improvement, and cement might be usefnl.

Of Mr. Frost's Tubulated Floors and Flat Roofs.

(228) This ingenious gentleman contrived a sort of square earthenware tubes, measuring 2½ inches square externally, and made by a machine in lengths of 10 feet, but cut previously to being baked, into pieces only 1 foot long. Two courses of these laid in pure cement, at right angles to each other, form a floor or roof 8 or 10 feet wide, of which specimens may be seen at his late artificial cement works now the property of Messrs. Bazeley and White. Mr. White has a very favourable opinion of this contrivance, which he considers a safe floor for any number of men to walk upon at the above width. In consequence of the tubulated form of its component parts, a floor or roof of this construction may be considered stronger than one made of plain tiles, or any other sort of flat tiles, of equal weight. The following figure represents a portion of one of Frost's tubulated floors in section, showing at one end its connection with the wall.

* Plain tiles are nearly 10½ inches long, 6¼ inches broad, and ⅝ths of an inch thick, and weigh about 2½ lbs. I believe that it was first proposed to use them in several courses laid in cement, for the covering of flat roofs, by the late ingenious Mr. Smart, who formed a pair of very flat rafters and a tie beam out of a single piece of scantling, by the addition of a small oak *King Block*, as it may be termed, to connect the heads of the two rafters. The very flat *Trusses* thus formed being placed at moderate intervals apart, he covered the whole with two or three courses of plain tiles laid in cement. Mr. Hight a builder of Dover constructed several roofs of this description, which he showed and explained to me about ten years ago. In his own house the bearings between the walls were 18 feet, for which he used fir scantling of 2½ by 9 inches, formed into flat trusses on Mr. Smart's principle, and placed at intervals of 2 feet apart. Over these he laid transverse battens of 2½ by 3 inches, at central intervals of about 11 inches, to support the joints of the lowest course of plain tiles, of which he used two courses for a common roof, but 3 courses for a Terrace over a Library in Dover, upon which people were to walk. These battens, which I suspect were also used by Mr. Smart in London, have been proved to be unnecessary, by the experience of Mr. Fowler's improved construction. This gentleman justly reprobates wood for the supports of a flat roof, as the alterations of form to which it is liable by shrinking, bending, or warping, produce very troublesome leaks, as is well known in Malta, Bengal and other countries, where flat roofs are common.

The tubes are so arranged as to break joint with one another in both courses, and a coat of cement stucco is applied both above and below this sort of floor.

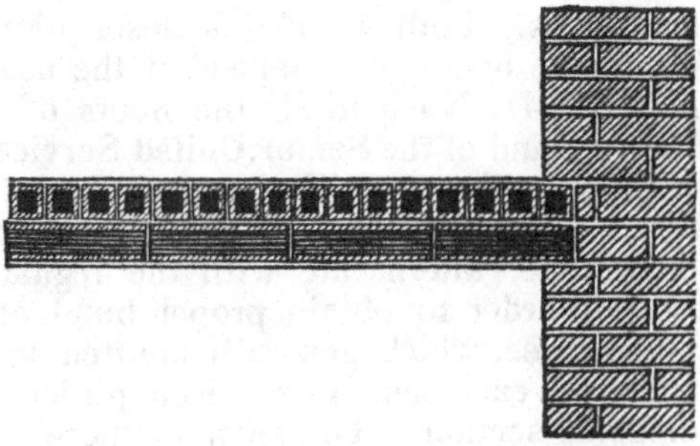

Section of Frost's Tubulated Floor.

(229) *Fire-proof Ceilings or Roofs.* These may be made of flat brick arches in cement, between iron girders, and to save the weight of solid bricks, similar arches have sometimes been formed with *Arch Pots,* or *Hollow Cones,* as they are more usually but I think improperly termed, in the manner represented in Section in the annexed figure.

Section of a Portion of Fire-proof Ceiling.

These arch pots, at least such as have been made in England, for other sorts may have been used in France and other countries, are square at top and round at bottom, the side of the upper square and diameter of the lower circle being exactly equal. Those which I have seen are of two sizes, one 8 inches high and about 4¾ inches square at top, the other 5¾ inches high and about 4¼ inches square at top,

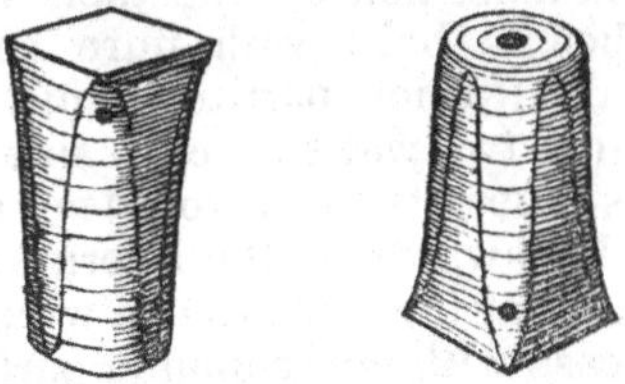

the former weighing about 4¼ lbs, and the latter only half as much. The first of the annexed figures represents one of these hollow cones seen in the position in which it is used with the square end upwards, whilst the second represents it turned upside down. The sides and bottom are scored, and a small hole is usually made in each, previously to these pots being baked, in order to afford a key for the cement or mortar used in constructing the arch, as well as for the plaster, when the ceiling is not an ornamental one, but formed by plastering the under surface of the pots themselves. Ceilings of this description were formed by Mr. Soane over the basement of the new Treasury buildings, by Mr. Nash in all the floors of new Buckingham Palace,* and of the Senior United Service Club, and by Mr. Wilkins in the new National Gallery. Various expedients, such as pieces of brick, or half pots broken down the middle, alternating with the regular arch pots were used, in order to obtain proper bond at the springing of these arches, which generally abutted against stone skewbacks on each side of the iron girders, as shown in the foregoing section. The span of these arches was usually about 6 feet, seldom more than 7½ and never less than 4½ feet, and their rise seldom exceeded 6 inches.

(230) *Fire-proof Partitions.* As upper apartments are usually smaller, especially in the attic story, than those below, partitions in those stories are not always supported throughout their whole length, by the internal walls of the building. They usually consist of a wooden framework lathed and plastered on both sides, and so braced as to press lightly on the floors below, when there is no wall beneath them, and to resist any action from the floor above, if there be one. Sometimes the interstices of a partition are filled with brickwork, termed ***Brick Nogging***, a sort of construction often observable in the walls of cottages, and never more than half brick thick, 4 inches being the usual thickness also of a wooden partition, exclusive of the laths and plastering afterwards added. The experiment of the summer-house, together with the properties of brick beams built with pure cement sufficiently prove, that a half-brick partition built with pure cement, and bonded into two walls on each side also with cement, which would support its ends, would not exert any unfavourable pressure downwards, in the intermediate space between those walls; and at the same time it would be fire-proof, which is not the case with any common partition.

(231) *Of the Advantages of using Cement for Chimneys.* The *Withes* or partitions between the flues of chimneys, as well as the chimney breasts, and frequently the boundary

* The roof of the Palace being intended to be flat, and the iron girders very deep, Mr. Nash employed two tiers of arches, each 8 inches deep, the uppermost of which was strengthened by bricks or tiles where necessary, until the general level for the proposed roof was obtained. This was coated with about one inch in thickness of a composition for terraces or flat roofs, proposed by Lord Stanhope, consisting of Stockholm tar, pounded chalk, and sand boiled together, and afterwards covered with large substantial slates. Both tiers of arches spring from pieces of Bath stone fitted to the sides of the iron girders.

walls or external withes of chimney shafts, being only half brick thick, the superior strength that will be obtained from laying such thin brickwork in cement will be obvious. The accidents destructive to property and sometimes to life, from high chimney shafts being blown down in gales of wind, will be prevented or rendered much less frequent, and so far as regards chimney breasts, the propriety of coating them towards the apartments by vertical courses of plain tiles fixed in cement, and breaking joint with the brickwork, has been suggested, which will obviate the risk of flame penetrating through the joints of the half brick wall, and setting the house on fire, which has at times been known to happen.*

(232) *That small materials may be formed into Steps for Staircases by means of cement.*

In districts, where stone cannot be procured at a moderate expense, I conceive that good artificial stones for staircases may be formed with plain tiles and bricks cemented together, using one course of bricks in the center, with 2 courses of tiles below, and 2 above. But the bricks should be about 13 inches long, and the tiles also 13 inches long and 6¼ inches wide, which would render it practicable to make a step of 13 inches in extreme width, including the bearing for the next step above it, and about 6½ inches in height, and of any convenient length at pleasure, breaking joint in every part, without unnecessary cutting. In short, if the system of making artificial stones for the steps of stairs should ever be adopted, it would be proper to make bricks and plain tiles on purpose; and the same remark will also apply to Frost's earthenware tubes, if these should be approved for staircases, the present size being a little too short to form the proper width or tread, and rather too high for the rise of a convenient step, if cemented together in 3 courses, which I think would be better than any other number.

GENERAL OBSERVATIONS ON FIRE-PROOF CONSTRUCTIONS.

(233) Having attentively examined the effect of great fires on buildings, I have invariably found that good brickwork seems to be a perfectly fire-proof material, but from the injudicious system of supporting the upper stories of brick buildings by wooden bressummers and story posts, and of letting bond timbers and wall plates, and the ends

* This expedient was suggested to me in conversation by Mr. Baker of Montague Place Russell Square, and of Stangate Wharf Lambeth, whose skill and experience as a Builder are well known.

of wooden girders and joists, into the walls, it seldom happens that the wood work of a building in England is burned, without its becoming necessary to pull down the whole of the brickwork afterwards. The most eminent British Architects have recently adopted a more judicious system, by substituting iron for wood where practicable, disusing bond timbers, and supporting their wall plates on projecting iron shoes, stone corbels, or gathering courses of bricks, without inserting either these wooden wall plates, or the joists which rest upon them, into the substance of the wall. I have observed however, after several great fires, that strong iron pillars and bressummers have been split, by first being exposed to a red heat, and then suddenly cooled by streams of water from the fire-engines. A query therefore occurs, whether such ironwork ought not to be protected by a coating of tiles laid in cement, which might be attached to the surface by fashioning it with holes or roughnesses for the purpose. I have also observed, that large stones in buildings when exposed to a strong fire, are invariably split or defaced, by fragments detaching themselves from the surface. Hence when a warehouse is constructed with iron pillars and girders supporting stone floors, it cannot be considered fire-proof, unless when filled with incombustible goods; for if the contents of such a building should take fire and burn briskly, the heat will break the stones of the floor, and may even crack or split the ironwork. I am not sure about the ceilings of arch pots before described, but I am persuaded that no fire that can occur, in a dwelling-house or warehouse, could possibly injure an arched ceiling of common bricks, though only half-brick thick, for as I said before all brickwork is fire-proof. I have observed, that the interior stucco on brick walls is usually partially detached, and so much injured by fire, that it is necessary to take it off and replace it afterwards; but in the mean time it guards the mortar or cement joints of the brickwork to which it was applied. If these should be exposed, which is always the case when the wall is not stuccod, I have observed that the lime in the calcareous mortar or cement of the joints is burned into quicklime, to a depth never exceeding one inch, but usually less; so that the greatest penetration of this process is far too insignificant to affect the solidity of the wall, which only requires the damaged mortar to be raked out of the joints, and to be replaced by pointing. Hence I would condemn stone floors for fire-proof buildings altogether, and recommend brick arches laid in cement, in preference, at the same time exposing the surface of iron when used, as little as possible.

(234) *That cement Bond may be used in Buildings, instead of timber Chain Bond, and to the disuse also of wooden Lintels.*

Sensible of the disadvantages of internal bond timbers, let into walls, flush with the surface, the most eminent British Architects have recently abandoned this construction altogether, substituting chain bond timbers instead of them, which being buried in the middle of walls, are not exposed to fire, and if they should decay, the wall will evidently be much less weakened, than by a like failure of the former. These timbers are generally so placed, as to form part of the lintels over the tops of rectangular doors and windows. This construction was considered very promising at the time, but I am informed, that dry rot has resulted from it in a few years, at least where fir or even English oak have been used. I believe that teak alone has remained sound in this confined situation, and probably Kyan's patent composition would render other sorts of wood no less durable; notwithstanding which it appears to me, that four or five courses of cement bond, that is of bricks laid in pure cement immediately over the apertures of doors and windows, strengthened by hoop iron in the joints, and continued entirely round the walls of a building at the same level, will have all the advantages of the common lintels and of chain bond, without any of the disadvantages of woodwork. The superiority of this sort of bond did not suggest itself until after Mr. Brunel had tried his memorable experiment of the brick beam, though I knew that this sort of bond had been previously used by Sir Robert Smirke, in several great buildings executed by his direction, but in which he did not use hoop iron, nor dispense entirely with chain bond or lintels, as I now propose. Sir Robert Smirke did however use hoop iron bond in some of his first buildings, but I believe in combination with mortar only.

GENERAL REMARKS ON THE SUPERIORITY OF CEMENT OVER LIME MORTAR.

(235) From the various examples of the wonderful power of cement, recorded or alluded to in the preceding Articles, as well as in former parts of this Treatise, it must be allowed, that the use of this admirable material renders many constructions both with stone and brickwork, not merely possible but easy, which with common mortar are

absolutely impracticable. It not only does away the thrust of arches, and renders centering of any magnitude unnecessary,* but what is more, it even dispenses with the arch-like form altogether, as was proved by the small summerhouse (223), and brick beams before described (225 and 226); so much so, that I am confident that the present model room in the Royal Military Repository at Woolwich, which is in the form of a round marquee or large tent, might be replaced entirely by brickwork and cement, without any central pillar, such as that which now supports the wooden and canvas roof of that large circular building. In short, there is scarcely any form of covering for the roofs of buildings, that may not easily be executed with brickwork and cement alone, from the stateliest Dome of European Architecture, to the fantastic Pinnacles of the Oriental Nations.

EXPERIMENTS ON THE COMPARATIVE STRENGTH OF VARIOUS SORTS OF CEMENT, NATURAL AND ARTIFICIAL.

(236) At the time when I decided upon erecting the small experimental summer-house before described, it was not only my own but the general opinion of the Officers of Engineers and others, who had witnessed the experiments previously tried with my Artificial Cement, that it was stronger than the best Natural English Cements sold in the London Market, because mixture C4 B5, the very first time it was tried, had set out 31 bricks from a wall (178), whereas the Natural Cements, though no doubt frequently tried, had not been known to set out more than 33 bricks in the course of 40 years (132); and as mixture C 4 B 5·5 proved afterwards to be stronger than C 4 B 5, it seemed only reasonable to suppose, that this stronger artificial cement mixture, if tried repeatedly, might set out 33 bricks, or more.

* The masons of Minorca who are distinguished by their skill, build all their arches of the handsome free stone of that island, in a very ingenious manner, by means of a sort of gypsum, called Gueesh, which sets as quickly and I believe with nearly as much strength as our English cement, but which is only fit for inside work, because it is injured and finally ruined by wet (18). In building a semicircular or elliptical vault, they begin at one end, working upwards from the spring of the arch on both sides, and supporting every stone by a stout reed, until they introduce the keystone at the crown of the arch, and thus finish a small portion of their proposed vault, after which they remove their reeds, and use them for another similar portion in continuation of the former, the stones of these two and of all the other successive portions being toothed into each other, to obtain proper bond. The Minorquin masons also make partitions, sometimes only 4 inches thick, of flat stones laid on edge and cemented together by Gueesh, which are quite strong enough.

Being myself strongly impressed with this opinion, and the quantity of cement necessary for the execution of the proposed summer-house being greater, than my own small Establishment could supply in any reasonable time, I requested Messrs. Francis and Sons to send me some casks of their best English cement, absurdly termed Roman cement, for the carcase of this little building; whilst I ordered some casks of Frost's artificial cement from the manufactory established by him, and now in possession of Messrs. White, Bazeley and Son, for the stucco; and I wrote to both parties, to request that they would send their very best cement of each sort, being required for an interesting experiment, in which inferior or even middling cement would fail.* Francis's cement appeared not only to me but to the military workmen employed, to be the best English cement that we had seen, far superior in quality to what would be generally accepted as good marketable cement, and on trying it against my own mixture C4 B5, it proved superior in strength, but not to C4 B5·5, which had the superiority over this and all the other cements, which we compared in the manner that shall be described.

In preparing to try the comparative cohesive strength of the several cements now alluded to, we provided 25 pairs of stone bricks of Portland stone,† all measuring 4 inches by 10, on those sides that were to be united by the cement joints, and all dressed alike, and compared with a pattern stone beforehand, in the manner recommended and explained in Article 192. These stones were selected and placed in 5 parallel rows, as equally assorted in respect to the state of their surfaces as possible, and lots were drawn, to decide which row or set of stones should be allotted to each of the 5 sorts of cement, that were to be tried in competition with each other. The whole were cemented together by the same experienced workmen in the usual manner, with stiffish cement paste, and all subjected to the same breaking or rather tearing apparatus at 11 days old, except one sort, which we could not conveniently try on the 11th day, and therefore we broke down two blocks of it on the 10th day, and 3 on the 12th, which we considered equally

* Messrs. Francis, White and Francis dissolved partnership amicably, on the 1st of January 1837, and separated into two firms, viz. Messrs. Francis and Sons of Nine Elms, Vauxhall, and Messrs. Bazeley, White and Son of Millbank, the latter retaining Mr. Frost's manufactory of artificial cement in Swanscombe parish, Kent, near Ingress Park.

† Purchased at half the usual price, being part of the balustrade of Rochester Bridge, blown down by a hurricane in November, 1836.

fair. In these, and in all our other experiments on the comparative adhesiveness and resistance of cements, &c, we made a rule never to increase the weights on the scaleboard by more than 1 lb, at a time, towards the close of each experiment, until fracture took place; except when we knew beforehand, that more than 3000 lbs, would be required to produce this effect. The cements now experimented upon by this system were as follows:

1st. Francis's natural cement, composed of Sheppy and Harwich stone mixed, in what proportions the makers did not inform me.*

2dly. The natural cement made in Sheerness Dock-yard, also composed of Sheppy and Harwich stone mixed, in the proportion of one measure of the former to three of the latter.†

3dly. Frost's artificial cement from the manufactory of Messrs. Bazeley, White and Son.

4thly. My own artificial cement, mixture C4 B5·5.

5thly. Another artificial cement mixture, made also by my direction for the sake of experiment, consisting of 4 parts by weight of dry chalk powder, 5 parts of blue clay, and 1 part of pounded iron scales from the Anchorsmiths' Forge in Chatham Dock-yard, which mixture is denoted by the emblem C4 B5 I.

(237) On tearing asunder the cement joints of these little masses, my two artificial cements proved the strongest of the whole, the cohesive strength of each being measured by the under-mentioned breaking weights, viz. first C4 B5·5 by 1453 lbs, secondly C4 B5 I by 1337 lbs, thirdly Francis's natural cement by 1223 lbs, fourthly the Sheerness Dock-yard natural cement by 1220 lbs; and lastly Frost's artificial cement by 705 lbs.‡

* They stated their opinion that Sheppy stone alone would yield a stronger cement than this mixture, which I doubt, because the Sheppy stone burned by me produced a cement that set too quickly, and with too much heat (161).

† The cement kilns and grinding apparatus in Sheerness Dock-yard were first established by the late Mr. Rennie, for the use of the new works under his direction.

‡ I also burned some cement stone sent to me from Kingston in Upper Canada, the average adhesiveness of which to bricks, when tried in the usual manner, was only 565 lbs.

TABLE IX. COMPARATIVE ADHESIVENESS TO PORTLAND STONE, OF VARIOUS ARTIFICIAL CEMENTS, TRIED IN COMPETITION WITH THE BEST ENGLISH NATURAL CEMENTS.

Sort of Cement used.	Age in days.	Fracturing Weight in lbs, in several successive experiments.					Average fracturing weight, in lbs.
Artificial.							
C4 B5·5	11	1009	2717	1247	1640	652	1453
C4 B5 I	11	1041	1576	1037	1156	1877	1337
Frost's	11	799	589	953	484	701	705
Natural.							
Francis's........	11	841	1261	1205	1093	1716	1223
Sheerness	10	1156	1688	....	....	....	1220
	12	813	1541	904	...	....	

(238) We were rather surprised that the iron did not increase the strength of our own artificial cement, for on previously trying various proportions of chalk, blue clay and iron, and making small experimental balls with each in the usual manner (189), the mixture above described appeared to yield much stronger balls, than any of the natural or of the other artificial cements that had been experimented upon by us. But I found by experiments afterwards tried on the resistance of cements, that this property is not in proportion to their cohesiveness, the latter being the most important of the two. I consider the circumstance of the chalk and blue clay alone, forming a mixture superior to the chalk, blue clay and iron, as being of the greatest importance, because the addition of the last ingredient, if necessary, would add greatly to the trouble and expense of making an artificial water cement on a great scale.

(239) From those experiments, Frost's cement appeared to be so much inferior to the others, that I am convinced that my objections to his mode of washing which he copied from M. Vicat, and to his long drying, as being both very prejudicial to the strength of artificial cement, are well founded. In consequence of these processes having entirely spoiled the first, which I myself attempted to make on a great scale, in 1829 and 1830, as stated in Articles 110 and 111, I have from that time repeatedly recommended both Mr. Frost himself, and the gentlemen who purchased his business, to abstain from washing their ingredients, and afterwards drying them to excess, notwithstanding which, they still persevere in Mr. Frost's original

system of manipulation.* Frost's artificial cement which I received in casks, even when quite fresh, was evidently so much weaker, than either the natural English cements, or the artificial cement made by me, that after the first casks containing it were expended, which completed only a part of the stucco of the summer-house, I abstained from ordering any more, at the request of the workmen, communicated to me by Mr. Howe, as they did not consider it strong enough for the purpose, although it had been my intention to have stuccod the whole of that small building with it. This objection was made to it, when used fresh from the cask, but I must admit, that what we used in setting the five joints, the strength of which was experimented upon and recorded in the preceding Table, was taken from a cask that had been open some days, rejecting however the upper surface of the cement powder contained in it, and taking it from the extreme bottom. Hence it may have been a little staler than the other cements tried in competition with it, so that I have no objection to rate its average adhesiveness at 803 lbs, instead of 705.† This brings it to two thirds of the average cohesive strength of Francis's or of the Sheerness Dock-yard cement, which is rather more than either Mr. Howe or the workmen are willing, as a matter of opinion, to allow it. I do not know the proportions of chalk and clay now used by Mr. White the successor of Mr. Frost, in the management of the artificial cement works alluded to, but from its appearance I suspect that he adds some other ingredient in making it, as it is of a darker colour, than any that I have ever myself obtained from chalk and clay alone.‡

* Mr. White as well as Mr. Francis being manufacturers also of the best natural cement, recommended their customers always to buy that in preference to Frost's, for constructions of any importance, especially if exposed to water. They continued the manufacture and sale of Frost's artificial cement, merely because it had at length found its way into the London market, though with extreme difficulty, and as some builders insisted upon having it, from its being cheaper than the natural cement, they were obliged to provide accordingly. When I told Mr. White of the great improvement of quality, that would arise from following my system of manipulation, he observed, that all the apparatus used being suited to Mr. Frost's original system, it could not be altered without occasioning much extra expense.

† I ordered another small cask of Frost's cement for the purpose of giving it a second trial, but unfortunately a mistake was made in the number of days for breaking down the joints, which were allowed to become too old to afford a fair comparison.

‡ Mr. Frost at one time added a proportion of brown ochre to his chalk and clay, I believe under the same impression that induced me to use coal dust, in order to prevent my artificial cement mixtures from being spoiled by washing and drying, as stated in Article 89.

(240) From the above experiments, as recorded in Table IX, we considered our own artificial cement C4 B5·5 to be at least equal to the best natural cements of England,* and might even have considered it superior to them, but for other occasional experiments, which render this point doubtful, part of which have been already related, and one of which must now be mentioned, namely that Francis's cement when used to unite five pairs of stone bricks of Kentish Rag stone, required in those five trials an average weight of 1498 lbs, to tear it asunder,† and what is very curious, C4 B5·5, the strongest of my artificial cements, did not show so much adhesiveness to this peculiar sort of stone, as mixture C4 B5, which it had far exceeded, not only in the trials with Portland stone, in which it had also proved superior to the natural cements, but likewise in similar trials with common bricks.

(241) But as all the cements were used pure in those experiments, whereas they are almost always mixed with sand, in the practice of building; in order to render our comparison more perfect, we had the same Portland stone-bricks cleaned again, and having made additional ones according to the same pattern, all as nearly alike as possible, we cemented them together again in pairs, by each of those five sorts of natural and artificial cements, mixed first with an equal measure of washed siliceous sand from the parade, and secondly with double their own measure or volume of the same, this last proportion being the maximum of sand, that any water cement will bear without being spoiled. In those additional experiments on the comparative adhesiveness of cement mortars as they may be termed, it occurred to me, that the same artificial cement mixture C4 B5·5 which was the strongest in its pure or nett state,

* The experiments which appeared to establish this opinion, were tried in presence of most of the Officers attending the Royal Engineer Establishment, on the 24th of May 1837, the day on which our gracious Queen (then Princess) Victoria attained her legal majority. This coincidence, quite unforeseen, was noticed with pleasure by the military workmen, who took an interest in the artificial cement, which they had themselves made.

† The weights actually used to separate the joints of Francis's cement, in those 5 trials, were 1380, 2444, 1772, 708, and 1324 lbs, respectively, yielding an average of 1526 lbs, but from being hurried at the time a mistake was made by our assuming previously, that the probable average weight would be 3000 lbs, which was suitable to cement joints of 30 days instead of 11. Hence all the weights were added by 56 lbs, or half hundred weights, at a time, until fracture took place; in consequence of which mistake, it becomes necessary to deduct 28 or the half of 56 lbs, from the above, leaving the probable correct average of 1498 lbs, in order to obtain a fair comparison with all our other experiments of a similar nature, in which the weights were added by 1 lb, at a time only.

might not perhaps bear quite so much sand, as another mixture C4 B4·5, containing rather more of calcareous matter, in proportion to the clay, than the former. This was found afterwards to be the case, on tearing them asunder by the usual apparatus, inasmuch as the last-named artificial cement, with one measure of sand, proved stronger than the former, and both were rather stronger than the natural cements used in the same manner; but the natural cements proved the strongest, when two measures of sand to one of cement powder were used, which last proportion is however more than the most experienced builders allow, either for cement mortar, or for stucco. Upon the whole therefore, I consider myself warranted in asserting, as I have done in the title page of this Treatise, that the artificial cement made by me, is equal in efficiency to the best natural water cements of England. The results of these additional experiments are as follows.*

TABLE X. COMPARATIVE ADHESIVENESS OF VARIOUS ARTIFICIAL AND NATURAL CEMENT MORTARS, TO PORTLAND STONE.

Sort of Cement used, and its proportion to the sand by measure.	Age in days.	Fracturing weight in lbs, in several successive Experiments.					Average fracturing weight, in lbs.
Artificial.							
(C4 B4·5) 1, Sand 1..	11	584	1158	850	1713	1088	1079
" 1, " 2..	11	430	442	444	512	405	447
(C4 B5·5) 1, Sand 1..	12	730	939	965	952	1016	920
" 1, " 2..	12	392	349	338	343	198	329
	13	407	221	387	....	....	
Frost's.. 1, Sand 1..	13	333	582	129	....	....	348
" 1, " 2..	13	147	194	298	....	....	266
Natural.							
Francis's 1, Sand 1..	11	498	1002	....	....	....	872
	13	1181	540	1140	....	....	
" 1, " 2..	11	506	282	....	....	....	578
	13	799	607	600	....	....	
Sheerness 1, Sand 1..	11	289	319	1289	....	....	632
" 1, " 2..	11	646	576	333	....	....	519

(242) REMARKS. Comparing the above experiments with those contained in Table IV (193), it will be seen that

* Not having stone-bricks enough, to experiment upon the whole of these cement-mortars as often as I wished simultaneously, some of them had only three trials instead of five. To give those as many as the others would waste so much time, that I shall not attempt it. As Frost's cement was not tried from a fresh cask like the others, I consider it fair to add one third to its average fracturing weights as stated in Table X, which will render its fracturing weights with one measure of sand equal to 464 lbs, and with two measures of sand equal to 354 lbs, instead of 348 and 266 respectively.

the artificial cement C4 B5·5 mixed with an equal measure of sand, is four times as strong as C4 B5 proved to be, when used in the same manner. This does not convince me, that the latter cement is weaker than the former, in the same proportion. It must either have been an inferior batch of artificial cement, for the quality of cement burned in small quantities, and having several batches mixed, as was often necessary with us, is more precarious than that obtained from the large kiln of the manufacturer: or it may have been rather stale, for several experiments tried by me with the same fracturing apparatus, especially upon the Sheerness Dock-yard cement, which I could not always procure fresh from the cask, have convinced me, that cement is more injured by staleness, than I at first supposed. When I had no other means of judging, than by making experimental balls and breaking them, the opinion formed could scarcely be considered more than conjecture or probability; but when the more accurate mode of estimating the strength of cement joints, by tearing them asunder by weights, suggested itself, I found that the cohesive strength of cement when stale, seldom averaged two thirds of its cohesiveness when fresh from the cask, and often much less. Whether stalish cement may not in time attain the same strength, but of course much more slowly, than fresh cement, as I at one time supposed, may now be considered doubtful. I am however inclined to the contrary opinion: and at all events, since in every critical work, such as a Lighthouse exposed to the violence of the waves, or a Tunnel in doubtful soil, the cohesive strength of the cement used must often be put to the test at a very early period; it is of great importance, that none but the best and strongest cement should be used, perfectly well burned and fresh from the cask, and in no other state; which may be ascertained partly by its not effervescing in the diluted muriatic or nitric acids (45), and practically by the rule explained in Article 189. To make our course of experiments more conclusive, whilst those last described were in progress, we prepared small square prisms of all the cements and cement mortars experimented upon, in the same mould before described (137), and broke them down in the same manner, in which we had before broken down our small experimental stones (199), and thus we ascertained the average resistance of each, as recorded in the next Table (XI), in which to render the comparisons contained in it more complete, we have repeated their average adhesiveness also.

(243) REMARKS. From the contents of the following Table, it will be sufficiently evident, that the resistance and cohesive strength of cements and cement-mortars, the latter being estimated by their adhesiveness to Portland stone, are not in proportion to each other, of which a remarkable proof is observable in the circumstance that the resistance of Frost's artificial cement in its pure state was $3\frac{1}{2}$ times as great, as that of C4 B5·5 mixed with an equal volume of sand; whilst the adhesiveness of the former only proved equal to seven-ninths of the latter.

TABLE XI. COMPARATIVE RESISTANCE OF VARIOUS NATURAL AND ARTIFICIAL CEMENTS AND CEMENT-MORTARS, EXPERIMENTED UPON AS SQUARE PRISMS, WITH THEIR COMPARATIVE ADHESIVENESS ALSO, EXTRACTED FROM THE TWO PRECEDING TABLES.

Cement, or Cement-Mortar, S denoting the Sand.	Age in days.	Weight per cubic foot, in lbs.	Resistance estimated by the breaking weight, in lbs, in several successive Experiments.					Average resistance. in lbs.	Average adhesiveness, in lbs.
Artificial.									
C4 B5·5	15	83	388	365	325	387	383	370	1453
(C4 B5·5)1, S 1..	14	101	111	97	108	....	....	105	920
.. 1, S 2..	26	119	123	117	123	....	....	121	329
C4 B5 I	14	87	560	468	442	....	....	490	1337
(C4 B5 I) 1, S 1..	14	106	154	124	129	....	....	136	929
Frost's..........	15	83	405	326	321	....	....	351	705
Frost's ..1, S 1..	15	111	113	161	129	....	....	134	348
.. 1, S 2..	26	117	76	87	75	....	....	79	266
Natural.									
Francis's........	14	92	321	443	410	507	558	448	1223
Francis's 1, S 1..	15	116	200	171	234	....	....	201	872
.. 1, S 2..	26	122	103	91	95	....	....	96	578
Sheerness	15	104	489	641	599	....	....	580	1220
.. 1, S 1..	15	114	97	217	128	....	....	147	632
.. 1, S 2..	26	115	65	66	67	....	....	69	519

THAT THE PRACTICAL MODE OF ASCERTAINING WHETHER A CALCAREOUS STONE WAS A WATER CEMENT OR NOT, AS DESCRIBED IN FORMER PARTS OF THIS WORK, WAS THE BEST THAT COULD BE USED FOR THAT PURPOSE, BUT NOT SATISFACTORY AS APPLIED TO THE WATER LIMES.

(244) My own mode of ascertaining whether a calcareous stone was a water cement or not, was to pound it when burned, to make the calcined powder up into balls with a moderate quantity of water, and to put these balls into a basin of water almost immediately afterwards, when if the stone experimented upon was a good cement, they set almost immediately into a very hard substance, if not they

fell to pieces. Nothing can be better or more appropriate than this method which decides the question at once. See Articles 45, 46 and 47.)

But the same process, when applied to limes, in order to ascertain their water-setting properties, is by no means satisfactory: for even the blue lias lime, which is the most powerful water lime in England, and probably in Europe, either goes to pieces in time, or does not set in a satisfactory manner, under water, when thus treated, as was stated in Article 24. Smeaton's mode of experimenting upon such limes, to all the details of which I had not properly adverted, when I began to investigate the subject, was much better than mine; but I consider the following process to be better than either, which I have partly borrowed, that is with modifications, from M. Vicat and General Treussart, two of the most recent and approved French writers on the subject of common and hydraulic limes and mortars, of whose able and important researches I shall endeavour to give a synopsis in the Appendix.

RULE FOR ASCERTAINING WHETHER A LIME HAS HYDRAULIC PROPERTIES.

(245) Burn some of the lime stone in a fire-place for three or four hours, until it ceases to effervesce in acids, as directed in Article 45, or if it be a lime in common use, procure some of it well burned and fresh from the kiln, and slake it thoroughly, by first pounding it in a mortar, and afterwards working it up into a stiffish paste with water, upon a slab with a spatula. Whilst being thus mixed, it will throw out more or less heat according to its quality. Wait therefore until it becomes perfectly cool, and then put it into the bottom of a cup or small basin, pressing the upper surface of it into a level form, after which if it should be rather too moist, let it dry a little until it becomes a stiffish paste. If it should crack in this state, which can only proceed from the whole of the lime not having been thoroughly slaked, when put into the basin, a circumstance most likely to occur with very powerful hydraulic limes, which slake with difficulty, such as the blue lias,* it must be taken out of the basin and remixed with a moderate quantity of water, and then put back again; after which let water be poured into the basin, until it is nearly full, taking care not to disturb the surface of the lime paste, which will now be completely immersed.

* See Article 24, and the note to Article 26.

In this state, let the paste be examined daily, by pressing it with the finger, and if it should become so much harder as not to yield to the pressure, it is a hydraulic lime, and has just begun to set. If on the contrary the lime paste, immersed in a stiffish state as above directed, should become softer under water, it is a common lime having no hydraulic properties. In comparing several hydraulic limes together, that which sets the soonest after immersion may be considered the most powerful.

In experimenting with the blue lias lime of Lyme Regis, in this manner, we found that it set rather more quickly, when made into a stiffish paste with a moderate quantity of water, than it did when mixed with excess of water into an almost fluid state, but both of these sorts of blue lias lime paste, having been covered with water about an hour after they were mixed, attained such a consistency, that they resisted the action of the finger next day, and eventually set into a hard substance, especially the former, which had a decided superiority over the latter.* We had before ascertained that the same blue lias lime putty, when these precautions were so far neglected as to immerse it whilst still warmish, without waiting till it had become perfectly cool, would not set under water at all.

(246) On mixing one measure of the same blue lias lime with one measure of sand, and allowing a couple of hours before immersion, this mixture also set under water, though not into so hard a substance, nor so soon as the lime alone had done: but on mixing one measure of this lime with two measures of sand, and treating it in the same manner, it became softer after immersion, and would not set at all, at or near the surface; but at the end of two months, after removing about half an inch of the surface which was quite loose, the remainder of it was found to be setting in a slow but satisfactory manner.

REMARKS ON THE INTRODUCTION AND USE OF PUZZOLANA IN ENGLAND.

(247) About the middle of the last century, Trass appears to have been in common use throughout England for all

* The last of these, which was mixed in rather a fluid state, adhered so firmly to the cup, that it was impossible to separate them, without breaking the lime, which was not easy, as it had become exceeding tough in setting. All the other experimental mixtures tried in this manner, whether of nett lime or of lime mortar, after setting at the surface, always detached themselves from the cup, in cakes of the form of the plano-convex lenses of opticians.

works exposed to the action of water, in the proportion as Smeaton observes of 2 measures of slaked lime to 1 of that substance. In respect to Puzzolana, it appears that it had not been used at all in England, and that he only met with it by accident, a merchant having imported a cargo of it from Civita Vecchia, on speculation, which nobody would purchase or even try, till Smeaton experimented upon its qualities, from having read of it in Belidor's Architecture Hydraulique, and finding it superior to trass, and offered by the owner at a lower price, the whole cargo was purchased and employed by him in the building of the Edystone Lighthouse, in which it was mixed with blue lias lime from Watchet, a small port of Somersetshire in the Bristol channel.* The Watchet lime used by Smeaton was burned and also slaked on the spot, with just water enough to cause it to fall down into a fine powder, which was closely packed in water-tight casks, and sent to Plymouth, and from thence to the Edystone rock as required, in which state Smeaton observes, that it kept so well, that he used the surplus of the same lime, which had been provided in the year 1757, for the Edystone Lighthouse, no less than 7 years afterwards, in the most critical part of the works for improving the navigation of the River Calder. At this period he says, that he found the lime in the casks perfectly good, but that he took the precaution of bruising and sifting some parts of it, which had become rather lumpy. This mode of slaking lime in powder, is equivalent to what was afterwards recommended by M. De la Faye in France, and which recently has been very generally used in that country, as will be mentioned in the Appendix. Smeaton mixed his slaked lime powder and puzzolana for the construction of the Edystone Lighthouse, in equal volumes, throwing both into the measure by a shovel with some degree of force, but without pressing them down.

(248) I before mentioned in Article 20, that the natural water cements of England, since most improperly and absurdly termed *Roman Cement*, of which the properties were first discovered by Mr. Parker, who took up a patent in 1796, for the exclusive use of them, had superseded the Trass of the Rhine, vulgarly termed Dutch Terrace, and the Puzzolanas of Italy. This assertion is perfectly accurate, in respect to private works in the neighbourhood of

* Having explained that he actually procured his lime at Watchet, he afterwards invariably calls it Aberthaw lime, being of the same quality as the latter.

London, and perhaps throughout England generally, and also so far as regards the use of trass, but not so in respect to puzzolana; for I find on more minute inquiry, that this substance, in combination with some sort of hydraulic lime, has frequently been used and generally in preference to cement, from the time of Smeaton to the present day, in public works of importance, such as Lighthouses, Harbours, Docks, &c, where strong hydraulic mortar was considered necessary.

(249) It is well known that puzzolana in itself has no cementing properties, having been found when analyzed chemically, to consist of the same component parts as the coloured clays of nature, that is chiefly of silica and alumina, with a small proportion of the metallic oxides, and sometimes also with *a Trace* of lime; but when mixed with the weaker hydraulic limes, or even with common lime, it constitutes a hydraulic mortar, that sets under water, certainly much more slowly than either the natural or artificial water cements; but those who are advocates for the use of this substance, consider it not less efficient or powerful in the end. I must confess that I entertain some doubts of the accuracy of this opinion, but as nothing can decide the question, but a sufficient number of trials of the strength of cement joints, and of puzzolana joints, which would require several months or perhaps a year to accommodate the latter, being the slower setting substance of the two, it is useless to discuss it.

(250) The following is a statement of the various mortars used by Smeaton, in which puzzolana was one ingredient.

SMEATON'S HYDRAULIC MORTARS FOR VARIOUS PURPOSES, COMPOSED OF ABERTHAW OR OTHER WATER LIME, MIXED WITH PUZZOLANA, OR WITH PUZZOLANA AND SAND.

No.	Name of the Mortar.	Winchester bushels by stricken measure of			No. of cubic feet	Expense per cubic foot.
		Slaked Lime Powder.	Puzzolana.	Common Sand.		
						s. d.
1	Edystone Mortar	2	2	..	2·32	3 8
2	Stone Mortar..............	2	1	1	2·68	2 1½
3	.. 2nd sort	2	1	2	3·57	1 7½
4	Face Mortar	2	1	3	4·67	1 4
5	.. 2nd sort	2	½	3	4·17	1 1
6	Backing Mortar	2	¼	3	4·04	0 11

(251) Smeaton states, that according to the common mode of mixing hydraulic mortar in his time, which was

done by continually beating it, one whole day's work of an able-bodied labourer was required for incorporating any of the mixtures above described, or any other well proportioned mixture, containing 2 bushels of lime as one ingredient; but by adopting a more judicious mode of mixing the ingredients proposed by him, he states that the time necessary for beating them might be reduced one half. This process of beating he considered indispensable.

I do not understand, what he meant in the above Table, by *Stone Mortar*, as distinguished from *Face Mortar*, but since the proportion of lime is the same in all, and we know that he considered the Edystone mortar No. 1, consisting of lime and puzzolana alone, without any sand, as being the strongest hydraulic mortar that could be made; it follows that he must have considered No. 6, which contains the smallest proportion of puzzolana, and the greatest of sand, as the weakest of all; the others being of intermediate qualities, better or worse, as they approximated more nearly to the former or to the latter of those mixtures. It may also be inferred, that he did not think it prudent, in making hydraulic mortar, to use more than 2 measures of sand and puzzolana inclusive, to 1 measure of slaked lime powder.

(252) In building the Bell Rock Lighthouse, opposite to the mouth of the River Tay, Mr. Stevenson used the same sort of lime that Smeaton had done, the blue lias of Aberthaw, in the proportion of 1 measure of slaked lime powder, 1 measure of puzzolana, and 1 measure of sand, which he says that he considered equally good, and of course much cheaper as a hydraulic mortar, than the Aberthaw lime mixed with puzzolana alone would have been, the latter of which substances then bore an exorbitant price. He states that it was not his wish at first to have used Aberthaw lime, but he found on trying the best limes, which the neighbouring counties of Edinburgh, Haddington, Fife and Forfar, could produce, that the like mixtures of these limes, with puzzolana and sand would not set under water at all.

I am not aware, that any other Civil Engineers of this Country have ever dispensed with sand entirely, as Smeaton did in the construction of the Edystone Lighthouse. But in determining the proportions of their hydraulic mortars, I rather suspect, without being certain as to the fact, that some of them have mixed their lime and sand in the same proportions, that they considered proper for the mortar of buildings exposed to air only, to which they added the puzzolana, under an impression that since this substance improves lime, it may improve mortar also. Smeaton seems to have apprehended on the contrary, that the addition of puzzolana, which has nothing calcareous in its composition, over and above as much sand as any lime made up into mortar can well bear, might have the same injurious effect in weakening if not ruining that mortar, as if an additional quantity of sand had been used; and therefore he lays down a rule, that whatever proportion of sand is known to make good common mortar with any sort of lime, the quantity of puzzolana used to improve this mortar, and render it fit for hydraulic purposes, must be deducted from the said proportion of sand.* Puzzolana communicates to limes the property of gradually setting under water; but perhaps it may be considered sufficiently proved, that this property does not convey the power of bearing much sand also, by the circumstance, that the powerful water cements of England, possessing much stronger hydraulic properties than any limes we know, are deprived of their cohesiveness, by a smaller proportion of sand, than that which common lime will bear (see articles 56 and 57.) On this subject however, I speak with less confidence, than in treating of cement, because I have tried so very few experiments with puzzolana, that I do not feel myself competent to give a decided opinion; which I have no hesitation in doing, as far as the natural and artificial cements are concerned. I think it right however, to record a few experiments recently tried by me upon puzzolana of excellent quality, and on two sorts of artificial puzzolana, made in imitation of it.

REMARKS ON PUZZOLANA. EXPERIMENTS ON THIS SUBSTANCE TRIED AT CHATHAM IN 1829 AND 1837.

(253) It appears from Vitruvius, that the properties of puzzolana were well known, and that it was in common use amongst the Romans in his time.† He describes it as a substance in the state of dust or fine powder found in the neighbourhood of Báiæ and in the environs of Mount Vesu-

* See the Article on puzzolana mortar, at the end of the 3rd volume of Smeaton's Tracts published in 1812.

† In book ii, chap. 6, in which he treats "de Pulvere Puteolano et ejus usu," he commences thus.

"Est et genus pulveris, quod efficit naturaliter res admirandas. Nascitur "in regionibus Baianis et in agris municipiorum, quœ sunt circa Vesuvium "montem, quod com mixtum cum calce et cæmento, non modo ceteris "ædificiis prœstat firmitates, sed etiam moles quœ construuntur in mari, "sub aquâ solidescunt." The word 'cæmentum' is here supposed to mean small stones. It certainly does not imply what we term cement.

vius, which, when mixed with lime, increased the strength of the mortar of common buildings, and caused that of moles or piers run out into the sea to indurate under water.

(254) After my experiments, tried with the view of obtaining an artificial cement, had first been attended with some success in 1828 and 1829, I procured some puzzolana, that had been provided for the extensive new works then in progress in His Majesty's Dock-yard at Sheerness, in order to experiment upon it, and compare its hydraulic powers with those of the Sheppy and Harwich cements, and with my own artificial cement. All my efforts at that time to produce a good hydraulic cement with puzzolana failed, in consequence of two errors in my mode of using it, which are injurious to its powers.

In the first place, I used the puzzolana in the same state in which I received it, that is with a very small portion of it only in dust, the rest being in lumps of various sizes, like a mixture of coarser and finer sands, in which state it does not combine properly with lime; for in order to produce that intimate union of the two substances, which alone can prevent the lime from going to pieces under water, it is absolutely necessary that the puzzolana should first be reduced to the state of an impalpable powder, which rule was not mentioned to me by the gentleman at Sheerness, from whom I received the article, and which possibly might not have been known, even to the persons employed there in making it into mortar.*

Secondly, I omitted to slake the lime completely, before I put it under water, instead of which I pounded the lime, but not the puzzolana, in a mortar, after which I mixed the quicklime powder and puzzolana into a paste with water, and made balls of the mixture, and put them under water immediately, which is a test that nothing but the strongest water cements will bear.†

* Because the puzzolana was thrown in its natural state into a mortar mill, worked by steam, and having two very heavy iron rollers, under which whilst being mixed with the lime and sand, this soft substance could not fail to be ground as fine as was necessary, to make it combine properly with the lime. But unless Mr. Rennie had explained, that not merely mixing but grinding was his object in using the mortar mill, this may have escaped the observation of the persons employed.

† In Smeaton's Observations on Water Cements, he does not lay down either of these rules pointedly; but his practice involved both, for he worked up his experimental mixtures so long, that the lime must have become thoroughly slaked, and in mixing his puzzolana mortar for practical purposes, the continued beating could scarcely have failed to reduce the puzzolana, if coarse, to a very fine state. I suspect that this effect was the chief advantage derived from the beating, and that the reason for it assigned by Smeaton, in his Account of the Edystone Lighthouse, was of very secondary importance.

(255) In my recent experiments with the same puzzolana, which had remained in my possession since 1829, I reduced it to the state of impalpable powder, and then mixed it with chalk lime perfectly slaked, in the proportion of 2 measures of puzzolana to 1 measure of chalk lime putty. I then put this mixture into a small basin, of which it occupied the bottom, its own upper surface being pressed down level. At the end of about an hour, I poured water gently into the basin, until it was nearly full. In this state the puzzolana mortar became harder every day, and on the fourth day after immersion it resisted the pressure of the finger, so that the surface was fairly set.

(256) I also mixed one measure of thoroughly slaked chalk lime putty, one measure of pulverized puzzolana, and one measure of sand together, and treating them precisely in the same manner, I found that the puzzolana still prevented the lime from going to pieces or dissolving under water, but that the sand injured the compound by retarding the setting of it, which after 11 days was only just commencing at the surface; but on handling it rather roughly in trying its hardness by pressing the finger into it, the sort of skin at the surface broke, and the mass below proved at the time to be in a very soft state, like wet clay. On breaking the former mixture of lime with puzzolana on the same day, to examine its comparative state, we found that it was formed into a cake hardest at the surface, but not soft in the center, for on breaking off and bruising a small piece, it was reduced to a dry powder.

OF ARTIFICIAL PUZZOLANA.

(257) The most ancient sort of artificial puzzolana is brick dust or tile dust, or fragments of pottery ground to powder, which being mixed with lime produce a sort of cement, much more impervious to water, than common mortar of lime and sand, and which in all probability was in common use in the time of the Romans, as may be inferred from some passages of Vitruvius.* It is not to be won-

* In book vii, chap. 4, entitled "de Politionibus in humidis locis," he commences thus.

"Quibus rationibus siccis locis tectoria oporteat fieri dixi, nunc quemad-
"modum humidis locis politiones expediantur, ut permanere possint sive
"vitiis, exponam. Et primum conclavibus, quæ plano pede fuerint ab imo
"pavimento alté circiter pedibus tribus, pro arenato testa trullissetur, et
"dirigatur, uti eæ partes tectoriorum ab humore ne vitientur."

The word *Testa* is translated by M. Rondelet as *ciment des tuileaux*, and I think he is quite right in this interpretation, for it is evidently contrasted with *Arenatum*, which is mortar composed of lime and sand, and therefore *testa* must mean mortar composed of lime and tiles or pottery, reduced by grinding to at least as fine a state as sand, if not to impalpable powder.

dered at, that men previously acquainted with the use of puzzolana, should have had recourse to this expedient, or that it should often have succeeded, since some of the natural puzzolanas and brick dust not only resemble each other strikingly in their appearance, but modern chemistry has enabled us to ascertain, that they contain the same component parts, both being originally clays, and both moderately calcined or burned, the one by volcanic agency, and the other in a kiln or clamp.

This primitive sort of artificial puzzolana has been in constant use from time immemorial in the South of Europe, and especially in France, where the word *Ciment* has been applied to it, which wherever it occurs in any French writer on Practical Architecture, without farther specification or explanation, invariably and exclusively means brick dust or tile dust proposed to be mixed with lime, for the purpose of forming a hydraulic mortar;* and I believe that the same practice has been followed in most of the other countries which formed a part of the Roman Empire.†

(258) Instead of confining themselves to the use of pounded bricks or tiles, the Dutch, to whom hydraulic mortars must have been of peculiar importance, have long been in the habit of burning a clay found under the sea on their coast, for the purpose of forming an artificial puzzolana, which is said to have been so good an imitation of the natural trass of Andernach on the Rhine, that it has been sold for such, and probably some of the Dutch tarras imported into England may have been of this description. More recently, since chemistry became generally known, a number of men of science, especially in France, have in like manner attempted to form more perfect artificial puzzolanas, than the indiscriminate use of tiles or bricks was likely to afford, by burning various natural substances, such as clays, basalts, schists, &c. for this special purpose, of

* This term will probably become obsolete in France, as M. Vicat, who has already improved the French nomenclature of calcareous cements by introducing the term hydraulic lime, instead of a less appropriate expression formerly in use, generally employs the phrase artificial puzzolana in preference to *Ciment*.

† The first works of which I had the entire personal superintendence, instead of attending by routine of duty, were two martello towers built at Sanitje and Adaya, on the coast of Minorca, in 1801. The small cisterns of these towers, and the larger cistern for the signal station of Mount Toro, which was also under my charge, were all lined by the native masons employed in building them, with a mixture of common lime and tile dust, according to the custom of that island, handed down to them no doubt from the time of the Romans, which was suggested by them, and approved by the Commanding Engineer, my Superior Officer.

which Mr. Baggé of Gottenburgh is said to have been the first to set the example in Sweden, about the middle of the last century. In this important research M. Vicat, who also made an artificial hydraulic lime, and General Treussart have exerted themselves with the greatest zeal.

(259) The expedient of using pounded bricks or tiles as an artificial puzzolana, was not unknown in England, as appears by a treatise on brickwork published by Mr. Batty Langley about the middle of the last century, but seems to have been little used. About 12 years ago, Mr. John White recommended the use of pounded bricks burned to a state of incipient vitrification, usually termed *Clinkers,* under the title of British Puzzolana, which does not appear appropriate, when applied to an expedient, that had been in general use in Europe for many centuries. His proceedings, which attracted a good deal of notice in London at the time, will be described in the Appendix. The system suggested by him has not however been adopted in practice, being infinitely inferior to cement for difficult and critical works, and the extra expense of his pounded clinker bricks not being considered sufficiently compensated in domestic buildings, by its superiority over common mortar, made of the same limes mixed with sand alone.

EXPERIMENTS ON ARTIFICIAL PUZZOLANA TRIED AT CHATHAM IN 1837.

(260) These were few in number and tried very recently, when I caused some blue alluvial clay of the Medway, and some of the brown pit clay of Upnor, which had both been used by me as ingredients for my artificial cements, to be made up into balls, and moderately burned; so that when reduced to impalpable powder, and mixed with water, they would not readily fall to pieces, though they had lost their original plasticity, which degree of calcination appeared to assimilate them as nearly as possible, to the natural puzzolana then in my possession. On mixing each of these two calcined clays, after pounding them in a mortar, with common chalk lime perfectly slaked and made up into a paste with water, in the proportion of 2 measures of calcined clay powder to 1 of lime putty, the mixture being worked up into a stiffish paste; we found that the artificial puzzolana, produced from the blue clay, set nearly but not quite as quickly under water, as the natural puzzolana; but that produced from the brown pit clay set much more slowly. We also mixed each of these two artificial puzzolanas with

common chalk lime putty, and sand, in equal proportions of all by measure. And as these mixtures were made up on the same day, with the like mixtures of the natural puzzolana, before described, we compared them all daily, and found that they set as follows, first the natural puzzolana and lime, secondly the calcined blue clay and lime, thirdly the calcined brown pit clay and lime, fourthly puzzolana, lime and sand, fifthly the calcined blue clay, lime and sand, and lastly, the calcined brown clay, lime and sand. The second mixture, namely, the calcined blue clay and lime was setting like the natural puzzolana in the most promising manner, though more slowly, forming like it a cake hard at the surface, but on breaking it when 11 days old, to compare it with the former, I found it rather soft in the center. The cake formed of the calcined brown clay was softer than either of these two, not being quite hard even at the outside, but was superior to all those mixtures, even to that of the natural puzzolana, in which sand as well as lime had formed a part. These last had scarcely begun to set, even at the surface, on the 11th day, when on breaking them, the inside of the best of them was like moist clay.

(261) From these experiments, I think it probable, that the same sort of clay, which makes the best ingredient for an artificial cement, when mixed with chalk previously to calcination, will also make the best artificial puzzolana, when calcined separately; and therefore that pounded bricks, which are always the produce of the coarsest clays, must be much inferior to any of the finer clays for this purpose.

(262) So far as regards the water setting properties of hydraulic mortars, experiments on a small scale, such as have been related in the preceding articles, are only decisive, in respect to the mortar in and near the outside of walls exposed to immediate and continued immersion. For experience has proved that hydraulic mortars, not capable of resisting the immediate and constant action of water in mass, will set in the center of walls whose surfaces are exposed to it, provided that these surfaces be protected by cement. Thus for example it was found necessary to protect the puzzolana mortar of the Edystone Lighthouse, and of the Bell Rock Lighthouse, by pointing the external joints of the former with gypsum as a temporary expedient, until the mortar should set, and of the latter with cement, these being exposed to the direct wash of the sea; whilst it was also found necessary to protect the blue lias lime mortar of the lower part of the piers of the

Menai Bridge, and of the Steam Packet Landing Wharf at Hobbs' Point Milford Haven, by cement. In short no substance with which I am acquainted, excepting cement, is capable of resisting the violence of the waves or of running water; but both puzzolana mortar as well as blue lias mortar without puzzolana, are capable of setting under water in a more tranquil state, but not without the gradual degradation of all the external joints, to a certain distance in; as was exemplified in the experiment with the blue lias mortar recorded in Article 246, of which the inside gradually set, whilst the surface gave way; and from my own experiments with small specimens of the Rosehill lime, that I have seen, and from Smeaton's favourable opinion of the Barrow lime, I consider it probable that the mortars of these two limes also, mixed with a moderate proportion of sand alone, will set in still water, like the blue lias; but so far as regards the weaker hydraulic limes, such as the Dorking, Halling, &c, I do not think that the mortars of any of these limes will set in a satisfactory manner under water in less than 20 years, unless they either be improved by a proportion of puzzolana, or confined to the middle of the wall only, the two sides being built with cement, or with some much stronger hydraulic mortar to protect the center.

DESCRIPTION OF THE MORTARS USED IN THE CONSTRUCTION OF THE DOCKS OF KINGSTON-UPON-HULL, AND OF THE STATE WHICH THESE MORTARS HAD ATTAINED, WHEN EXAMINED BY MR. TIMPERLEY, ABOUT HALF A CENTURY AFTER THE FIRST OF THEM, THEY HAVING ALL BEEN BUILT AT DIFFERENT PERIODS, WAS FINISHED.

(263) My doubts on this point have been caused by reading an able and interesting account of the Harbour and Docks of Kingston-upon-Hull, written by Mr. Timperley, and published in the first volume of the Transactions of the Institution of Civil Engineers, from which I shall extract some facts relating to the state of the various sorts of mortar, used in building the walls of the basins and locks of that port.

In building the Junction Dock at Hull, which was commenced in 1826, under the direction of Mr. Walker, the present President of the above Institution, it became necessary to pull down a part of the walls of two former docks, in order to form a communication or junction between them by means of this new dock, as its name imports; and at this period Mr. Timperley the resident or executive Engineer at that port had an opportunity of examining the mortar of those walls. Of the two original docks, one was called the Old Dock, communicating with the river Hull, which was planned by Mr. Grundy and finished in 1778, after four years labour; whilst the other, called the Humber dock, communicating with the harbour on the river Humber, was planned by Mr. Rennie and Mr. William Chapman of Newcastle-upon-Tyne, jointly in 1802, commenced next year, filled with water in December 1808, and opened for business in about six months afterwards.

(264) The basin walls of the Old Dock were built entirely of brickwork with a stone coping, laid in mortar compound of Warmsworth lime and sand, part of which was fresh water sand, the rest selected from the excavation.

The basin walls of the Humber Dock were also built of bricks, with the exception of a stone *thorough Course* at the bottom of the vertical fender piles, three courses of stone on the level of an average tide, the lowest of which was a thorough course also, and the coping. The mortar was made of Warmsworth blue lime, and sharp fresh water sand only. The lime having been ground in its dry state in a mill worked by a steam engine, was mixed with two parts of sand for the front work, and water having been added the whole was ground again, and the mortar used immediately afterwards, whilst hot and fresh. The backing mortar was compound of 1 part of unslaked lime to 3 parts of sand measured and tempered in the usual manner. The brickwork of the front and back was laid in mortar, the rest grouted in every course. Part of these walls being built a little before winter, the front mortar was affected by the frost, but the joints were afterwards raked out, and pointed with puzzolana mortar.

The locks in both of these docks as being the most important parts of the work and more exposed than the basins, were entirely faced with stone, and fronted with puzzolana mortar, but backed with lime mortar of the quality before described. The front mortar for the Humber Dock was compound of 1 measure of ground Warmsworth quicklime, 2 thirds of a measure of ground puzzolana, and $1\frac{2}{3}$ measure of sharp fresh water sand.

On pulling down part of the Humber Dock at the period before alluded to, Mr. Timperley observes 'that all the 'puzzolana mortar was found to be exceedingly hard, being 'both in colour and hardness like a well burned brick, but 'the whole of the other mortar of that dock with the excep-

'tion of a very small portion under an inverted arch was 'still perfectly soft, and might have been worked up again, 'without adding a drop of water.'

In respect to the puzzolana mortar in the Lock, he observes that it appeared to adhere much better to the brickwork, than to the stones, which he seems to consider a defect, but which as I myself have found greater adhesiveness to stone in puzzolana than in common mortar, I ascribe to the inexperience of the workmen in using it. Now as the water was not let into this dock or great basin, in building which no puzzolana had been used, until about 6 years after the work was commenced, which must have been of the greatest use to the mortar, by allowing ample time for a great part of it to set in air, previously to immersion, notwithstanding which it appears that even at the end of 18 or 20 years more, scarcely any part of this mortar had set at all,* I conceive that this proves in the strongest light, the propriety of never using any weak water lime mixed with sand alone, for the purposes of hydraulic Architecture; and the hint was not lost upon Mr. Walker, as will presently be mentioned.

(265) In respect to the great basin of the old dock, part of which was pulled down at the same period, 'the front 'of the wall below an average tide, for about 8 or 9 feet 'under the coping had but an indifferent appearance, the 'bricks being in many places much decayed and rubbed by 'the vessels, and the mortar washed out of the joints, but 'below this, the bricks were generally in a much better 'state, and the pointing nearly entire. It has been before 'observed that the mortar for this wall was made partly 'from sand dug out of the dock, which was far from being 'of the best quality; the interior of the wall was grouted, 'and not very sparingly, as in some places the mortar was 'found nearly as thick as the bricks. The mortar in the 'inside of the wall varied very much in quality, according 'to circumstances. Where the wall was solid and undis-

* Mr. Timperley explains that the mortar in the front of the wall, which was every where soft, was much out of the joints for 9 or 10 feet from the top. Below this the joints were not wasted, but had thrown out a sort of stalactite or calcareous incrustation that entirely covered the face of the wall. As it appears that these different actions on the external surface of the joints took place, the former above, the latter below the level of the high water of neap tides in the basin, I conceive that the moderate action below may have proceeded from the water in the lower part of the basin not being in motion or continually changed, as in river wharfs, or in piers carried out into the sea, which produces a violent action upon common or inferior mortars. (See Articles 10, 11 and 12.)

'turbed, it was very hard, requiring picks, and in many 'places, sledges and wedges, to take it down; but where 'the wall had given way, or been otherwise disturbed, and 'cracks and cavities thus caused in the inside, the mortar 'was in general very soft. This was observed in a variety 'of places, and it was not uncommon to see the mortar in 'one part of the wall exceedingly hard and good, and 'within a few inches from it, where the wall was open, and 'the water had found its way, quite soft and bad, or but 'little harder than when first built.'

(266) The author observes, that such works were almost in their infancy in England at the time when this dock was built, and therefore he considers it upon the whole to be creditable to the Engineer who designed it, notwithstanding several defects, which he has mentioned, such as 'the insufficiency of the piling, and the foundation, which 'was only level with the bottom of the dock, not being low 'enough, in consequence of which the walls have subsided, 'and been forced forward in several places by the pressure 'of the earth behind, &c.' In the front of the wall, the section of which was in the form of a common sloping revetment, an error was committed, by building the courses of brickwork at right angles to the slope, for about 14 inches, or a brick and a half inwards, beyond which the whole remaining brickwork of the wall was built in horizontal courses. Thus there was one unbroken joint nearly vertical, from top to bottom, throughout the whole extent of the dock wall, than which one can scarcely conceive any thing more injudicious. Owing to this defect, in the original construction, Mr. Timperley remarks that 'the parts of the 'wall were not only unconnected, but in many places 'entirely separate, so that a rod may be thrust down many 'feet between them. It was observed also, that where the 'wall had given way, it was completely separated from the 'counterforts, to the extent of one or two feet or more in 'the worst places, whereby the strength of the wall had 'been greatly reduced.*

* The wall was 34 feet 6 inches high, including the coping, with an exterior slope of 5 feet, built on a horizontal piled foundation. It was 10 feet thick at bottom, and 4 feet 6 inches thick at top just below the coping, with 4 small offsets in rear, the first pair at the height of 10 feet, the second pair at 15 feet up. The counterforts were at clear intervals of 35 feet apart, and of the uniform width of 9 feet throughout. Their length, measured from the back of the wall, was 8 feet at bottom, but only 3 feet 6 inches at the top, which terminated 4 feet below the coping. This diminution was not effected by a slope, but by 3 steps of equal height and width nearly, at the back of each counterfort. I am of opinion, that no sloping revetment with a

(267) I shall here remark, that though the mortar in the center of the undisturbed portions of this wall had set into a respectable hardness in about half a century, yet from the state of the same sort of mortar in the Humber Dock, it is probable that the whole of it must have remained soft for at least 20 years, and to this soft imperfect state, I ascribe the separations that have been mentioned; for with all the disadvantages of an unbroken vertical joint, which is extremely prejudicial in building with common mortar, the front would never have separated from the body of the wall from this cause, if cement had been used; of which the Thames Tunnel, consisting of many hundreds of vertical joints of this description, affords the strongest proof. (See the Note to Article 58.) Nor could the dock walls possibly have been forced forward, if laid in cement, without drawing the counterforts along with them, with which in that case they would have composed one homogeneous mass, as incapable of separation as solid rock.

(268) The walls of the basin of the Junction Dock, the most recent of the whole, were also built with brickwork, but with a facing of 10 courses of stone immediately below the coping. Mr. Timperley observes, that the basin had been begun, and the East wall in particular, had been finished as high as the stone work, with mortar composed entirely of unslaked blue Warmsworth or Weldon lime, and clean sharp fresh water sand, in the proportion of 2½ measures of sand to 1 of lime for the facing, and 3½ measures of sand to 1 of lime for the backing. Trusting to the high reputation of these limes, which were considered good water limes in the neighbourhood, Mr. Walker had proposed to build the whole of the basin walls in the same manner; but on breaking into the walls of the Humber Dock, as before mentioned, which brought the striking inferiority of that sort of mortar to light, he directed some experiments to be tried on various mortars, in consequence

horizontal foundation resting entirely upon vertical piles in soft soil like the above, can be considered secure against being forced forwards at top; but I shall not enlarge upon this point, which would lead me away from calcareous cements to another branch of the extensive subject of Practical Architecture.

The lock of this old dock must have been the most defective part of the original design, because the walls had yielded so much in the course of seven or eight years after it was built, as to require to be taken down about 12 feet from the top. One side was rebuilt in 1785, and the other the following year. But it does not appear that these repairs were effectual, for it afterwards proved necessary to pull down the entrance lock and basin entirely, in consequence of their very ruinous state in 1814, and to rebuild both, according to a new design and under the direction of Mr. Rennie.

of which, he ordered the whole front of the remaining part of the Junction Dock walls to be faced with puzzolana mortar, composed of 1 measure of unslaked blue Warmsworth or Weldon lime, ½ measure of puzzolana powder, and 2 measures of sharp fresh water sand; but he directed the hollow quoins of the Lock-gates, as being of more importance, to be faced with 1 measure of Halling lime, 1 measure of puzzolana, and 2 measures of sand.* For the backing of his walls, he still continued to use the same sort of mortar, with which he had commenced.

(269) In the lower part of the East side of the basin of this dock, where no puzzolana had been used, an important circumstance, not noticed by any former writer, or at least new to me, was observed by Mr. Timperley, who states that 'notwithstanding the thickness and solidity of 'the walls, the water in wet weather found its way through, 'so that they were exceedingly damp even in front, and in 'several places the water literally ran down the face of the 'wall. This was ascribed to the mortar and grout not 'hardening sufficiently, as in all cases where the front was 'set in puzzolana mortar, although the walls were a little 'damp in places, the water never penetrated through.'

In respect to the puzzolana mortar of this dock, which Mr. Timperley had an opportunity of examining, when part of the wall was pulled down and rebuilt, in consequence of having been injured by shipping, he remarks, 'that it was 'in a good state, although not so hard in the interior as in 'front, the mortar in the beds of the stone work also being 'more indurated, and for the most part adhering better than 'in the vertical joints,' which I ascribe to the former having been stiffer than the latter, which must have been poured into these joints in the state of grout. He also remarks, 'that this puzzolana mortar, in the front of the walls, was 'in general hard and good, even before the water was let 'in, the only defective part being in the West end of the 'dock, where the wall was damp in consequence of its

* Mr. Timperley states that in the course of the experiments before alluded to, he found that the Halling lime was stronger than the Warmsworth, which he describes as a magnesian limestone, and also stronger than the Weldon or Fairburn limes, none of which when made into flat cakes with sand, and allowed some days to dry previously to immersion, would ever set under water at all, but dissolved in a few weeks. Hence these limes must be very weak, for the Halling lime itself possesses very moderate hydraulic powers. But the addition of the half measure of puzzolana to 1 measure of those limes having rendered them rather superior to the Halling lime, and as this mixture with a moderate quantity of sand was not much more expensive than Halling lime mortar made with sand alone would have been, it was used in preference.

P 2

'being backed with wet soft earth. Some part of the mor-'tar also, being used late in the year, was a little perished 'by the frost and required fresh pointing.'

(270) A similar fact was remarked by the same intelligent observer, on breaking into the walls of the old Fortifications of the town of Hull, supposed to have been built with stone in the reign of Edward the Second, and repaired and strengthened with bricks in that of Richard the Second. 'The mortar was of two kinds, one composed 'of lime and sand only, the other of lime and powdered 'tiles or bricks, with very little sand; both were with a very 'few exceptions very hard, the latter being the more so. 'The mortar appeared to have been used in a very soft 'state, or as grout, but by no means well tempered, small 'lumps of pure lime resembling hard tallow, being inter-'spersed in great abundance. In three or four of the bot-'tom courses, and 9 to 18 inches in width of the back of the 'wall, where it was in a damp state, it had not set in the 'least, and at the bottom in particular appeared like pure 'sand, whilst the neighbouring parts being dry, were par-'ticularly hard and united like a rock.'

THAT GOOD HYDRAULIC MORTAR SHOULD BE USED NOT ONLY FOR THE FRONT, BUT FOR THE BACK ALSO, OF THE WALL OF A DOCK, BASIN OR WHARF, WHEN INFERIOR MORTAR IS USED FOR THE CENTER, BUT THAT IT IS BEST TO REJECT INFERIOR MORTAR ALTOGETHER, AND TO USE GOOD HYDRAULIC MORTAR FOR THE WHOLE WALL.

(271) The propriety of never using common lime mortar, or even the weaker hydraulic limes and sand alone, for the whole backing of the walls of a dock, basin or wharf, appears to me to be sufficiently proved by some of the facts quoted from Mr. Timperley, which I shall here recapitulate, first by the penetration of water from the rear, through the entire thickness of the East wall of the Junction Dock whilst in progress, secondly by the imperfect state even of the puzzolana mortar in front of the West end of the dock from the same cause, and thirdly by the mortar in the back of the wall of the Fortifications of Hull, not having yet begun to set, though part of it must be at least 500 years old.

(272) Mr. Timperley's description of the state of induration, in which the various sorts of mortar used in the walls of the docks of Kingston-upon-Hull were found, when examined by him at various periods, from one or two years

to nearly half a century after they were built, may be considered a record of experiments on a great scale, and of the most convincing nature, upon the fitness or unfitness of these mortars for the purposes of Hydraulic Architecture: and the same most liberal and useful Institution, which has been the medium of laying so much valuable information before the public, will probably be the channel of future communications of the same interesting nature, respecting the state of similar works in other parts of the United Kingdom, which will afford far more satisfactory data for future hydraulic constructions, than any experiments on a smaller scale can possibly do. One thing only is wanting to render such observations complete, as far as they regard the mortar of new wharf or dock walls. That is to ascertain the time in which each sort of mortar sets, in the center of such walls, at a certain distance below the usual level of the mass of water acting on the front of them. For this purpose it appears to me, that the only satisfactory mode will be to cut a small opening into the back of the wall for about 3 feet in, by working from the bottom of a shaft sunk in rear of it. Let this be done in about a year after the dock has been filled with water, or the wharf finished; and when the state of the mortar has thereby been examined, let the opening be filled with brickwork and cement. If the mortar should not have set at this time, similar small openings may be made afterwards for the same purpose, at successive periods of not less than 6 nor more than 12 months, until it shall have set.

(273) In objecting to the use of mortar made of any of the weak hydraulic limes, such as the Warmsworth lime, which appears not to be much better than chalk lime, for the whole backing of walls exposed to water in front, and to damp earth in rear, let it be fully understood, that I do not speak in the spirit of criticism, or in depreciation of the eminent Engineers, who have occasionally adopted this practice; for I should have done the same myself in their place, never having seen any objection to it, until after the perusal of Mr. Timperley's valuable, but very recent paper. And I am even aware, that in one instance at least, namely in the wall of the London Docks, the backing of which was built with mortar said to be composed of 2½ measures of sharp sand to 1 measure of Dorking lime, this sort of mortar proved to be in a very hard and satisfactory state, when part of the wall was pulled down in 1828, of which I was myself a witness. But as these Docks were built about

25 years before this partial demolition took place, we do not know how long the mortar may have taken, to acquire the respectable state of induration, in which it was found at the period alluded to.*

(274) But if the practice, which I consider as having been proved to be generally objectionable, be set aside for the future, and the system of protecting not only the front but the back also of every such wall with strong hydraulic mortar, such as cement or puzzolana, be considered essential, it may be remarked that the curved leaning revetments, usually adopted by the Civil Engineers of this country, for the profiles of their dock walls, have either been made of the uniform thickness of 6 feet from the bottom upwards, or seldom exceeding 8 feet near the bottom, and 6 feet at top, and the counterforts usually placed at clear intervals of 15 feet apart, have sometimes been made 3 feet, and never more than about 4½ feet square.† Hence if we suppose

* The facing mortar of this dock, which was I believe planned by Mr. Rennie, consisted of 1 measure of Aberthaw lime, ½ measure of puzzolana, and 1½ measure of sand, being precisely the same as Face mortar No. 4, proposed by Smeaton in his table of mortars (250). I had this information from Mr. Palmer, Engineer of the London Dock Company in 1828, who was then building a new basin, in preparing for which, part of the old wall alluded to was taken down.

† For example, the Humber Dock wall, planned by Mr. Rennie, is 32 feet high, including the stone coping, with a concave exterior slope described by an arc of a circle, of which the greatest horizontal ordinate to the said absciss of 32 feet is 6 feet 8 inches. The foundation of the wall and of its counterforts was laid out in an inclined plane, higher in front than in rear, which in profile is represented by a line converging towards the center of the above-mentioned arc. This plane is at right angles to the sloping piles previously driven, consisting of a line of sheeting piles in front, and of two rows of common piles parallel to these, at central intervals of about 3 feet apart, except where there are counterforts, under each of which there are 2 more piles forming a fourth row at unequal intervals. A longitudinal waling is spiked along the heads of the sheeting piles, and sleepers upon the heads of the others, supporting transverse close planking, upon which the foundation of the brickwork was commenced.

This profile proved scarcely strong enough, inasmuch as Mr. Timperley observes, that even before the walls were raised to their full height, it was found that they 'had been forced forward on the east and west sides near 'the middle, 2 feet from a straight line, carrying the foundation piling along 'with them.' These appear by the plan to be the longest sides of the Dock. 'As a security, a quantity of earth about 10 feet high in the center, 'diminishing gradually to 6 feet at each end, was immediately laid in front, 'where it still remains. A length of the upper part of each wall was also 'taken down and rebuilt in a straight line. Some time after the Dock was 'finished, the water having been drawn down to within 13 feet of the bottom, 'for the purpose of making a level bed for the counterbalance weight of the 'gale chains, the east wall again gave way a little, but the movement 'ceased on the rising of the tide. The circumstance operated as a warning 'not to draw the water so low in future.' I conceive it probable, that if the walls had been built with cement, they might have resisted the pressure of earth, to which with common lime mortar, they were evidently unequal, except when that pressure was counteracted by the water with which the that 2 feet of the wall in front, and 2 feet in rear, including the back and sides of the counterforts are to be built with strong hydraulic mortar, there will remain so small a space in the center of the wall for the inferior mortar, that a query naturally suggests itself whether it is worth while to use it in such works at all. I think not: for although there can scarcely be a doubt, that inferior hydraulic mortar will set in time in the center of a wall, when protected on each side, yet there are objections to this construction; partly from the difficulty of bonding the work properly in those parts, where the different sorts of mortar meet, and partly from the probability of unequal settlements taking place, in a wall built of inferior and consequently very slow setting mortar in the center, and of superior and very quick setting mortar on each side.

(275) Upon the whole it appears to me, that it is a much sounder principle, to build the entire mass of every dock or wharf wall with one and the same sort of hydraulic mortar throughout, which may either consist of cement powder measured lightly and sand in equal measures, or of blue lias lime as it comes from the kiln and sand, in the proportion of 2 measures of the latter to 1 of the former; or if any inferior lime be used, let it be ground to powder, and mixed with puzzolana also ground, in equal quantities of 1 measure of each, both measured lightly, to which let

dock was filled. A similar change of form, but of less importance, took place in the center of one of the long sides of the great basin in Sheerness Dockyard from the same cause, on letting out the water, after the basin had been kept for some time full.

The wall of the Junction Dock planned by Mr. Walker, is also a curved leaning revetment, but of a much stronger profile than the above, though some feet lower, having its height or abscisss equal to 27 feet 6 inches, and its greatest ordinate or extreme slope 5 feet. It is 10 feet thick at bottom from whence it is reduced to 8 feet 7½ inches by footings or offsets at the height of 2 feet; and from thence to 6 feet at top, partly by gradual diminution, and partly by 4 small offsets, the 2 lowest of which are placed at about two thirds of the height of the wall, and the others 3 feet higher. The counterforts, parallel in profile to the back of wall, and with similar offsets opposite to the former, are 4 feet 6 inches square above the height of 2 feet. The foundations are oblique and supported by inclined piles driven at a slope of 2½ inches per foot. There are four rows of common or bearing piles, the first not far from the front of the wall, the fourth under the backs of the counterforts, with the intermediate rows at equidistant transverse central intervals of 4½ feet apart, but longitudinally they are only 3 feet apart. The fourth is not a continued row like the others, there being only 3 piles under each counterfort. Before the first row of those piles and connected to it by an intermediate waling, a line of sheeting piles has been driven under the whole front of the foundation. Transverse half timbers and over them longitudinal planks are fixed over the heads of the piles, forming a sort of grating, below which 18 inches of brick rubbish grouted was laid in the intervals between the piles, as the commencement of the foundation, and a similar concrete was laid at the foot of the wall and covered with earth.

1 or at the utmost 2 measures of sand be added.* Except in working with cement mortar, the outer joints will in process of time be washed out or deteriorated for some little way in, which must happen in wharfs executed by tidework. In this case it is best to wait till the defect gradually takes place, at which period the wall will also have settled its utmost, and then let the damaged mortar be raked out, and the joints pointed with cement. In similar dock walls built by means of cofferdams, let the pointing of the external joints with cement be executed just before these dams are removed, at which period in all probability the full or final settlement of walls, built as supposed with blue lias lime mortar, or with the mortars of weaker limes and puzzolana, will have taken place; because the quick setting properties of these mortars put an end to this process much sooner than in common mortars, composed of weak hydraulic limes or of common chalk lime and sand alone.

(276) For my part, I consider good cement as affording the most perfect sort of mortar for all walls of the description now under discussion, and this opinion is borne out by Mr. Timperley's description of the only part of the works of Hull where cement was used, namely in a slip for repairing the mud boats and the lock gates, which was built in 1829 on the west side of the entrance basin, abutting upon the Humber. He remarks that 'the coping and front 'brickwork of this slip were set with Parker's cement and 'sharp fresh water sand in equal proportions, and although 'exposed to the waves and swell of the Humber, have stood 'hitherto with scarcely a failing joint.'

HASTY EXPERIMENTS ON VARIOUS LIMES, LIME MORTARS, AND PUZZOLANA MORTARS.

(277) After having become convinced of the superior accuracy of experimenting upon the strength of cement by

* Mr. Stevenson was induced from his experiments on a small scale, to form an opinion that no mixture of puzzolana and sand, with any common lime whatever, would make a good hydraulic cement, and therefore he used Aberthaw lime which was much stronger, and which if protected at the surface by cement, would have set without any puzzolana at all. I think it probable, that his mode of experimenting, like that which I adopted myself at first, must have been rather too severe a trial of the hydraulic properties of mortars, for my own more recent experiments at Chatham, tried in a different and less rigorous mode, establish the fact of puzzolana communicating the important property of setting under water to common chalk lime, and Mr. Timperley's reports upon the works at Hull, prove in a more decisive manner, that a smaller proportion of puzzolana than I would myself recommend, imparts the same valuable property to the Warmsworth lime, which from his description I consider to be very little superior to chalk lime.

the double process of breaking it down in square prisms, and of tearing it asunder in the joints of stone bricks united by it in pairs, I became anxious to apply the same system to limes also, and accordingly caused a similar series of experiments to be tried upon the blue lias lime and common chalk lime, both in the state of hydrates or putties, and as mortars, with and without puzzolana; and also upon the Halling lime and its mortars, but not with puzzolana. Not being able to afford the time usually supposed necessary, for such experiments, for which a period of 12 months has generally been allowed, I should have suppressed them, if they had not developed some remarkable circumstances respecting the properties of limes and mortars, which I could not have anticipated, either from my own former observations, or from those of other writers.

TABLE XII. COMPARATIVE RESISTANCE OF VARIOUS LIMES, LIME MORTARS AND PUZZOLANA MORTARS, FORMED INTO SQUARE PRISMS, OF THE SAME DIMENSIONS AS THE SMALL ARTIFICIAL STONES BEFORE EXPERIMENTED UPON.

Lime measured in Paste, P denoting Puzzolana in Powder, and S Sand.	Age in days.	Weight per cubic foot.	Resistance estimated by the breaking weight, in lbs, in several successive Experiments.					Average resistance, in lbs.
Blue Lias (Hydrate)..	14	....	328	....	....	....	..	142
	15	77	116	86	119	....	..	
	19	64	128	77	....	....	..	
Blue Lias 1, S 2	15	124	209	175	189	....	..	157
	19	107	115	96	....	....	..	
.. 1, S 3	16	101	117	117	161	123	142	132
.. 1, P 1	15	100	109	100	140	....	..	103
	19	80	75	91	....	....	..	
.. 1, P 2	15	94	159	200	179	133	136	161
.. 1, P 1, S 1	15	114	168	172	184	....	..	141
	19	90	59	124	....	....	..	
.. 1, P 1, S 2	15	109	133	105	109	127	117	118
.. 1, P 1, S 3	15	108	103	102	107	123	147	116
Halling Lime (Hydrate)	14	63	35	51	49	....	..	45
	15	71	47	57	42	35	45	
Halling Lime 1, S 2 ..	15	111	29	35	36	21	21	45
	15	118	48	46	127	....	..	
.. 1, S 3 ..	15	98	64	51	98	....	..	62
	15	118	59	56	63	50	56	
Chalk Lime (Hydrate)	14	63	43	21	35	....	..	34
	34	....	52	20	....	....	..	
Chalk Lime 1, S 2	14	106	42	39	43	97	76	59
.. 1, S 3	15	120	49	85	71	61	39	61
.. 1, P 1	15	84	219	270	186	123	179	195
.. 1, P 2	15	92	180	142	152	130	179	157
.. 1, P 1, S 1	15	107	214	215	188	180	189	197
.. 1, P 1, S 2	15	111	109	88	83	91	102	93
.. 1, P 1, S 3	15	104	75	68	107	65	56	74

TABLE XIII. COMPARATIVE ADHESIVENESS OF VARIOUS LIMES. LIME MORTARS AND PUZZOLANA MORTARS TO PORTLAND STONE, WITH THEIR COMPARATIVE RESISTANCE ALSO, REPEATED FROM THE PRECEDING TABLE.

Lime measured in Paste, P denoting Puzzolana in Powder, and S Sand.	Age in days.	Adhesiveness estimated by the fracturing weight, in lbs, in several successive Experiments.					Average adhesiveness, in lbs.	Average resistance, in lbs
Blue Lias (Hydrate)..	3	481	....	....	....	..	460	142
	15	655	419	456	413	..		
	16	338	....	....	....	..		
Blue Lias 1, S 2	15	506	534	....	....	..	499	157
	16	330	473	454	....	..		
.. 1, S 3	15	448	557	422	419	557	481	132
.. 1, P 1	15	506	654	490	579	..	519	103
	16	364	....	....	....	..		
.. 1, P 2	15	607	988	685	316	..	596	161
	16	384	....	....	....	..		
.. 1, P 1, S 1	15	845	....	....	....	..	530	141
	16	280	221	678	627	..		
.. 1, P 1, S 2	16	499	445	835	406	443	526	118
.. 1, P 1, S 3	15	652	637	627	629	686	646	116
Halling Lime (Hydrate)	16	333	109	442	436	221	308	45
Halling Lime 1, S 2 ..	16	333	247	288	284	377	306	45
.. 1, S 3 ..	16	303	358	352	379	277	334	62
Chalk Lime (Hydrate)	17	400	221	757	275	275	386	34
Chalk Lime 1, S 2	17	386	238	305	330	221	308	59
.. 1, S 3	17	260	317	363	202	275	283	61
.. 1, P 1	17	497	277	450	329	309	372	195
.. 1, P 2	17	365	450	301	629	366	422	157
.. 1, P 1, S 1	17	310	611	254	644	477	459	197
.. 1, P 1, S 2	16	410	221	545	433	501	422	93
.. 1, P 1; S 3	16	295	359	359	250	406	334	74

(278) From the above experiments on lime and lime mortars combined with those previously tried on cement, the following inferences may be drawn, which generally confirm but occasionally disprove the opinions before advanced by me, from less copious and direct evidence, and which also differ in several points from received opinions.

First. Of the effects of Sand & of Puzzolana on Cement.

Sand in all cases diminishes the strength of cement, whether as estimated by its adhesiveness, its resistance, or its water setting powers. Puzzolana is still more injurious to it than sand, as we found by a series of experiments tried by us to ascertain this point, which will afterwards be more particularly noticed. In short every extraneous substance, excepting the carbonate of magnesia, which is far too expensive to use for building purposes in this country,* injures, and when added in a certain excess, entirely ruins, cement.

(279) *Secondly. Of the effects of Sand on common Chalk Lime or weak Hydraulic Limes, such as Halling Lime.*

Sand appears rather to diminish the adhesiveness of Chalk lime, but slightly to increase that of Halling lime, in either case the difference being insignificant. Its effect upon their resistance is more marked, especially upon that of chalk lime, which it nearly doubles, whilst it only increases that of Halling lime by about one half.

Thirdly. Of the effects of Sand on strong Hydraulic Limes, such as the Blue Lias.

It appears to increase both their adhesiveness and their resistance in a slight degree, and the blue lias does not seem to be materially injured by mixing it with more than twice its volume of sand, contrary to our former opinion (26) which was chiefly derived from the apparent shortness of the mortar made with 3 measures of sand to 1 of lime, which mode of judging by the eye alone was mere guess work, compared with the accuracy of those experiments in Table XIII, by which it has been so far disproved, that although a mixture of 2 of sand to 1 of blue lias still appears to make the best mortar, yet 3 to 1 does not spoil it, as we formerly believed.

Fourthly. Remarks on the use of Sand generally.

Though sand deteriorates cement in every respect, yet a moderate proportion of it is always used for the sake of economy, except under critical circumstances in very important works; and as sand is much cheaper even than lime, it is proper always to build with mortar, not with nett lime, for the same reason; since although it produces very little effect upon the adhesiveness of limes, yet it generally adds to their resistance, which is a very important property.

(280) *Fifthly. Of the effects of Puzzolana on common Chalk Lime.*

Puzzolana communicates to common chalk lime the important property of setting under water (255). It also appears to increase the adhesiveness of common chalk lime, but very moderately, when mixed with it in the pro-

* In one part of India it is so abundant and consequently cheap, that Mr. J. Macleod, a Surgeon in the Hon. East India Company's Service, who had discovered its cement properties in India as I did in this country (106), unknown to each other, recommended the Government of Madras to use it as a cement in their public works (See the Appendix).

portion of 2 measures of puzzolana powder to 1 of lime paste: but it increases the resistance of this lime in a most extraordinary degree nearly sixfold, rendering it in 17 days nearly double of the resistance of the best concretes and mortars made of the same lime mixed with gravel and sand alone, though from 8 to 12 months old or upwards, as will be seen by comparing our last experiments recorded in Tables XII and XIII, with those of Tables VI and VII. I suspect that the resistance of limes and mortars, which depends upon their hardness, and on the tenacity of their particles or component parts, is at the same time a test of their water setting properties, and therefore that the remarkable inferiority of the resistance of common chalk lime, or of any pure lime, is the cause of these limes going to pieces so much sooner under water than any other, when exposed to immediate immersion. I suspect also that chalk lime mortar, which will never set under water, rather dries than indurates in air; for the experiments recorded in Table XIII, compared with those of Table IV (193), appear to prove the very extraordinary fact, that chalk lime mortar, at least in small joints like those of single bricks, acquires as much adhesiveness in 2 or 3 weeks, as it is capable of attaining in 20 or 30 years. However paradoxical this may appear, such is the inference, that may be drawn from those experiments; and if it be erroneous, there is no mode of accounting for it, otherwise than by supposing that the old mortar may have been shaken in the joints in cutting it out of the wall, or in forming the mortises for the nippers, by which it was torn asunder. I apprehend, however, that the maximum strength of chalk lime mortar or of any common lime mortar is very moderate indeed, and that the superior hardness generally ascribed to old mortars, is more often a skin at the surface, than uniform throughout.*

Sixthly. Of the effects of Puzzolana on common Chalk Lime mixed with Sand.

Puzzolana being much more expensive than lime, the greatest proportion of it used in making mortar is 1 measure of puzzolana powder to 1 of lime, combined with various

* There is an old Priory at Dover, now in ruins, supposed to have been built in the reign of Henry the 1st, the mortar of which appeared to me to be much stronger than any I had ever seen, for I was obliged to borrow blacksmiths' tools from a forge opposite, to break the surface. Afterwards I requested Colonel Arnold to have a compact piece of about a cubic foot of this old masonry cut off, and sent to me by coach, on receiving which, I found that the inside of it had no pretensions to any very extraordinary hardness. The lime used for this mortar must probably have been the Kentish Rag stone, as it seems superior to Chalk lime.

proportions of sand, for which Smeaton's rule that the latter shall not exceed 2 measures, when the lime alone without puzzolana will bear 3 measures of sand, is probably the most judicious. By our experiments it appears that 1 measure of sand added to 1 of puzzolana powder and 1 of chalk lime paste, produces the strongest puzzolana mortar with that species of lime; that one more measure of sand in addition diminishes the adhesiveness of this mortar, in a moderate degree not exceeding 10 per cent, but that it diminishes its resistance by 60 per cent, which may be very injurious to the mortar, especially if used under water; and therefore, so far as our present hasty and by no means numerous series of experiments may be relied upon, this increase of sand appears by no means advisable.

Seventhly. Effects of Puzzolana upon the weaker Hydraulic Limes, such as the Halling Lime, &c.

Puzzolana must no doubt produce the same favourable effect upon these limes as it does upon chalk lime, and in building under water with any of them, it would not be prudent to dispense with the use of it, as has been proved by the example of the mortar of the Docks at Hull, which was bad or imperfect, except in those parts where puzzolana had been added to it (261). The Halling lime of this neighbourhood, though superior to any of the limes used at Hull, is not capable of resisting immediate immersion, as may be seen in the wharf of the limeworks at Halling, and in the wharf wall adjacent to, and connected with the new Baths at Rochester, in both of which, being exposed to the Medway during part of every tide, the mortar has been washed out of the external joints for about an inch in depth, in the course of a few years.

Eighthly. Effects of Puzzolana upon a strong Hydraulic Lime, such as the Blue Lias.

Puzzolana appears to increase the adhesiveness of blue lias lime, and when added in the proportion of 2 measures of puzzolana powder to 1 of lime, it increases its resistance also, but for the reason before stated, 1 of puzzolana to 1 of lime would be the utmost proportion of the former ingredient, that could be recommended in practice. When to the last quantities of these two important ingredients are added 1, 2 and even 3 of sand, the adhesiveness is still increased, apparently in contradiction to Smeaton's rule, for as our experiments with blue lias lime and sand, recorded in the same table XIII, proved that this lime makes good mortar with 3 measures of sand, it ought according to Smeaton when mixed with 1 measure of puzzolana to bear only 2 measures of sand. Upon the whole puzzolana appears to increase both the adhesiveness and resistance of the blue lias lime, but in such a moderate degree, as to render it doubtful whether it is worth while to use it at all, with any lime possessing such very powerful hydraulic properties, unless its price should become as low as that of the lime itself, which can never be the case in this country.

(281) *General Remarks on the above.* Having had sufficient experience of the inexpediency of forming definitive conclusions from a limited number of experiments, I think it possible that the average results of those contained in Tables XII and XIII, might be altered or modified by trying a much greater number, especially with a different mode of measuring the lime, for I found reason soon afterwards to doubt, whether the measurement in paste, which I thought I had good reason for adopting, and which had also been previously used by M. Vicat and General Treussart in France for experimental purposes, was quite so accurate as I had imagined. This doubt occurred in the course of a new series of experiments next tried by me, in order to compare the above with three other modes of measuring lime, that have been used for practical purposes in this country, in hopes of fixing something definite in respect to lime measure ; it being evident, that the produce of the same quantity of quicklime from the kiln, may contract or dilate into much smaller or greater apparent quantities, if afterwards measured by discordant modes and in different states. This comparison, if attainable, appeared to be of great importance, for without some satisfactory mode of distinguishing real from nominal quantity, the experiments or practice of one Engineer or Architect cannot be of any use to, nor even understood by another, who measures his lime by a different mode, and in a different state.

OF THE UNCERTAINTY OF MEASURE, AS AN ESTIMATE OF ACTUAL QUANTITY, IN SUCH SUBSTANCES AS COALS AND LIMES.

(282) About six years ago, whilst investigating the subject of measures and weights, I ascertained that the measurement of all solid substances sold in lumps, easily broken, and not spoiled by breaking, such as coals and quicklime is liable to considerable uncertainty. For example, I found by experiment, that when one cubic foot of solid coal was broken

into 30 pieces, and these, together with the rubbish and dust produced in the breaking, were thrown into a cubic foot measure, the same quantity in this state occupied 2¼ cubic feet of fair level measure. On breaking the pieces into still smaller ones 300 in number, which produced a greater proportion of rubbish, they filled 1¾ cubic foot: but on pounding them entirely into rubbish and dust, they only filled 1½ cubic foot. In all these states the real quantity of coal was the same, though the estimates of it by measure were so very different; besides which I also found, that a compact 10-cubic-feet measure contained a greater quantity by about 5 per cent, than ten times the contents of a cubic foot measure of coals, in pieces of the same average size. Upon the whole, although the practice of coal measure then sanctioned by law, for all England except the metropolis and its vicinity, did not admit of the apparent quantity purchased at one place, being doubled by breaking when sold at another, as was the case in those experiments, which I purposely tried under extreme circumstances; yet as the merchants purchased their coals in very large lumps, by the North Country chaldron of 53 cwt, which was estimated at 2 legal imperial chaldrons, though in reality rather more; and as these were usually broken smaller in the voyage for the purpose of increasing their apparent quantity, when measured by so small a vessel as the heaped imperial bushel, which was the only mode of measurement then used in the sale of coals to the public; I felt myself warranted in stating after having taken every pains to investigate the subject, that the quantity of coals, weighed when shipped, and entered officially at the Custom House of Newcastle, as 6 North Country chaldrons, and estimated by law as being equivalent to 12 imperial chaldrons, might afterwards be measured and sold in the South of England for 15 or 16 imperial chaldrons.*

* Accordingly I suggested in a little work published by me in 1834, entitled 'Observations on the Expediency and Practicability of simplifying 'and improving the Measures, Weights and Money used in this Country, 'without materially altering the present Standards,' that the sale of coals by measure should be prohibited throughout the whole British Empire, as being liable to great uncertainty and consequently to error and frauds, to the equal prejudice both of the public and of the conscientious coal merchant; and that they should be sold exclusively by weight, according to the practice which had some years before been established by law, I believe at first as an experimental measure only, for London and its immediate vicinity, in consequence of the recommendation of a Parliamentary Committee, who had previously investigated this important subject, before I took it into consideration at all; so that I claim no further credit from my own experiments on the measurement of Coals, than having confirmed the judicious views of that Committee. This beneficial measure was afterwards extended by Act

(283) The same uncertainty of measure applies to lime, but not in so great a degree, for the lumps of lime from the same kiln are always of one uniform size nearly, it being necessary for the conveniency of burning, that all large masses of lime stone must previously be broken into pieces not larger than 3 or 4 to the cubic foot, before they are put into a flame kiln, and into pieces of about 200 to the cubic foot for a common kiln. Originally lime was always measured in baskets, and sold by a quantity termed *the Hundred*, probably meaning 100 pecks, as I ascertained, not without some trouble, that the contents of 25 bushel baskets of Winchester measure not heaped were formerly served out for that quantity, which it has since become more usual to measure in a larger basket of 18 to the hundred.* But to save the trouble of measuring by one basket full at a time, the principal builders of the metropolis, rejecting this tedious process, have for many years past agreed with the lime dealers, to accept the contents of a cubic yard measure raised one inch higher than the top, as an equivalent for the hundred of lime: and by degrees this unintelligible term is wearing out, the cubic yard adopted in lieu of it being now more generally used, not only for the measure but to designate quantity.

(284) If lime like coals were a necessary of life, and consequently sold every day by retail, the same quantity if purchased by a large measure such as the cubic yard, when afterwards sold by a much smaller measure, such as the

of Parliament to the whole kingdom, though strongly opposed by persons ignorant of the subject, or having an interest opposed to change, who in their evidence before successive Parliamentary Committees on Weights and Measures, pretended that the weight of coals might be so much increased by wetting them, that this mode of sale would be more uncertain and more liable to fraud, than the system of measurement. On the contrary, other experiments also tried by me with the greatest care, and recorded in the same treatise, had fully proved the entire fallacy of this objection; for I found that coals are not only less capable of imbibing, but also that they are less retentive of moisture, as they shake it off sooner, than almost any other substance. When kept 72 hours under water and weighed immediately, coals of a marketable quality and size, only gained 4 per cent of additional weight, though dripping wet at the time, in which state no person who had to pay by weight would purchase them at all. But when moistened to such a moderate degree, as would not strike the eye at once and disgust a customer, coals of the same description only gained from 1 to 2 per cent of additional weight, which is a mere trifle compared with the abuses to which measurement was liable, for as I said before, coals purchased in lumps of one size by one measure, and sold in lumps of a smaller size by a much smaller measure, may gain 25 or even 30 per cent of apparent quantity.

* This larger basket was afterwards adopted, for the conveniency of carrying the proper contents of the heaped Imperial bushel, when it was made the only legal measure of coals, without the risk of a part falling out.

last mentioned basket containing 1½ cubic foot, would dilate into a larger apparent quantity, by filling 19 or 20 such baskets, to the prejudice of the retail purchaser, who would only be allowed 18 to the cubic yard, this being the just proportion, which the capacity of the smaller actually bears to that of the larger of those two measures. But as lime in small quantities is so seldom required, that the sale or measurement of it by retail is not an object of any public importance, it may be observed, that the mode of measurement by the cubic yard, now in general use in wholesale dealings, affords a sufficiently fair average estimate of actual quantity, and is liable to as little uncertainty as can reasonably be expected; for in using so large a measure, there is no temptation for the lime-burner to break even his flame-burned lime into smaller pieces. On the contrary it has been found, that by so doing he would not increase but diminish the apparent quantity, to his own loss. Accordingly the builder, in purchasing his lime in lumps by the cubic yard, may depend upon receiving nearly the same average quantity of the same sort of lime from the same kiln. In purchasing different sorts of lime burned in different kilns, the quantities of each contained in a cubic yard will not always be equal, but they will be proportional, and I believe that the difference will scarcely exceed 10 per cent, the excess of real quantity being in favour of the small pieces or kiln-burned lime, about 9 cubic yards of which will contain as much lime as 10 cubic yards of the same sort of lime, when burned in a flame kiln.

FOUR MODES OF MEASURING LIME. EXPERIMENTS TO ASCERTAIN THE SPACES, OCCUPIED BY THE SAME QUANTITY OF LIME FROM THE KILN, MEASURED IN THE FOUR DIFFERENT STATES, TO WHICH THOSE MODES APPLY.

(285) *First mode, in Lumps as it comes from the Kiln.* This is the usual mode of measuring lime, which if any large compact measure such as one containing a cubic yard, or even 10 cubic feet only be used, will, as has just been explained, always afford a tolerably fair estimate of quantity or of proportional quantity; not so, if a much smaller measure, such as any of the common lime baskets be used. The former being the mode of measurement always understood, if nothing be said to the contrary (9), whenever a builder proposes to tender for executing a work by contract, in which the specification requires the mortar to be made of 3 parts of sand to 1 of lime, he calculates the whole quan-

tity necessary, and estimates the expense of it, in the proportion of 3 cubic yards of sand to each cubic yard of lime, according to the current prices of these articles at the time being, as delivered on the spot; and afterwards in the execution of the work, it is fully understood, that these ingredients are to be mixed in the above proportion as nearly as possible, in making the mortar.

(286) *Second mode, in Slaked lime Powder.*

This mode, first adopted by Smeaton for the mortar of the Edystone Lighthouse, and recommended by him for all hydraulic mortars, applies to lime broken small and slaked by a moderate quantity of water, sprinkled over it by a watering pot, after which it should be covered up until it falls down into a powder, for which more or less time will be required according to the quality of the lime. If slaked in the afternoon of one day, it will be ready to use sometime next day, for even the blue lias which is the slowest in slaking of all our English limes, from its having the strongest hydraulic properties, does not usually require more than 18 hours for this purpose (see the note to Article 26). I am not aware that this system, though recommended by such high authority, has recently been used in England.

(287) *Third mode, in Quick lime Powder.*

When the Architect or Engineer requires his mortar to be made with lime fresh from the kiln, and reduced to fine powder, it is either pounded by manual labour, or if the work be of considerable magnitude, the contractor will find it worth his while to provide himself with a proper mill for grinding it. It must be sifted before it is used, and the coarser parts, if lime, submitted again to the same pulverizing process; if core, thrown away. This mode was adopted by Mr. Stevenson in Scotland, who reduced the Aberthaw lime, used by him in making the mortar for the Bell Rock Lighthouse, to the state of impalpable powder without slaking it.* The same has often been done in England by order of some of our most eminent Civil Engineers and Architects, for works or buildings of importance; but I believe that this practice has recently been entirely disused, excepting for concrete foundations, and by Mr. Ranger for his artificial stone.

(288) *Fourth mode, in Slaked lime Putty or Paste.*

This mode, adopted by us for our most recent experiments at Chatham, in consequence of the third mode having been found objectionable (210), applies to quicklime fresh

* Said to be slaked lime powder by mistake, in Article 252.

from the kiln, pounded in a mortar and afterwards thoroughly slaked, by mixing it with a moderate quantity of water, not applied all at once but gradually, until after throwing out more or less heat, it shall become quite cool, and then remixing it with more water, to bring it into the state of a stiffish paste, in which it is to be measured. In order to define this state more accurately, let it be understood, that the lime paste, used by us in the experiments recorded in Tables XII & XIII (277), was as nearly as possible of an intermediate consistency between that of glaziers' putty, and of common mortar prepared for use, being a little thinner than the former, but not quite so moist as the latter; so that in all those experiments, in which we mixed our lime paste in this state with sand, it was necessary to add some more water afterwards, in proportion to the quantity of sand;* especially in those experiments, in which the mixture thus made was to be used for uniting our small Portland stone-bricks together in pairs, for which purpose the consistency of good common mortar was of course requisite. The small experimental prisms, being used independently, were generally made a little stiffer than the same mixtures, when applied to joints.

Lime in lumps in the first state is always measured by fair level measure, rather full than otherwise, that is with the uppermost lumps projecting a little above the brim of the vessel as was before explained (283). Lime in powder in the second and third states has always been measured by stricken measure, both for practical and for merely experimental purposes, in the manner described in Article 65. In the fourth state of putty or paste it was also measured by stricken measure, with the precautions mentioned in the same Article.

(289) This being premised, our proceedings were as follows. Having selected an equal quantity of the best blue lias, Halling and chalk limes, at their respective kilns, all in lumps, and extremely well burned, we broke them into small pieces as nearly as possible of the same average size, and all capable of passing through a $2\frac{1}{2}$ inch ring, after which we measured them in a 1000-cubic-inch measure, and weighed the contents of each, when their weights proved to be $28\frac{3}{8}$, $21\frac{1}{2}$ and $18\frac{1}{4}$ lbs, respectively. Hence the average weights of well burned blue lias, Halling and chalk limes, in the state that has

* We measured our sand dry. Had it been wet, more water would not have been always required to make it into mortar, in mixing it with the lime paste.

been described, that is in rather small pieces suited to the common lime kiln, may be estimated at 49, 37 and 31½ lbs, per cubic foot respectively.* What difference may be produced by supposing them of larger pieces, and consequently to have been flame burned, I shall not pretend to say, but in this state also, their respective weights must be nearly proportional to the above. If part were burned in the flame kiln and part in the common kiln, their weights per cubic foot might cease to be in direct proportion to these numbers, for the reasons explained in the preceding article. In fact there is no possibility of attaining a correct estimate of the real quantity of any sort of lime, by measure, which is only an approximation. Weight on the contrary affords a most accurate estimate of quantity, provided that the lime be well burned and fresh from the kiln; notwithstanding which, I was always of opinion until very lately, that it would be highly improper to legalize the sale of lime by weight; because I supposed that in that case lime imperfectly burned, or lime originally well burned but afterwards allowed to become stale by exposure to air, neither of which are fit for use, would bear a higher price than fresh well-burned lime in its most perfect and serviceable state, in which it is well known that it weighs much less than in any other.

(290) Knowing however that weight afforded the only means of estimating the actual quantity of lime, with sufficient accuracy for our proposed comparisons, we next weighed several portions of all those three sorts of lime, in equal quantities of 2 lbs. each, one half of which we reduced to quick lime powder, by pounding them in a mortar, after which we measured them lightly, by throwing or dropping them into the measure and striking the contents, without shaking or compressing them. We next made each portion up into a paste, by mixing its produce in quick lime powder with water, gradually added as long as the lime seemed to require it, during which the process of slaking commenced with such violent heat in the chalk lime, that the quick lime powder, though well wetted, expanded into slaked lime powder, before the whole of the water necessary for obtaining the paste could be measured and added. Not so with the other limes, which threw out more moderate heat, espe-

* The blue lias lime used for the experiments which are now about to be described, was supplied by Mr. Gladdish of Northfleet; but what we had previously used, in all our former experiments on the adhesiveness and resistance of that excellent lime, was burned in our own little kiln, from stone procured at my request by Captain Savage from Lyme Regis, from whence Mr. Gladdish also imports his blue lias lime stone which he burns in his own kilns at Northfleet.

cially the blue lias, and therefore did not resume the dry form, during the measurement of the water. The paste thus formed, apparently with a sufficient quantity of water, was then made up into a lump and put by, in which state the slaking process accompanied by heat still continued for some time, splitting and cracking each of these lumps more or less in the act of partially setting. When we considered that all action of this sort had subsided, for which we always allowed several hours, and sometimes a whole night, if it suited the men's working hours better, we remixed each of those lumps with an additional quantity of water, at which period they could not fail to be thoroughly slaked, in order to make them again into paste of the proper consistency, which we then finally measured, and recorded as the slaked lime paste produced from the quick lime powder; and it is rather curious that the total quantity of water used in this change, sometimes exceeded the actual quantity of slaked lime paste, produced from the quick lime to which it had been added, proving that part of the water had evaporated, or been carried off in the hot fumes proceeding from the lime in the act of slaking.

The remaining half of the portions of quick lime before mentioned were sprinkled with water, and then covered up with cloths, and allowed to fall down into the state of slaked lime powder. Not knowing before hand what quantity of water would be required to effect this object, we tried the Halling lime with two different quantities of water, namely 24 and 32 cubic inches to the 2 lbs; and as the smaller quantity of the two seemed sufficient, we used the same for the other limes also. But in order that this process might be equally complete in all, we allowed two nights and a day for this purpose to the blue lias, as the slowest slaking lime of the three, and one night for the Halling lime, but only 4 or 5 hours to the chalk lime, after which we measured them lightly, and noted the produce of each portion in the state of slaked lime powder. Finally we added a sufficient quantity of water to form each of them into a paste of the desired consistency, which we measured and recorded as slaked lime paste produced from the slaked lime powder.

Whilst these experiments were in progress, the results of which will be recorded in the next Table (No. XIV), we tried others connected with the same subject of investigation, which shall now be stated, as they could not conveniently be included in the same Table.

(291) *First. Of the Increase of Weight in the slaking of Quick lime to a Powder by watering it. That very little more than one half of the water used is solidified.*

After slaking five 2 lb. portions of our chalk lime to an impalpable powder with water, the spaces occupied by which will appear in the Table, we weighed each portion in this state abont 5 hours after they had been wetted, and found that they had acquired an increase of weight of 3015, 3890, 3565, 4705 and 3523 grains respectively, making a mean increase of 3340 grains in each portion, which is rather less than one fourth of the original weight of 2 lbs, (or 14000 grains). Hence it appears, that only about 11-20ths, or little more than one half of the 24 cubic inches of water used, which must have weighed at least 6100 grains, were solidified in this process.

Secondly. Of the Circumstances, which may render the measurement of Lime in the state of fine Powder uncertain.

(292) It having been stated by Smeaton, that the lime used by him for the Edystone Lighthouse was slaked into powder and sifted, and the fine powder only received, which was rammed into water-tight casks, after which it was measured for use in the manner particularly described by him, which may be considered equivalent to what is called measuring lightly, it appeared desirable to try what difference, if any, may have been produced by this process.

We therefore took one of our 2lb. portions of the Halling lime, which by the addition of 24 cubic inches of water had been slaked into fine powder, occupying 240 inches of space, when measured lightly: we compressed it in a cubical vessel, by forcing it down with a wooden plug nearly fitting the top of the vessel, and then remeasured it, when it occupied only 112 cubic inches. Afterwards we shook it out of the vessel, and measured it lightly, when it occupied 155 cubic inches. Thus slaked lime powder will by compression occupy less than one half of its original space, and when measured after temporary compression, without loosening the particles, it may occupy less than two thirds of its original space. Hence it is possible that Smeaton's Edystone mortar, said by him to consist of 1 measure of puzzolana powder, mixed with 1 measure of slaked lime powder, if adopted by another Engineer, who did not ram his slaked lime powder before he measured it, might be equivalent to a mixture of 1 measure of slaked lime powder with 1½ measure of puzzolana.

(293) Again, Mr. Stevenson states that he used 'puz-'zolana and lime in the state of a dry impalpable powder, 'and clean sharp sand, in equal proportions by measure,' for the mortar of the Bell Rock Lighthouse; and he informs us that he ordered these ingredients to be packed in separate casks, and sent out to the rock to be mixed; and therefore the workmen would no doubt empty three casks, 1 of lime, 1 of puzzolana and 1 of sand, at the same time, and mix their contents together, instead of remeasuring the ingredients. If therefore the lime was compressed in these casks, which I consider probable, though not stated, another Engineer, who believed that he was following the same proportions for a similar work, but who did not direct his lime to be rammed before he measured it, would in reality use a different mixture consisting of 1 measure of quick lime powder, mixed with $1\frac{2}{3}$ measure of puzzolana and $1\frac{2}{3}$ measure of sand. For in order to ascertain, what difference different modes of packing the lime might have produced in Mr. Stevenson's proportions, we compressed three several 2 lb. portions of our Halling quick lime, which when pulverized and measured lightly, had occupied 97, 95 and 98 cubic inches respectively, and found that after this process they only occupied 57, 58 and 57 cubic inches. Thus the unslaked powder of lime of this description is capable of being compressed into 6 tenths only of its original space.

Of the Proportional Spaces occupied by the same quantities of Puzzolana, and of Sand, in different states.

(294) Apprehending that some uncertainty, in respect to the composition of puzzolana mortars, might occur from the want of precision in the writings of authors upon this subject, none of whom have stated whether they measured their puzzolana powder lightly or under compression, moist or dry, excepting Smeaton, who particularly mentions that he always dried it for the purpose of grinding it, and who also probably, though it is not stated, must have measured it in the same mode, in which he measured his lime powder, that is rather lightly; it appeared desirable to lay down more accurate data for future measurements and specifications. For this purpose, our stock from Sheerness having been exhausted in the experiments already recorded or alluded to, we procured some puzzolana imported from Civita Vecchia by Messrs. Francis and Sons, and having reduced it to a perfectly dry impalpable powder, we ascertained that the same quantity, which in this state weighed 45·43 lbs, occupied 1st, 1000 cubic inches when measured

lightly; 2dly, when taken out of the same cubical box and put in again by small portions at a time, each of which was forced down by means of a square board fitting the inside of the measure, with the full strength of an able-bodied man, the same dry impalpable powder thus compressed measured 862 cubic inches; and 3dly, when again taken out and mixed in successive portions, with quantities of water, amounting to 416 cubic inches in all, into a stiffish paste, it occupied 950 cubic inches of space.*

Thus the measurement of puzzolana is liable to much less uncertainty than that of lime, for on reducing those proportions to a simpler form, it appears that 10 measures of puzzolana measured lightly as a dry impalpable powder, contain the same quantity of that earth nearly as 9½ measures of stiffish impalpable puzzolana paste, and as 8½ measures nearly of dry impalpable puzzolana powder after being well compressed.

(295) In respect to the measurement of sand for the mortar of brickwork or of regular masonry, in which the thickness of the joints is usually limited, never exceeding half an inch, it is liable to less uncertainty than that of lime, the coarser parts, such as come under the denomination of *small Gravel,* being necessarily rejected, by sifting it through a screen, usually having openings not exceeding one sixth part of an inch in the clear. The chief difference will arise from the state of the sand itself at the period of measurement, which according to our former trials occupied, when wet, only about four fifths of the space that it did in its usual rather moist state. (See the Note to Article 9). On repeating these measurements recently, modified in such a manner as to obtain more precise results, we found that the same quantity of sharp clean river sand, which in its moist state stood 12 inches high in a regular cubic foot measure, and therefore completely filled it, only stood 10⅜ inches high in the same measure when perfectly dried; and very little more than 8½ inches high when thoroughly wetted, not by merely pouring water over it in the measure, as we had done before, but by mixing every part of the sand with water. Hence instead of occupying only 4 fifths or 8 tenths of its original space, the wet sand now occupied very little more

* Though no more water was used than was necessary to give, to each portion of paste, an intermediate consistency between glazier's putty and common mortar prepared for use, yet when the whole of them were collected and levelled in the cubical box, part of the water rose to the top in a sheet about one eighth of an inch in depth, below which the moist paste was of the regular consistency, that has been described.

than 7 tenths of the space, which it had previously filled in its usual moist state. In all our experiments at Chatham, we measured our sand rather drier than usual, which system having been adopted at first by accident, it was proper to persevere in it in order to obtain results consistent with each other. I believe that this is the only circumstance that has not before been explained, and it is so far of importance, that it should be understood, inasmuch as it must have insured in all our experiments rather a fuller proportion of sand, than would be obtained by measuring it in the moister state in which it is commonly supplied and used by builders.

(296) In respect to solid substances, which may be reduced from a fine powder to a paste of a given consistency or specific gravity, by mixing them with water, the measurement of them in the latter state yields a much more accurate estimate of real quantity than in the former, provided that no chemical change be produced by the addition of the water. I was in hopes therefore, that the measurement of all thoroughly slaked limes in the state of paste, would have yielded a sufficiently just estimate of the proportional quantity of quick lime, from which a given volume of the paste was formed; as I believed that after thorough slaking, change of weight might take place, but no material change of bulk. I was however disappointed in this expectation, as will be seen by a careful examination of the results of the experiments recorded in the following Table, which had they been more satisfactory, would not have required so long a preamble, upon the uncertainty of the measurement of solid bodies broken into irregular and unequal fragments, and afterwards reduced to a dry powder or to a stiffish paste.

(297) To render the following Table more available I have subjoined an Abstract of it, obtained by adding the produce of five 2 lb. portions of each lime together, in their four several states, which gives the sum and mean of the results at the same time, with double quantities for the slaked lime paste, because it was obtained both from the quick lime powder, and from the slaked lime powder; and where more quantities than one appear, their mean or average is entered opposite in the same column. We tried the Halling lime twice as often as the two others, because the results of our first experiments with this lime appeared rather incongruous.

TABLE XIV. COMPARATIVE SPACES OCCUPIED BY THE SAME QUANTITY OF LIME, IN DIFFERENT STATES.

Limes fresh from the kiln, but broken into small equal lumps, with their weight per cubic foot, in this state.	Weight of each of the 5 portions of lime, in lbs.	Water used and spaces occupied by the Lime, in cubic inches.					
		Proportional space occupied by it in small lumps.	Quick lime powder, or slaked lime powder produced.	Water used for slaking to a powder, or to the first paste.	More water added gradually for the final paste.	Total quantity of water used.	Paste produced from the quick lime powder, or slaked lime powder.
			Quick.				
			66	29	40	69	81
			67	29	37	66	80
Blue Lias ..49·032 lbs.	2	70·48	67	31	32	63	79
			72	31	37	68	81
			70	31	39	70	81
			Slaked.				
			114	24	33	57	70
			118	24	30	54	76
Blue Lias ..49·032 lbs.	2	70·48	118	24	28	52	64
			116	24	31	55	69
			120	24	31	55	78
			Quick.				
			97	54	29	83	84
			95	48	29	77	88
Halling ....37·152 lbs.	2	93	96	48	30	78	84
			97	52	30	82	85
			98	48	31	79	93
			Slaked.				
			240	24	61	85	91
			230	24	62	86	91
Halling ....37·152 lbs.	2	93	229	24	62	86	93
			227	24	61	85	92
			226	24	62	86	95
			Quick.				
			80	51	6	57	73
			95	64	6	70	77
Halling ....37·152 lbs.	2	93	102	56	8	64	77
			98	61	8	69	80
			98	64	3	67	78
			Slaked.				
			201	32	54	86	87
			186	32	48	80	90
Halling ....37·152 lbs.	2	93	195	32	44	76	85
			188	32	45	77	87
			200	32	46	78	90
			Quick.				
			85	37	67	104	105
			84	33	71	104	115
Chalk Lime..31·536 lbs.	2	110	83	36	68	104	106
			88	40	79	119	115
			87	40	70	110	106
			Slaked.				
			191	24	75	99	105
			203	24	74	98	110
Chalk Lime..31·536 lbs.	2	110	202	24	72	96	117
			206	24	80	104	115
			180	24	62	86	95

ABSTRACT OF THE RESULTS CONTAINED IN TABLE XIV.

Mean quantity by weight, and name of each sort of Lime experimented upon.	Space, or Spaces and their Mean, in cubic inches, occupied by the same quantity of Lime in different States.			
	In Lumps.	In Quick lime Powder.	In Slaked lime Powder.	In Slaked lime Paste.
10 lbs. of Blue Lias Lime.	352	 342	 586	402, 357 } 380
10 lbs. of Halling Lime.	465	483, 473 } 478	1152, 970 } 1071	431, 462, 385, 439 } 429
10 lbs. of Chalk Lime.	548	 427	 982	547, 542 } 545

The above results are by no means satisfactory, for Halling lime is of intermediate quality between the blue lias and chalk; but its results are not intermediate as they ought to have been, and as I have no doubt they would be if more numerous experiments were tried, and if lime measure were susceptible of greater accuracy. The slaked lime powder and the slaked lime paste appear to be the most unsatisfactory states for measurement, there being a difference of about one fifth by measure in the apparent quantity of each, as produced from the same real quantity of Halling lime; and if the blue lias and chalk limes had been tried as often as the former, it is probable that similar discrepancies might have been observable.

Such being the uncertainty of the measurement of lime, even in its most perfect state of well burned quick lime, I became desirous of observing the progressive changes produced upon it by the action of the atmosphere, for the purpose of ascertaining whether its proportional bulk or its proportional weight are most affected by this action; for which purpose I used quick lime powder, in order to come sooner to a conclusion, although lime is never pounded or ground for practical purposes, until required for immediate use, and therefore it can very seldom remain long, probably not more than one night at the utmost, in this state.*

(298) *Of the gradual Increase of the Weight and Bulk of Quick Lime, caused by Exposure to Air.*

To ascertain this, we measured 1000 cubic inches of well burned fresh chalk lime, previously broken into small lumps of the same average size, and capable of passing

* Unless a surplus remained on leaving off work on a Saturday night, which would be exposed to air, until the Monday morning following.

through a 2½ inch ring, as before. This quantity according to our former experiments, ought to have weighed only 18¼ lbs, whereas it weighed 20$\frac{4}{9}$ lbs, being at the rate of 35 lbs. nearly to the cubic foot. When reduced to quick lime powder it scarcely filled 4 fifths of its original space. Adding more pounded quick lime therefore to make up the deficiency, we found that 1000 cubic inches weighed 24·81 lbs, whereas according to our former experiments, this weight of quick lime powder ought to have filled about 1365 cubic inches; which discrepancies may be considered additional proofs of the uncertainty of measure.

The quick lime powder, of which the bulk and weight have just been stated, was kept in a dry apartment in a compact box open at top, exposing rather more than 1 superficial foot to the action of the air; and was taken out and weighed, and measured again, twice every week. Twelve examinations of this sort have taken place since it was pulverized, the results of which are as follows.

TABLE XV. EFFECTS OF AIR ON QUICK LIME POWDER.

Number of Days after the first Observation.	Measurements in Cubic Inches.	Successive Differences.	Weights in lbs, and decimal parts.	Successive Differences.	Increase per Cent at each period.	
					In Bulk.	In Weight.
Same day.	1000	0	24·81	0·00	0	0
3	1138	138	25·49	0·68	14	2¾
7	1251	113	26·08	0·59	25	5
10	1347	96	26·41	0·33	35	6½
14	1437	90	26·80	0·39	44	8
17	1538	101	27·14	0·34	54	9¼
21	1634	96	27·42	0·28	63	10½
24	1691	57	27·68	0·26	69	11½
28	1780	89	27·95	0·27	78	12½
31	1798	18	28·11	0·16	80	13¼
35	1819	21	28·26	0·15	82	13¾
38	1932	113	28·43	0·17	93	14½
42	2010	78	28·56	0·13	101	15

Thus the action of the atmosphere, which increases both the weight and the bulk of lime by gradually slaking it even in a dry apartment, produces a much greater proportional augmentation of the latter than of the former, in the same period, inasmuch as the bulk of our quick lime powder had actually doubled, but its weight had only obtained an increase of not much more than one 7th part in 42 days.

(299) *Of the rapid Increase in the Weight of Quick Lime, when converted into a Hydrate by mixing it with Water.*

The weight of chalk lime hydrate 14 days old was 63 lbs,

to the cubic foot (see Table XII, Article 277), but by a reference to the abstract of Table XIV, it will be seen, that a given quantity of chalk lime, measured in lumps from the kiln, yields as nearly as possible an equal volume of slaked lime paste. Hence one cubic foot of chalk lime in lumps, which in two trials weighed 31½ lbs; but in a third 35 lbs, as has just been stated, ought to produce one cubic foot of chalk lime paste. Supposing the last estimate to be the most accurate of the two, it therefore appears, that the quick lime after being changed into a hydrate, must have gained in weight in the proportion of from 35 to 63, or from 5 to 9 nearly, in the short space of 14 days. It is rather curious, that these numbers are the precise proportionals, between any given quantity of pure quick lime, and the carbonate or natural stone of the same lime, from which it must have been produced. Thus the lime used in those experiments appears to have recovered in 14 days, the whole weight of carbonic acid gas driven off from it in the kiln, or at least the greater part of it; for we may suppose a small portion of the weight of each of our experimental prisms to have consisted of moisture not yet evaporated. That the proportion of moisture must have been inconsiderable, I infer from the dry appearance of the inside of those very small prisms, when broken, by the weights used in experimenting upon their resistance.

THAT WEIGHT AFFORDS A MUCH MORE ACCURATE ESTIMATE OF THE QUANTITY OF LIME, THAN MEASUREMENT.

(300) The above will be evident from considering a circumstance, which seems hitherto to have been overlooked by writers on this subject. Carrara marble and common chalk, both being pure carbonates of lime, yield when burned quicklime of precisely the same quality; but any given measure of the former, as being a more compact stone and of much greater specific gravity, contains a greater quantity in proportion, not only of the carbonate but of quick lime after calcination, than the same measure of the latter.* Hence the mortar made by adding 3 cubic feet of sand to 1 cubic foot of chalk lime, would be a very different mixture, from that made by adding 3 cubic feet of sand to 1 of white-marble-lime. But if 3 cubic feet of sand be added to 31½ lbs. of well-burned chalk lime, which is the average quan-

* We found that the chalk of Chatham, when perfectly dry, as it ought to be for a fair comparison, weighed from 90 lbs (164), to 94·99 lbs, per cubic foot (Table VIII, Article 212). Hence the specific gravity of this chalk varies from 1440 to about 1520, that of water being 1000. The specific gravity of white marble is usually stated at about 2700.

tity contained in 1 cubic foot of that lime, it will form precisely the same sort of mortar, as may be made by adding 3 cubic feet of sand to 31½ lbs of well-burned white-marble-lime, although this quantity of the last-mentioned lime will not occupy much more than half a cubic foot of average space. Weighing therefore affords by far the most accurate estimate of the real quantity either of any limestone, or of the well-burned quicklime produced from it, whose weights though very different are always proportional to each other; and it appears from the experiments, which have just been related, that weighing affords also a much more accurate estimate than measurement, of the original quantity of lime, if through neglect or unavoidable circumstances, it has been allowed to become stale, that is in a state of transition from perfect quicklime towards a thoroughly slaked lime powder. In short, since I tried these last experiments, I see very strong reasons in favour of the sale of lime by weight, which before occurred to me as quite inadmissible.* If the conditions of the agreement state that the lime, supplied to the builder, shall be well-burned and fresh from the kiln, which is to be known either by its slaking as quickly and readily as the sort of lime ordered usually does, or to a still greater degree of certainty, by requiring that it shall not effervesce in acids when delivered, there can be no reasonable apprehension of the purchaser being defrauded. Upon the whole, the most accurate mode of determining the proportions of the ingredients of mortar would be by weight of quicklime, and by measure of sand, and although in practice, it is not likely that this nicety of determining quantity will ever be adopted; yet I conceive, that it ought to be used in preference to measurement, but not without reference to measure, in future experiments for determining the best proportions of lime for common and hydraulic mortars: and even in designing any great work or building, the Engineer or Architect should have weight so far in view, in combination with measure, that he should order the proportion of sand to the lime by measure to be greater, when using lime produced from a very heavy compact stone, than when using lime of the same chemical quality produced from a lighter and more porous stone. I am not aware that any former author or practical man ever weighed

* The sale of lime by weight was and probably may still be customary in some parts of Ireland, which until I tried these last experiments appeared to me extremely objectionable; and I reprobated it accordingly in my Treatise on Measures, Weights, &c, before quoted; but now I consider it a much better system than that which prevails in England.

his lime systematically, except Dr. Higgins, who weighed not only that, but his other ingredients also, including even the sand; in which latter part of his system I see no advantage, for the proportion of sand, necessary to form good mortar with any given quantity of lime, does not depend on its mean specific gravity, but on the magnitude and average space occupied by the particles, of which measure affords quite as accurate an estimate as weight.*

THAT IT IS DESIRABLE TO SPECIFY THE PROPORTIONS OF THE INGREDIENTS OF MORTARS WITH MORE PRECISION, IN DESCRIBING IMPORTANT WORKS, THAN HAS HITHERTO BEEN USUAL. SUGGESTIONS FOR THIS PURPOSE.

(301) In the actual construction of great works, no misunderstanding can occur, because the Engineer or Architect, who himself determines the proportions that shall be used by measure, is often present and sees the state and manner in which the ingredients are measured, and if he should not approve of either, he has it in his power to apply the remedy; and whatever uncertainty there may be in any description of his mortar, he himself is never at a loss, knowing how every thing was done, and indeed excepting under peculiar circumstances, actual measurement is either not employed at all, the proportions of the ingredients being left to the tact or sagacity of the laborers who make the mortar (9), or if measure be used the rudest sort only is employed, that is shovelfuls or barrowfuls.

But from the observations contained in several of the preceding articles, it appears to me that in describing the mortar of any very important work, it is desirable that not only the state in which such ingredients were measured, but also the peculiar mode of measurement employed should be specified or recorded, with greater precision than has hitherto been usual.

For this purpose, I beg to suggest, that whenever the common mode of measuring lime in lumps from the kiln has been intended and used, this shall be particularly specified, and that the average weight per cubic foot of the lime in this state, estimated however not from the contents of a

* A reader dipping carelessly into Dr. Higgins's Treatise on Calcareous Cements, published in 1780, may be astonished at finding him recommend 6 parts of sand to 1 of lime as a good proportion for mortar, for after explaining that he determined his proportions by weight, he does not always repeat the same thing in the course of his work. His proceedings and the sort of cement, or rather improved mortar and stucco, formed by him, for which he took out a patent, will be noticed in the Appendix.

single cubic foot measure, but from that of some larger measure, which need not exceed 10, and should not be less than 5 cubic feet shall also be recorded.* In fact, supposing it required to mix 3 measures of sand with 1 measure of Halling lime, it would afford much greater accuracy and uniformity in the quality of the mortar, to direct 3 cubic feet of sand to be mixed with 37 lbs. of quick lime fresh from the kiln, or as a 10 feet-measure is most convenient, let it be stated that the mortar shall be made in the proportion of 10 cubic feet of sand to about 124 lbs. of lime, which is nearly equivalent to the former; and the mode of measuring the lime from the kiln should also be described or specified, because although it may be presumed that fair and full level measure is intended or has been used, it is best to leave nothing doubtful.

When the lime from the kiln is directed to be ground to a fine powder, it is still more essential to define the mode in which it is to be or has been measured, whether lightly, or after temporary compression only, or under actual compression.

In respect to sand, the custom is to serve it in by stricken measure, in whatever state it may be at the period of sale, which may vary between more or less dry or wet, according to circumstances, known to the persons who use it, but not to others unless explained, amongst which circumstances the state of the weather has its influence; for sand is not kept under cover, but laid out in masses in the open air. Hence in order that we may have any precise knowledge of the real proportion, which the sand bears to the lime, in the mortar of any work of importance, the person who describes such mortar, ought to specify particularly the state in which the sand was measured, whether absolutely dry, or damp, or wet; because the actual quantity of

* Whilst investigating the subject of Measures and Weights, I found by repeated trials, that a 10-cubic-feet measure made of two rectangular wooden cases, open at top and bottom, and each measuring 2 feet by 2½ feet square in the clear and 1 foot high, either to be used separately as 2 five-cubic-feet measures, or jointly by placing one upon the other, was a more convenient arrangement for the measurement of dry materials, as well as for calculation, than the cubic yard measure in common use, which is also usually made in two parts, each composed of a similar case 3 feet square in the clear and 18 inches high. In small buildings where only few masons or bricklayers are employed, the half of the 10-cubic-feet measure may be the most convenient, as small quantities of mortar only are required to be mixed at a time. But to use a smaller measure than this would not afford a satisfactory estimate of average quantity, because the cubic foot of lime in lumps, measured singly in a one-cubic-foot-measure, is not equal to one tenth part of the contents of a ten-cubic-feet measure (282), nor to one fifth part of the contents of a five-cubic-feet measure.

sand obtained by the same measure in these three states, varies considerably between the second, which is the minimum, and the latter, which is the maximum of quantity. But the sand used for building in this country, is scarcely ever in either of these two extremes of perfectly dry or wet. It generally varies only between more or less damp, and probably the difference in real quantity, between equal measures of it in those two states, does not exceed one tenth, in the practice of building at any one place. To describe accurately the state, in which it has generally been used for the mortar of any important work, the author should specify not only the mean space occupied by it in that state, but also the spaces which the same quantity of the same sand is capable of occupying when perfectly dry, and when thoroughly wet, stating also its weight per cubic foot when perfectly dry, there being no certainty as to the weight of sand in any other state. Moreover to enable a person, who does not know the sort of sand obtained from a particular locality, to understand the nature of it, the size of the particles should be described in the way that has been done by M. Vicat, by stating the diameter of the smallest and of the largest grains composing it, the latter of which may be sufficiently defined by describing the sort of screen through which even very fine sand is almost always sifted, to exclude pieces of wood or other extraneous substances generally found in it. When sand and gravel are to be mixed together, in any given proportion, the size of both should also be described in the same manner; and even in using some natural mixture of these ingredients, such as Thames ballast, in the mortar of any important work, it is desirable that it should be defined in the same manner, for the use not only of foreigners, but of our own countrymen, in those parts of the United Kingdom where it is not used, although those who are accustomed to the daily use of it will of course need no such description.

In respect to puzzolana powder, let it be stated whether it is to be or has been measured, in a perfectly dry impalpable state, loosely and lightly, or under compression. I find on inquiry that it is generally understood, that puzzolana ought not to be used in the coarse state in which it is imported; but the necessity of grinding it very fine or to an impalpable powder is so little known, that the Engineer or Architect, who proposes to use it, should never omit to make this one of the conditions of his specification.* I mention this, because when Mr. John White† attempted to introduce his artificial puzzolana formed of pounded bricks as was before mentioned (259), he states particularly that they should not be ground into a fine powder, but left in a much coarser state, which from his description exactly resembles that of the natural puzzolana when imported from Italy.

Experiments before alluded to, in order to ascertion the effect produced by Puzzolana, and by Artificial Puzzolana upon Cement.

(302) In consequence of puzzolana having so much improved the strength of limes and lime mortars, but especially of chalk lime, as we had found by our last experiments (277), it appeared desirable to ascertain also what effect it might produce upon cement; and in order to obtain more conclusive results, I directed not only the natural puzzolana, but also the artificial puzzolana, which we had made from the blue clay of the Medway in the manner before described in Article 260, both reduced to impalpable powder, to be tried in competition with each other; mixing first each of these puzzolanas with cement in equal measures; secondly, 2 measures of each puzzolana to 1 of cement; and afterwards mixing 1 measure of each puzzolana and 1 of cement, with 1 measure, with 2 measures, and with 3 measures of sand, successively. The results of these experiments, in which all those mixtures were employed, both for uniting Portland stone-bricks in pairs, and as square prisms in the usual manner, are contained in the following Tables (XVI and XVII), on examining which, and comparing the latter in particular with the corresponding parts of the contents of Table XI (243), it will be seen that both the natural

* Not having been used in this country, except on special occasions, and never generally for private works, there are very few persons who deal in puzzolana, and the sale of it has not yet been arranged into a regular system, the dealers sometimes selling it to the builders in its original coarse state, and sometimes previously grinding it, but not to an impalpable powder, as it ought to be. Although either the dealers in this article or the builders may grind it, it appears to me that it will be more convenient for both parties, that the former, who are usually cement manufacturers also, and who consequently must have a grinding apparatus in their possession, should execute this process, than the latter, who have no use at all for such an apparatus in the common course of their business. Notwithstanding the authority of Smeaton, I think it probable, that puzzolana may be ground more conveniently and economically in a wet state with considerable excess of water, than dry.

† Not Mr. John Bazley White, the cement manufacturer. I am informed that Mr. John White who proposed the British puzzolana was originally a Builder, and has since been employed as a Civil Engineer.

and artificial puzzolana were injurious to the adhesiveness and resistance of cement, which they generally diminished even more than the like proportions of sand alone had done; but that of the two, the artificial puzzolana injured the cement rather less than the natural puzzolana, under the like circumstances.

At the same time in order to ascertain what effect, the addition of 1, 2, 3, 4 and 5 measures of each of those two puzzolanas to 1 measure of Francis's cement, would have upon the hydraulic powers of that cement, we made several balls of all those mixtures, and immersed one of each sort under water on the same day on which they were made, and did the same on each successive day for three days afterwards, considering those which set the first day, better than those which would not set until the second or third day, and so on.

Of the mixtures with the natural puzzolana, F1 P1, and F2 P2, and of those with the artificial puzzolana F1 AP) 1, and F2 (AP) 2, set, or rather did not fall to pieces, when immersed on the same day, about 3 hours after they were made. Of the other mixtures with the natural puzzolana F1 P3, F1 P4, and F1 P5, all set, when immersed 24 hours after they were made, or at 1 day old, but none of the corresponding mixtures with the artificial puzzolana F1 (AP) 3, F1 (AP) 4, and F1 (AP) 5, would set until they were 3 days old.*

After being immersed at those periods, the whole of them were kept under water until the tenth day after they were made, when on splitting them by a chisel they were all found soft not only inside but outside. In short puzzolana, which so much improves the hydraulic powers of common lime, spoils cement, and of the two it appears that the artificial is more injurious to its water setting properties than the natural puzzolana; which however is of very little importance, as no one would ever mix these two valuable substances, cement and puzzolana together, to make not only a much more expensive compound, but a very bad one, whereas if used separately both are good.

(303) The unexpected superiority of adhesiveness of our artificial puzzolana, when mixed with cement, induced us to try further experiments, for the purpose of ascert-

* We tried similar proportions of sand with the same cement in the same manner, all of which retarded and injured its water setting powers, but not quite so much as the puzzolana had done, because the balls of cement mixed with sand were not so soft, when broken after the same periods of immersion.

aining its effects upon chalk lime and its mortars, in comparison with those which we had already tried with the natural puzzolana, and the same lime and mortars; and at the same time, in order to enable us to judge whether the chalk lime, about to be used with the artificial, was equal to that which had previously been used with the natural puzzolana, we experimented also upon the strength of this lime, and of the mortars made by mixing it with sand alone, as we had done with part of the chalk lime used in our former experiments. The results of these new experiments contained in the 8 lowest lines of the two following tables, when carefully collated with the corresponding experiments contained in the lowest lines of Tables XII and XIII, (277), afforded what we considered a very fair comparison between the natural and artificial puzzolanas, which gave us a much more favourable opinion of the strength of the latter, than we could possibly have anticipated.

TABLE XVI. COMPARATIVE RESISTANCE OF VARIOUS MIXTURES OF CEMENT, WITH NATURAL AND ARTIFICIAL PUZZOLANA, WITH AND WITHOUT SAND, AND ALSO OF CHALK LIME AND ITS MORTARS, WITH AND WITHOUT PUZZOLANA, FORMED INTO SMALL SQUARE PRISMS OF THE SAME SIZE AS THOSE BEFORE EXPERIMENTED UPON.

Cement Powder or Lime Paste, mixed with P, Puzzolana Powder, (AP), Artificial Puzzolana Powder, and S, Sand.	Age in days.	Weight per Cubic foot in lbs.	Resistance estimated by the breaking weights in lbs., in several successive Experiments.					Average resistance in lbs.
Francis's 1, P1	15	96	103	95	71	96	76	88
" 1, P2	15	101	59	59	67	50	61	59
" 1, P1, S1	15	112	49	58	70	75	68	64
" 1, P1, S2	15	110	28	33	47	39	33	36
" 1, P1, S3	15	116	33	42	35	44	49	41
Francis's 1, (AP) 1	15	96	91	82	79	89	140	96
" 1, (AP) 2	15	99	117	112	112	96	95	106
" 1, (AP) 1, S1 ...	15	116	67	72	74	77	75	73
" 1, (AP) 1, S2 ...	15	114	44	49	46	56	39	47
" 1, (AP) 1, S3 ...	15	116	44	47	37	30	30	38
Chalk Lime (Hydrate)	15	65	39	25	35	32	37	34
Chalk Lime 1, S2	15	107	39	39	26	28	40	34
" 1, S3	15	112	35	32	26	26	57	35
" 1, (AP) 1	15	82	108	116	206	157	173	152
" 1, (AP) 2	16	89	264	105	77	228	171	169
" 1, (AP) 1, S1 .	16	98	123	100	117	112	124	115
" 1, (AP) 1, S2 .	15	105	75	54	60	82	..	68
" 1, (AP) 1, S3 .	15	114	65	40	46	53	49	51

TABLE XVII. COMPARATIVE ADHESIVENESS TO PORTLAND STONE, OF VARIOUS MIXTURES OF CEMENT, WITH NATURAL AND ARTIFICIAL PUZZOLANA, WITH AND WITHOUT SAND, AND ALSO OF CHALK LIME AND ITS MORTARS, WITH AND WITHOUT PUZZOLANA, TOGETHER WITH THE RESISTANCE OF THE SAME MIXTURES REPEATED FROM THE PRECEDING TABLE.

Cement Powder or Lime Paste, mixed with P, Puzzolana Powder, (AP), Artificial Puzzolana Powder, and S, Sand.	Age in days.	Adhesiveness estimated by the breaking weight in lbs., in several successive Experiments.					Average Adhesiveness in lbs.	Average Resistance in lbs.
Francis's 1, P1	11	1250	423	255	441	561	586	88
„ 1, P2	11	199	305	303	221	379	281	59
„ 1, P1, S1	11	652	641	597	557	277	545	64
„ 1, P1, S2	11	202	359	309	272	165	261	36
„ 1, P1, S3	11	240	233	247	190	179	218	41
Francis's 1, (AP) 1	11	949	593	748	622	571	697	96
„ 1, (AP) 2	11	531	409	484	440	398	452	106
„ 1, (AP) 1, S1....	11	398	320	566	562	1000	567	73
„ 1, (AP) 1, S2....	11	468	134	526	414	162	341	47
„ 1, (AP) 1, S3....	11	387	316	281	253	268	301	38
Chalk Lime (Hydrate)....	17	92	239	219	125	..	169	34
Chalk Lime 1, S2	17	393	288	329	358	277	329	34
„ 1, S3	17	235	237	272	170	..	229	35
„ 1, (AP) 1	17	513	613	165	457	373	424	152
„ 1, (AP) 2	17	684	838	1105	1177	1021	965	169
„ 1, (AP) 1, S1 .	17	354	712	602	590	628	575	115
„ 1, (AP) 1, S2 .	17	663	596	322	267	369	443	68
„ 1, (AP) 1, S3 .	17	395	404	320	348	362	366	51

(304) Having thus tried the comparative adhesiveness and resistance, both of the natural and of our artificial puzzolana mixed with chalk lime and the mortars of that lime in various proportions, we next experimented upon the same mixtures, in order to ascertain their comparative hydraulic powers, in the same manner in which we had experimented upon the various mixtures of those puzzolanas, combined with cement, that is by making several balls of all those mixtures, and putting one of each under water on the same day that they were made; and also 1, 2, 3 and 4 days afterwards.

No mixtures, either of the natural or artificial puzzolana, would set under water on the same day on which they were made, nor even on the day following when one day old; but the whole of them went to pieces on those days.

The mixtures denoted by C1 P2 and C1 (AP) 2, each composed of 1 measure of chalk lime mixed with 2 measures of the natural or of the artificial puzzolana, set under water at 2 days old, and when taken out and broken 5 days afterwards they were both hard inside, but the latter seemed rather the hardest of the two.

The mixtures of the natural puzzolana denoted by C1 P1, C1 P1 S1, and C1 P1 S2, and those of the artificial puzzolana denoted by C1 (AP) 1, and C1 (AP) 1 S2, set under water at 3 days old, and when taken out and broken 4 days afterwards, were soft inside.

(305) *That the Artificial Puzzolana made by calcining the Blue Clay of the Medway, is not quite equal in its water-setting Properties to the Puzzolana of Italy, and also rather inferior to it in Resistance, but that it is much superior to it in Adhesiveness.*

First, in regard to their comparative hydraulic powers, the experiments which have just been described, would seem to prove, that there is very little difference between the two, as the artificial was only inferior to the natural in one trial out of five; but on referring back to our former experiments also (260), in which the latter had rather the superiority, it appears proper to admit, that the mortars made with our artificial puzzolana may be expected to set very nearly, but not quite as soon, as the like mortars made with the natural puzzolanas of Italy.

Secondly, in regard to their resistance, by comparing the results recorded in Table XII (277), and in Table XVI, in the last 5 lines of each, it must be admitted, that the artificial is generally inferior to the natural puzzolana in this quality, since only one in five of the artificial puzzolana mortars evinced greater resistance, than the corresponding mortars of the natural puzzolana.

Thirdly, in regard to their adhesiveness, by comparing in like manner the results recorded in Table XIII (277), and in Table XVII, in the last five lines of each, it will be seen that the artificial puzzolana is much superior to the natural in this important quality.

For example, the mixture C1 (AP)2, composed of 2 measures of artificial puzzolana combined with 1 measure of chalk lime paste, not only possessed twice as much adhesiveness as the corresponding mixture of natural puzzolana with the same lime; but at 17 days old it even proved superior in that quality to all but one of the cement mortars, that we had made by mixing the best artificial and natural cements with sand in equal measures of each, when these last mixtures were from 11 to 13 days old, as will be seen by referring back to Table X (241).

Thus the blue mud or alluvial clay of the Medway affords us the means of rivalling by very simple processes, not only the most powerful natural cements of England, but also the puzzolanas of Italy and the trass of the Rhine, neither of which valuable properties would probably have come under my notice, if it had not previously been constantly used by the Soldiers under my command at Chatham, by their own choice and not by my desire, for a very different purpose (2).

(306) I shall here remark, that puzzolana, and trass, as well as the coloured clays of nature, together with several basalts and slates or other schistous rocks, all agree in consisting principally of silica and alumina with a proportion of the oxide of iron, the two former substances usually amounting at least to two thirds and sometimes to three fourths of the solid matter in the whole compound, and the latter varying from perhaps one 12th or one 10th to one 5th of the same, besides which other component parts of less importance such as carbonate of lime, magnesia, and the alkaline salts are also found, usually in smaller proportions than the iron, and not alike in all. But though there is thus a considerable similarity in their character, as ascertained by chemical analysis,† and though my experiments with the blue clay of the Medway induce me to place great confidence in the artificial puzzolanas, that may be formed by the moderate calcination of any very fine impalpable clay, I doubt whether any of the basalts or schists are equally suitable for that purpose, as has been asserted; because the basalts have a degree of hardness equal to that of overburned bricks, whilst the best puzzolanas are in the state of moderately burned bricks; and I apprehend, that even the schists are generally in too hard a state to yield satisfactory results. If however on a fair and ample trial of their comparative strengths, they should not prove so much inferior to the fine clays of nature, as I anticipate, they may be made available as artificial puzzolanas, in those districts, in

* If artificial puzzolana made from this clay should ever be manufactured for sale, in England, which is not improbable, it will, however useful and efficient it may prove, have no pretension to a new discovery or improvement. (See Article 258, and Articles LXX and XCIII, in the Appendix.)

† See the analysis of trass and puzzolana in the Appendix, and the analysis of slates, schistous rocks and basalts in any approved work on mineralogy.

which they are abundant but where little or no clay is to be found, which is the case in many parts of Scotland. Compact clays not soft enough to be plastic, and yet not hard enough to furnish stones fit for the walls or roofs of buildings, such as are occasionally to be met with, as if I recollect rightly in the island of Heligoland, would probably form a useful substitute for puzzolana, by merely pulverizing them.

FARTHER EXPERIMENTS ON THE STRENGTH OF CEMENT JOINTS, FOR CONNECTING LARGE STONES.

(307) The experiments recorded in from Article 214, to 218 inclusive, for ascertaining the adhesiveness of the natural and artificial cements to large Bramley-fall stones, did not lead to any decisive result, as to the comparative superiority of either, owing to repeated breakages, occasioned by the joints of both cements proving stronger, and the apparatus for trying them, much weaker, than had been anticipated; but they sufficiently demonstrated what had hitherto been considered doubtful, that cement may be used with advantage for masonry composed of the largest and heaviest stones.

In continuation of the experiments in Article 218, we connected the same two Bramley-fall stones there described, by a joint of our artificial cement mixture C4 B5·5, and used the common mason's lewis for suspending them, which was done by letting it about 5½ inches into the uppermost stone. On applying weights in the usual manner, the joint was torn asunder by 8459 lbs, but on inspection, it was found to be an imperfect one, part of the space between the two stones being without any cement for about one sixth part of the whole extent, so that from this defect in the center of it, the real surface of cement joint acted upon was only about 830 superficial inches, instead of 1000, which it ought to have been.

At the time when this fracture took place, the lewis was gradually being drawn out of the upper stone, which had been splintered in consequence, and would certainly have given way before the joint, had the latter been perfect.

(308) We therefore renewed the same experiment, after cleaning the stones, taking every pains to make our new cement joint of 1000 superficial inches quite perfect, and now instead of a lewis, we used an iron plug, secured into the whole depth of the upper stone nearly by melted lead, and so formed that it could not possibly be drawn out without breaking the stone.

After applying 7110 lbs , the bolt of the shackle of this iron plug broke, and both stones, with the weights bearing on the lower one, fell to the ground.

Having replaced the iron bolt by a piece of steel of greater strength, and submitted the same joint which appeared uninjured to a second trial, the plug itself broke off close to the upper stone, after 11,825 lbs, were applied, and the stones and weights again fell to the ground.

Next day, as the joint, which was now 17 days old, still appeared uninjured, we cut 2 mortises into the upper stone, each 5 inches deep, similar to others that had been cut before, when this stone formed part of a larger one, used in a former experiment. Into these mortises, we inserted 4 iron bearers, on opposite sides of the stone, and secured them with melted lead; and having again suspended our two stones by these bearers, we applied weights to the lower stone as before, and gradually increased them, until the joint was torn asunder, which was effected by a total weight of 12,982 lbs., being rather more than we had previously calculated upon (218.) The cement in this joint happened to be unusually thick, being 7-sixteenths of an inch in mean thickness, so that some parts of it were more than half an inch thick, which seems to prove that thick joints are not so very prejudicial to the strength of brickwork or masonry laid in cement, as we had before supposed.

Considering that the joint might have been shaken by the two falls, as well as by the act of cutting the mortises that have been mentioned, into the upper stone, the great weight required to tear it asunder, after those circumstances, is a proof that our artificial cement possesses not only great adhesiveness, but considerable toughness, in which last and very important quality we had before remarked, that it appeared to excel the natural cement. (218.)

EXPERIMENTS ON THE RESISTANCE OF BRICK BEAMS BUILT WITH PURE OR NETT CEMENT, WITH AND WITHOUT HOOP IRON BOND. GREAT IMPORTANCE OF THIS BOND.

(309) In order to ascertain how much of the extraordinary resistance of brick beams built with cement, such as were before described in Articles 225 and 226, might be owing to the hoop iron bond, the following experiments were tried at my request, by special permission and authority of the Honourable Board of Ordnance.

1st. We built a brick beam with pure cement, two bricks wide and four courses of bricks high, upon two brick piers built with cement and sand mixed in equal measures, each of which was 2 feet 6 inches high and 2 bricks square above the level of the ground, below which there was one footing or foundation course $2\frac{1}{2}$ bricks square, bedded in common mortar, with the earth well rammed.* The beam which was $18\frac{3}{4}$ inches wide and 12 inches high, measured 13 feet 1 inch in extreme length, with a clear interval of 10 feet between the piers, to which it was united by joints of pure cement, as represented in the annexed figure, which shows also the apparatus afterwards used for breaking it down, consisting of a frame and planks for carrying the weights, suspended from a round bar of wood, resting on the center of the top of the beam.

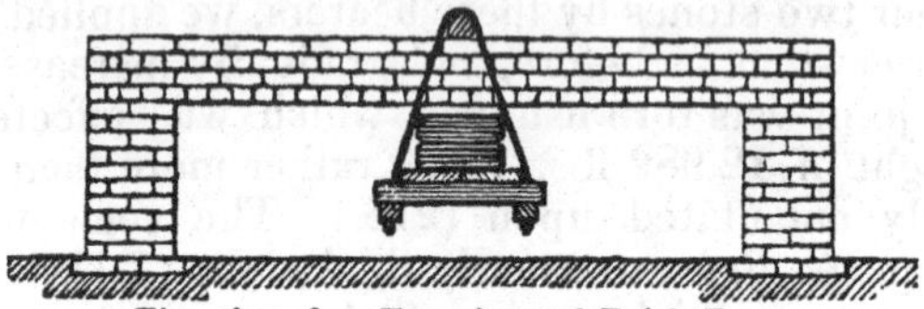

Elevation of an Experimental Brick Beam.

2dly. We built another brick beam precisely similar to the above, but with the addition of 5 longitudinal pieces of hoop iron, extending the whole length of the brickwork, 2 of which were placed in the lower joint, 1 in the center joint, and two in the upper joint of the beam, as represented in section, in the following figure, which also shows the breaking apparatus before mentioned.

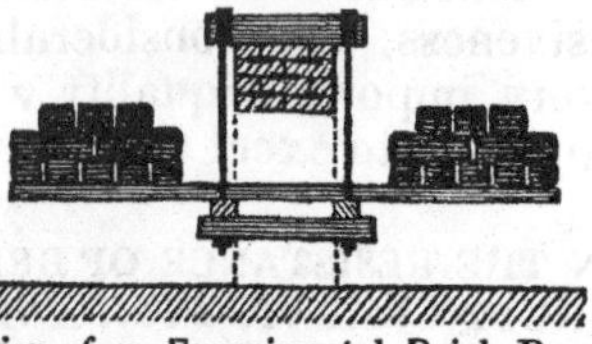

Section of an Experimental Brick Beam.

* In the 2 piers we used 189 bricks, 2·8 cubic feet of cement powder and 2·8 cubic feet of sand. Thus the quantity of cement was rather less than 26 cubic inches per brick, and was in the proportion of 1-5th nearly of the whole mass of brickwork in those piers, the sand and cement together measuring originally about 2-fifths of the said mass, of which they afterwards formed a part. In the brick beam we used 272 bricks, and 7·11 cubic feet of cement powder, which was therefore at the rate of rather more than 45 cubic inches per brick, and was in the proportion of rather more than one third of the whole mass of brickwork afterwards composing the beam, which was finished in 8 hours by 2 bricklayers and 2 labourers.

3dly. We constructed a third beam similar to the last in every respect, excepting that the brickwork was built with mortar, composed of 3 measures of sand, mixed with 1 measure of well-slaked Halling lime putty, of a stiffish consistency.*

We removed the wood work upon which the 1st and 2d beams had been built, in 10 and in 7 days respectively, after they were finished, and at the end of 50 days we applied weights to break them down, which were laid upon strong planks placed on a sort of scaleboard below the beam, suspended by means of 2 bent pieces of iron passing over both ends of a transverse bar of wood, resting on the center of the top of the beam. The planks, by projecting considerably beyond the sides of the piers and beam at both ends, allowed the iron weights to be applied without interfering with the brickwork.

(310) The first beam, which was built with pure cement, without the aid of the hoop iron bond, gave way under a total weight of 498 lbs., only, including the apparatus which has just been described. But the two lower joints had been cracked entirely across the soffit, before any weight was laid upon the beam, which therefore was not a perfect one. This crack, which was very fine, had not been observed until a day or two before the experiment, which consequently was not a fair one; but the circumstance sufficiently proves that the centering was removed too soon, and that a brick beam of the above dimensions, built with pure cement, is precarious. In falling, this beam broke into three pieces, of which the centre one was not quite 3 feet long, and the other two differed in length by about 10 inches. One of the two fractures agreed with the original crack that has been mentioned, and was about 1 foot from the center of the beam.

(311) The second beam, built also with pure cement, but having the advantage of the five longitudinal bond irons, showed a deflection of 1 tenth of an inch when loaded with a total weight of 2537 lbs., which gradually increased to 2 tenths under a weight of 3718 lbs., and at this time there was a crack across the two lower courses of the beam.

* In this beam were used 272 bricks, 7·11 cubic feet of sand, and 2·37 cubic feet of Halling lime putty, so that the sand alone measured rather more than one third, and the lime putty rather more than one ninth of the whole mass of brickwork of the beam. The ingredients composing the mortar therefore originally measured rather more than 4 ninths of the said mass, of which they afterwards formed a part. This beam was built by the same 2 bricklayers and 2 labourers in 10½ hours.

When 3945 lbs., were laid on, the crack extended to the 3d course of the beam, and the deflection had increased to 5 tenths. With a weight of 4308 lbs.; the deflection was equal to 1 inch, but with a weight of 4523 lbs., the crack extended entirely across the beam, throughout its whole depth, and the scale board supporting the planks on which the weights, consisting of iron ballast, were applied, fell to the ground. On removing these weights and the whole apparatus, the beam appeared in elevation, as shown in the following figure, with its center 4 inches lower than the ends. In consequence of the beam being bedded with pure cement on the tops of the piers, it tore off the uppermost course of brickwork of each pier, which adhered to the ends of it, separating themselves entirely from the remainder of their respective piers.

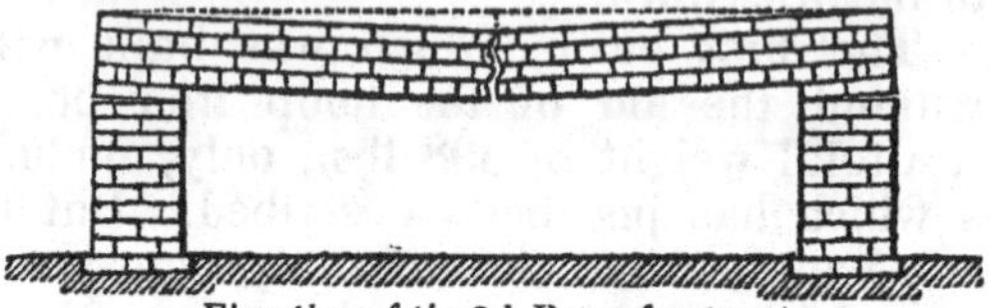

Elevation of the 2d Beam fractured.

We next raised the apparatus for the weights higher, by supporting the bar of wood upon two pieces of iron ballast, laid on each side of the fracture at the top of the beam. This enabled us to apply weights again, which rather proved the strength of the hoop iron than of the brick and cement work, the latter having been already completely fractured, and on increasing our weights to 4314 lbs, the beam again broke down, and fell as low as the scale-board beneath it would permit, and at this period the center of the beam was 22½ inches lower than its ends, as shown in the elevation in the next figure, which represents its state, after the weights and all the breaking apparatus, except the scaleboard, which prevented it from touching the ground, had been removed.

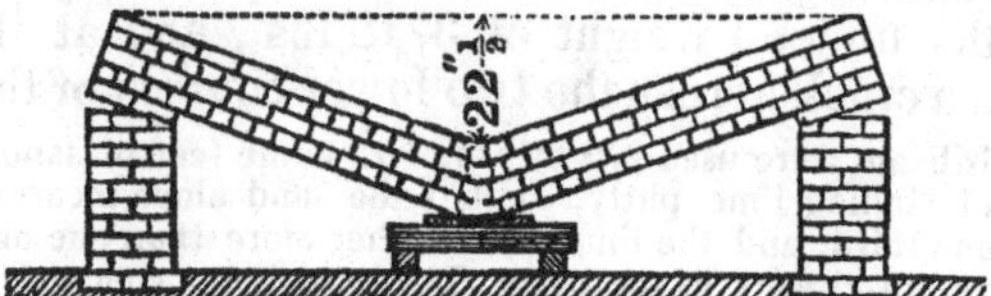

Elevation of the 2d Beam entirely broken down.

On examining it, we found that the center iron was bent, and the parts of the brickwork there though fractured were in contact or nearly so; but the two lower irons had been broken, and the brickwork there had opened out considerably, and two small irregular vertical cracks showed themselves, one on each side of the principal fracture, and a few of the lower course of bricks had been broken off in this part, probably by the violent action of the two irons when broken. In the upper part of the beam, which had been subject to compression, the bricks in the center were forced violently against each other and crushed, and the two upper irons were buckled but not broken.

Thus the first brick beam, built with pure cement appears to have been scarcely capable of supporting its own weight, from having been deprived of its centering, when only 10 days old; for being thereby injured, when its strength was tried after the cement was 50 days old, it broke down under a weight of 498 lbs: whilst the second beam strengthened by the hoop iron bonds required nearly 9 times that weight before it was fractured, and even then only fell 4 inches in the center, being held together by the hoop iron.

(312) In trying these experiments we did not wish to produce any extraordinary results, but to ascertain a doubtful point of some importance, and therefore we did not give to any of our beams such dimensions as would have required very enormous weights to break them down, like those made by Mr. Brunel and Messrs. Francis. Insignificant, however, as the weights that cracked and finally fractured our second brick beam, may appear, it was not in reality weaker, in proportion to its dimensions, than either of the above.*

(313) The woodwork supporting the third beam was removed in 42 days after the brickwork had been finished, but, observing next day, that a perceptible though small deflection had taken place, we replaced it, until the mortar was 50 days old, when we finally removed it, and subjected this beam to the same breaking apparatus, as the others. The original deflection at this time, was nearly a quarter of an inch, which on loading the beam with 700 lbs. increased to 3 tenths of an inch, and on adding 42 lbs. more, the

* The comparative resistance of beams of the same materials is stated by a popular rule, to be inversely as their lengths and directly as their widths and as the squares of their depths. Hence if the above rule be considered sufficiently correct to afford a rough estimate approximating to the truth, the comparative strength of Mr. Brunel's beam ought to have been 8 times and of Messrs. Francis's 11 times greater than mine.

beam broke, and the loaded scale board came in contact with the ground. On removing the weights and all the apparatus for supporting them; although there was now a complete fracture through the center of the beam, it had no tendency to fall lower, being held together by the hoop iron, but the center of it had sunk 6 inches lower than the ends, so that it was apparently nearly in the same state as the second beam after its first fracture; but on examination, we found that the whole of the irons had been drawn inwards, from their original position on one side of the beam along the joints of the mortar, to the depth of more than an inch and a half from the end of the beam on that side, with which they had originally been flush. In this state we loaded the beam with pig iron ballast piled over its center, until it broke down under a weight of 361 lbs., when that half of the beam in which the irons had been drawn from their original position fell almost all to pieces, every joint of the greater part of it being broken or injured, whilst the other half of the beam in which the irons had not been moved, held together in one compact mass, of which the greater part was apparently uninjured. None of the irons were broken in this experiment, but those belonging to the weaker end of the beam were violently bent and contorted. The annexed figure is a view of the beam, when thus finally broken down. One of the piers was moved out of its original perpendicular position, and separated a little from the ground on one side of it by the action of the beam in falling.

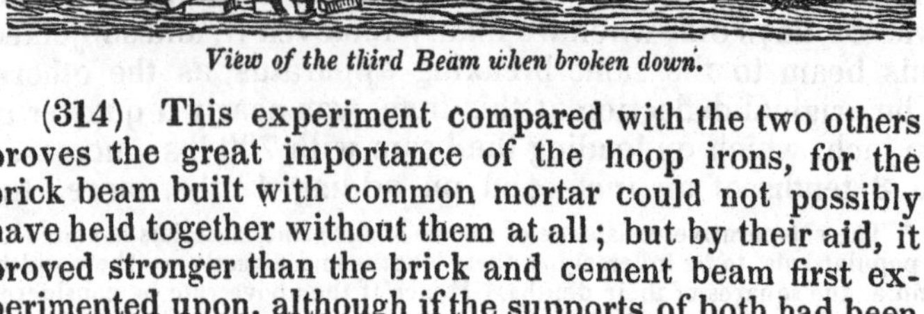

View of the third Beam when broken down.

(314) This experiment compared with the two others proves the great importance of the hoop irons, for the brick beam built with common mortar could not possibly have held together without them at all; but by their aid, it proved stronger than the brick and cement beam first experimented upon, although if the supports of both had been kept standing for the same period, the latter would no

doubt have had the superiority. But although of some value even with common mortar, the circumstance of such mortar not having tenacity enough to prevent the iron bars from being drawn along the joints, when acted upon by a moderate force; after which they cease to be an efficient bond, is an objection to this combination. But when hoop irons are used to strengthen a brick beam built with pure cement, the mutual adhesion of the cement, and of the irons is so perfect, that no force can separate them without producing the complete fracture of the brickwork, which is thus resisted by all the tenacity of the iron.

(315) The conclusion, that may be drawn from the above, is that cement bond, consisting of 4 or 5 courses of brickwork laid in pure cement, if strengthened by longitudinal pieces of hoop iron in all the joints, may be used to supersede not only the wooden lintels of doors and windows, but all timber bond generally in the walls of buildings, as suggested in Article 234, which was written before we had tried these last experiments. In using hoop iron bond in walls, the irons should extend if possible the whole length of each wall in one piece; but if a break be necessary, the adjoining ends need not be united together by the blacksmith, but turned down at right angles into one of the vertical joints of the wall by the bricklayers themselves. Without hoop iron bond, on the contrary, the additional strength communicated by cement alone would not suffice in difficult cases.

It is to be observed, however, that a continued string of 4 or 5 courses of cement and hoop iron bond, in the walls of a building, would not be exposed by any means to the same strain, as our experimental brick beams; for it would not have to bear much more than its own weight in all the unsupported parts over a door or a window, there being other windows above those, and in all the intermediate portions of the wall corresponding with the ends of our experimental brick beams, the courses of cement bond alluded to would not only be supported from below, but their strength would be greatly increased by the weight of the solid parts of the wall above, it being well known that all beams have a much greater resistance, when firmly fixed, than when merely supported at their ends, which Mr. Barlow in his able and useful Treatise on the strength and stress of timber estimates from his own experiments, as being in proportion to the numbers 3 and 2. Besides which, 10 feet between the bearings is a much greater width than would be given in practice to the windows, or even to the doors of the largest building, unless the latter were carriage-gateways, which are more usually covered by semicircular or elliptical arches, than by flat arches or straight lintels.

It only remains further to remark, that the flattest and thinest brick and cement arch has sufficient power to resist great pressure, in openings of 10 or even 15 feet, as was proved by one of our former experiments (134); though a straight brick and cement beam is not to be recommended, over such openings, unless consolidated by hoop iron bond.

(316) We next experimented upon the most perfect fragment of the first brick and cement beam, by laying it upon two piers at the clear distance of only 3 feet apart, and applying the same apparatus, we loaded it in the usual manner, as represented in the annexed figure, when it broke under a weight of 2356 lbs.

(317) We experimented also in precisely the same manner upon the most perfect half of the third or brick and mortar beam, which we laid upon two piers at the clear distance of 2 feet 2½ inches apart, but not having room for our former apparatus in this narrow space, we laid one single piece of iron ballast transversely across the center of the beam, upon which we built others, projecting beyond it, in alternate courses longitudinal and transverse. On applying 4887 lbs, two bricks fell out from the bottom of the pier. When the weight was increased to 4993 lbs, a third brick also fell, and on increasing it to 5147 lbs. the beam fell to pieces. Thus this portion of the Halling lime mortar beam appears to have been stronger in proportion even than that of the pure cement beam, owing no doubt entirely to the advantage of being bonded by the hoop iron.

FINAL RESISTANCE OF MESSRS. FRANCIS'S EXPERIMENTAL BRICK BEAM, BEFORE DESCRIBED.

(318) Since the above was written, the experimental brick and cement beam with hoop iron bond, of Messrs. Francis and Sons at nine Elms Vauxhall, to which a weight of 24,000 lbs, had been permanently suspended, as described in Article 226, was broken down on the 14th of February 1838, in the presence of many scientific and professional men, whom they had invited to witness this interesting

experiment, the ground being required for the metropolitan terminus of the Southampton railway. Not being able to attend, as I wished, I learned on inquiry, that no crack was observed, before the entire fracture of the beam, which was suddenly effected, when the additional weights gradually laid on amounted to a total of 50,176 lbs, (or 22 tons 8 cwt). It was broken into three parts, two of which were very nearly equal, each consisting of almost one half of the beam, whilst the third near to the top was a very small irregular lump; the fracture which commenced vertically from the bottom upwards, nearly all along the middle of the beam, having, when very near to the top, diverged into two branches in the form of the capital letter Y, which I think must have been caused by two parallel cross-timbers near to the center of the top of the beam bearing equal portions of the weights, which were not suspended from one central piece of wood or of iron, as in my experiments.

The appearance of the brick beam after being broken down is represented in the following sketch. The apparatus for bearing the weights, on coming to the ground had prevented the entire downfal of the parts, and the small lump before-mentioned having got jammed in between the two large fragments prevented them from descending equally low. The piers were fractured and forced out of their perpendicular, and if the fall of the beam had not been arrested, they would no doubt have been pushed entirely over.

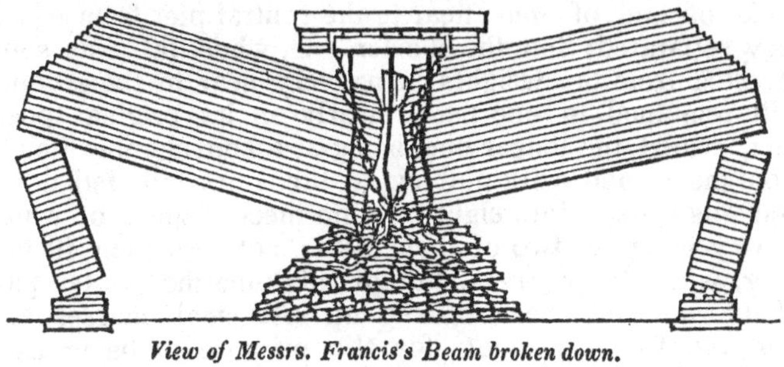

View of Messrs. Francis's Beam broken down.

The two large portions of the beam were perfectly sound, with the exception of one or two very small cracks near the edges of the great fracture; and no injury whatever had been done to any part of it by the frost.*

* This sketch does not exactly agree with the section and elevation of the beam in Article 226, in which the whole brickwork of the beam was correctly represented, but some small parts of the woodwork, &c, by which the weights were suspended, had been omitted by mistake.

THAT MR. BRUNEL'S SEMI-ARCHES BEFORE DESCRIBED FELL DURING THE LATE SEVERE WINTER, BUT NOT OWING TO THE FROST, WHICH HAD NO EFFECT ON THE CEMENT.

(319) These semi-arches having fallen down on the 31st of January, 1838, and this having been supposed at the time to have been in consequence of the frost, I took the first opportunity of inspecting the ruins, and of inquiring into the cause, which Mr. Brunel himself explained to me. Not having described this extraordinary portion of bridge with sufficient precision, in Article 59, in which an elevation of one side of it only was given, the construction of which was not as in ordinary bridges, common to the other side also, I refer for a more detailed and accurate description of it to to the Appendix. In the mean time, suffice it to say, that no part of the brick or cement work was injured in the slightest degree by the frost, and that the fall of the semi-arches was solely occasioned by the foundation of the central pier, by which they were supported, having been deranged by excavating near it to a much greater depth for a gasometer and its fitments, and also by the ground immediately in contact with the brick footings of the pier having been frozen hard, and then suddenly thrown into a soft wet state by a thaw. Owing to these circumstances, the pier moved, and the shorter of the two semi-arches descended, until the weights attached to it struck the ground, which caused a shock and a sort of vibration that broke off both of them, near to the central pier from which they separated: but the shorter one, whose fall was soon arrested from the weights being very near to the ground, still remained in one mass, nearly in its original position, whilst the longer one, being at a considerable height from the ground, with nothing to break its fall, fell down and was broken into eight or nine pieces, some of which were very large, two of them being not less than 40 feet long, when I saw them, but at this time the greater part of the ruins had been broken up into small masses, and removed from the spot, for the purpose of being used in the drains under the tunnel. Mr. Brunel informed me, that his semi-arches had been about four years in building, as he had from time to time added both to the length and width of brickwork first intended, and he observed that in falling, the fractures had generally taken place at those parts, where the new work had been added to the old, where he considered the cement joints to be less per-

fect, in consequence of the surface of the old brickwork, to which the new cement was attached, having got smoked and greasy. Mr. Francis's brick beam on the contrary, having been built without intermission, was one homogeneous and very perfect mass, and fell as a similar mass of solid rock of the same dimensions and of equal specific gravity and resistance would have done, if experimented upon in the same manner.

(320) As an additional example that cement is frost-proof, the brick and cement summer house, built by me for experiment, as described in Articles 223 and 224, has not been in the slightest degree affected by the extraordinary severity of the late winter of 1837-8.

(321) *Experiments tried at Chatham, on the Strength of Hoop Iron, in opposition to a Tensile Force.*

The extraordinary resistance communicated to brick beams by longitudinal hoop iron bond, the value of which far exceeded our expectations, induced us to try a few experiments on the actual strength of the hoop iron alone, in opposition to a force tending to tear it to pieces in the direction of its length, and acting from one end, the other being fixed.

First, with 1½ *inch hoop Iron.* This was the sort used by us, in the second and third brick beams, that have been described, its width being 1½ inch, and its thickness being what is described as No. 15 wire gage, or 1 sixteenth of an inch nearly, and its weight per foot being 2520 grains, or 3·6 lbs. Having fixed a piece of this hoop iron in a vertical position, and suspended weights to the lower end of it, which we gradually increased, it broke down under a total weight of 6163 lbs., after the iron had elongated 13 sixteenth parts of an inch in the length of about 18 inches.

Secondly, 1 *inch hoop Iron.* We next broke down a piece of hoop iron 1 inch wide, of No. 18 wire gage, weighing 1220 grains to the foot, by a total weight of 2961 lbs., which produced a previous elongation of half an inch.

Thirdly, three quarter inch hoop Iron. Finally we broke down a piece of hoop iron 3 quarters of an inch wide, of No. 20 wire gage, and weighing 640 grains to the foot, by a total weight of 1842 lbs., which produced a previous elongation of 1⅜ inch.

The results of these experiments agree very nearly with the statements of the strength of iron of this description in the works of Professor Barlow and Tredgold, and as the strength of iron has been fully investigated, and accurately stated by these and other Authors of reputation, we did not think it worth while to pursue this subject any farther.*

THAT COMMON BRICKS OF GOOD QUALITY, UNITED BY PURE CEMENT, ARE MUCH TOO WEAK FOR THE STEPS OF A GEOMETRICAL STAIRCASE, AS WAS BEFORE PROPOSED, AND THAT EVEN WITH THE AID OF HOOP IRON, THEY ARE SCARCELY STRONG ENOUGH FOR THIS PURPOSE.

(322) In article 232, I stated an opinion that good artificial stones for staircases might be formed with small materials, such as bricks or plain tiles, or hollow earthenware tubes, which I did with little hesitation, knowing that I would have an opportunity of proving the thing by direct experiment, before this treatise could be published, so as either to confirm this opinion if just, or to correct it if erroneous. For this purpose I caused an artificial stone, to represent one step of a geometrical staircase, to be built in 5 courses, of which the center one was composed of bricks, the others of plain tiles, all united by pure cement supplied by Messrs. Francis and Sons, and breaking joint with each other. When finished it was stuccod with a mixture of the same cement and sand in equal measures, after which it was 4 feet 9 inches long, 14½ inches wide, and 7¼ inches high. Eight days afterwards, we fixed one end of this step 9 inches deep into one of the inside walls of a building in the North stable yard, Brompton Barracks, bedding it in the same sort of cement mortar as the stucco, and pinning or wedging it well in, with fragments of plain tiles and pieces of hoop iron. Thirty two days afterwards, when the cement was 40 days old, we laid one piece of pig iron ballast transversely on the center of the step, and then laid others parallel, and close to it alternately on each side, until it was loaded with 5 such pieces, weighing in all 540 lbs, by which it was fractured close to the wall, and on falling broke into two pieces. Considering that the weight pressing upon one step of a geometrical staircase of the above dimensions might possibly, though not usually amount to 600 or 700 lbs, and that building materials ought not to break down under less than double the ordinary weight, to which they may be exposed, this experiment convinced us, that artificial steps made partly of com-

* Mr. Howe who suggested and prepared the apparatus for these experiments, is of opinion, that straps of 1½ inch hoop iron, would be strong enough for connecting the rafters and tie beams of roofs of moderate span, instead of ordering wrought iron straps for the purpose.

mon bricks, and united by nett cement, are far too weak for a geometrical staircase.

(323) To ascertain how far hoop iron bond might render steps of this sort efficient, we cut the most perfect fragment of our second brick beam (311) to the same length of 4 feet 9 inches, and fixed it in the same manner in the same wall, from which it projected 4 feet, representing one step of a geometrical staircase, 18¾ inches wide, and 12 inches high, which though inconveniently large for a real step, we could not diminish without spoiling it, and were unwilling to lose the benefit of one experiment.

Afterwards, when the cement was sixty eight days old, we applied iron ballast tranversely to this step, distributing it equally over its whole surface commencing from the center outwards, and on the application of 3560 lbs, it broke close to the wall, but was held so fast by the hoop iron bond, that it proved impossible to disengage it by a man jumping upon or even by battering the top of it with rammers, until all the pieces of hoop iron were cut off by a chisel at the wall. This proves that a step of this description 12 inches high, and 18 inches wide would be quite strong enough, but those dimensions are altogether too large for a staircase. If a step of the same materials had been built of the more convenient and usual dimensions of only 6 or 7 inches in height, by 14 or 15 inches in width, one cannot estimate its comparative resistance to a breaking weight at more than 720 lbs, which would scarcely be strong enough for the step of a geometrical staircase.

(324) Reflecting upon this disappointment, the defect appeared to rest rather with the bricks than with the cement, but it appeared probable that tiles, which are always made of more compact clay and better compressed in the moulds, might be strong enough for this object, to which the bricks were inadequate. We therefore made several small prisms 4 inches long and 2 inches square, each consisting of 5 pieces of plain tile, as shewn in the first, or of 2 pieces of paving tile, as shewn in the second of the annexed figures, which were sawed to the proper size by a piece of hoop iron, and then ground nicely, so that the parts stuck together like one piece.

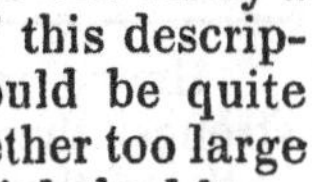

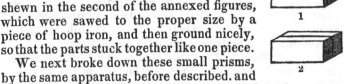

1

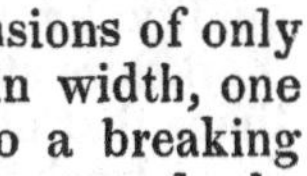

2

We next broke down these small prisms, by the same apparatus, before described. and in the same manner (199), in which we had broken down our small square prisms of the same size, whether of artificial or of natural stone, brick, &c, of which the results are inserted in the following Table; but to render it more complete, and to save the reader the trouble of referring back to former parts of this Treatise, for the sake of comparison, the resistances of Yorkshire stone and of bricks have been repeated from Table VIII (212).

TABLE XVIII. COMPARATIVE RESISTANCE OF YORKSHIRE STONE, TILES AND BRICKS, EXPERIMENTED UPON AS SQUARE PRISMS.

Description.	Weight of Prism in Troy Grains.	Weight per Cubic foot, in lbs.	Resistances or breaking weights in lbs, in several successive experiments.					Average Resistance in lbs.
Yorkshire Stone ..	9571	147·67	2976	2500	3185	....	....	2887
Plain Tiles	7154	110·38	1006	1658	764	1189	1225	1168
Paving Tiles	7308	112·75	1148	988	1073	1225	1188	1124
Superior Bricks ..	5944	91·71	{ 704	795	717	955	622 }	752
			{ 640	722	706	823	.... }	
Inferior Bricks ..	....	......	204	262	522	....	....	329

Hence we found, that the plain tiles seemed a little stronger than the paving tiles, and that the resistance of both of them exceeded that of common bricks of good quality by about one half. Also that the resistance of Yorkshire stone was about $2\frac{1}{2}$ times that of tiles, nearly 4 times that of superior bricks, and about 9 times that of inferior bricks.

THAT ARTIFICIAL STEPS OF PLAIN TILES OR OF PAVING TILES, UNITED BY PURE OR NETT CEMENT, WITH HOOP IRON BOND, ARE STRONG ENOUGH FOR A GEOMETRICAL STAIRCASE, THOUGH INFERIOR TO YORKSHIRE STONE IN RESISTANCE. THAT TILES RATHER LARGER THAN COMMON PAVING TILES, SHOULD BE PREPARED ON PURPOSE FOR SUCH STAIRCASES, IF ADOPTED.

(325) In order to ascertain the comparative resistance of natural and artificial stones, supposed to represent the steps of a geometrical staircase, we prepared three steps all 4 feet 9 inches long, each of which we fixed with cement 9 inches deep into the inside wall of the same building before alluded to, wedging them well up in their places.

First, a Step of Yorkshire Stone. This was a sound piece of good stone, 13 inches wide and $7\frac{1}{2}$ inches high, and of the length before stated; and under an apprehension that, on breaking it down, the end of it might tear out part of the wall, which was $1\frac{1}{2}$ brick thick and built with common lime mortar,* we let in three courses of brick headers over it, which we laid in nett cement to strengthen them. The annexed figure, marked 1, represents this step, fixed in the wall as above described, in longitudinal section.

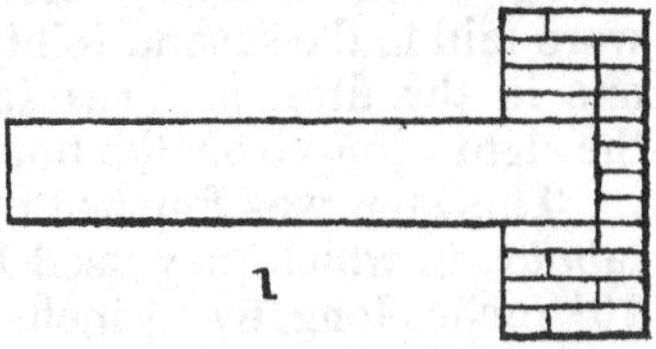

1

Having loaded this stone with pig iron ballast, laid transversely, the first piece over the center of the stone and the others alternately to the right and left of it, until the whole surface was covered, we gradually applied a second and a third course of iron ballast, and so on, commencing as before, from the center, until the stone suddenly broke down under a total weight of 3514 lbs, without giving warning by any previous perceptible deflection or crack. The fracture was very regular, and nearly but not exactly coinciding with the inside face of the brick wall, from which it projected rather less than half an inch, and which contrary to our expectation was not injured by the stone in the slightest degree, a circumstance which proves the efficacy of the common mode of fixing steps of this description in a wall.

(326) *Secondly, an Artificial Step of Plain Tiles and Cement, with Hoop Iron Bond.*

This step measured $16\frac{3}{4}$ inches in width and $7\frac{1}{4}$ inches in height, and was built in courses, each consisting of 9 tiles laid transversely in reference to the length of the step, and $5\frac{1}{2}$ laid longitudinally, alongside of the former, as represented in plan, in the second figure annexed, and so arranged as to break bond with each other in every two successive courses.

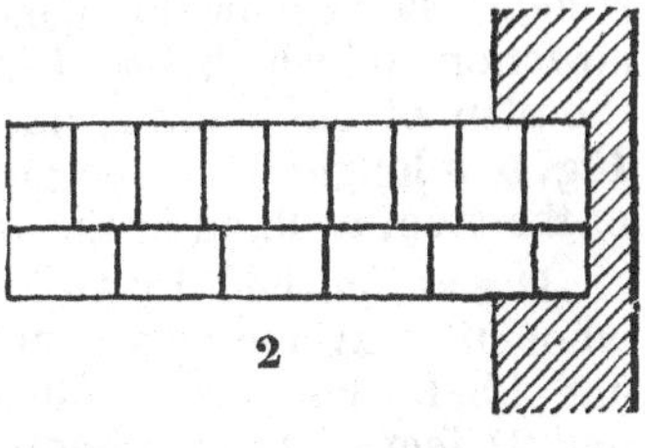

2

Nine such courses were required to complete this step, as shown in section in the third figure annexed, and

* This supposition was hastily adopted before we had considered the subject properly. For notwithstanding the great superiority of the resistance of Yorkshire stone to that of the best bricks, the weight capable of breaking the stone was far inferior to that which would have been necessary for crushing 117 superficial inches of solid brickwork; because the end of the stone acted upon a portion of the wall measuring 13 by 9 inches.

five longitudinal pieces of hoop iron were used to strengthen it, two of which were laid in the second joint, one in the fifth, and two in the eighth joint from the bottom.

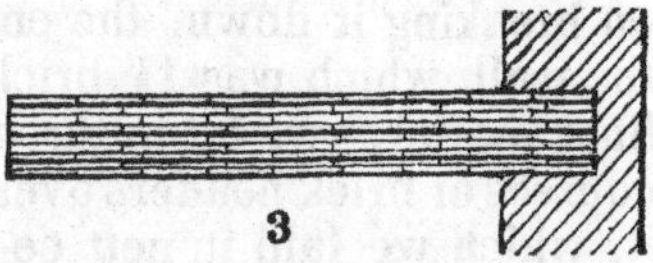

3

This step was finished by two men in $8\frac{1}{4}$ hours of actual labour, in which they used $130\frac{1}{2}$ plain tiles, each averaging $10\frac{1}{4}$ inches long by $5\frac{1}{4}$ inches wide and nearly $\frac{1}{2}$ inch thick, with 4725 cubic inches of cement powder, and 24 lineal feet of hoop iron $1\frac{1}{2}$ inch wide. It was let into the wall 7 days afterwards.

On applying pig iron ballast to this step in the same manner that has been described, it broke down under a total weight of 2396 lbs, but a very small crack near the top of the step had been observed, before the whole of this weight was laid on. The cement was at this period 105 days old.

(327) *Thirdly, an Artificial Step of Paving Tiles and Cement, strengthened by Hoop Iron Bond.*

This was made of four courses of paving tiles, and measured 12 inches wide and 7 inches high. Each course consisted of $4\frac{1}{2}$ tiles, laid so as to break joint in successive courses.

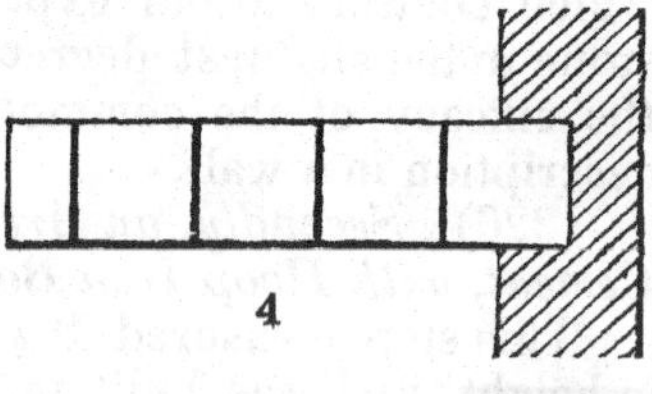

4

Five pieces of hoop iron were laid longitudinally in the joints, two in the first joint, one in the second, and two in the third joint.

The two figures annexed sufficiently explain this construction, of which Fig. 4 is the plan of one course, and Fig. 5, a longitudinal section of the step, as fixed in the wall.

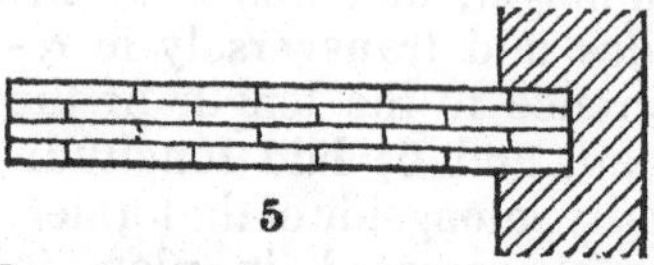

5

It was finished by two men in $4\frac{1}{2}$ hours, who used 16 whole tiles averaging 12 inches square and $1\frac{1}{2}$ inch thick, and 4 half tiles, with 1150 cubic inches of cement powder, and 24 feet of $1\frac{1}{2}$ inch hoop iron, and was let into the wall 2 days afterwards. It was broken down in the same manner and on the same day with the two former steps, by a total weight of 1263 lbs, the cement being at this period 107 days old.

REMARKS. In both of these artificial steps, the fracture took place near to the wall, but that part of each though entirely broken through was suspended by the irons, which did not break, but elongated or were drawn out from their original position within the wall, just enough to admit of the far end of each step striking the ground in falling. In reference to the consideration before stated (322), the stone step may be considered to possess a resistance of about 5 times, the plain tile step a resistance of about 3 times, and the paving tile step a resistance fully double of the greatest weight, ever likely to press upon one step of a geometrical staircase 4 feet wide; that is provided its width, which was only 12 inches, had been increased to 14 or 15 inches, which is the more usual width of the steps of such staircases, and which would of course increase its resistance in proportion. I shall observe also, that as the resistance of plain tiles and of paving tiles without cement was proved to be very nearly equal by our former experiments recorded in Table XVIII, (324), the marked superiority of resistance of the plain tile and cement step over the paving tile and cement step, though both formed of materials equally or nearly equally strong, may probably be ascribed in this experiment, to the former being composed of a much greater number of tiles, and therefore having a much greater number of cement joints than the latter. Notwithstanding, however, this inferiority of the paving tiles, I would reeommend their being used in preference to plain tiles, if the expedient of building staircases with artificial steps composed of tiles, with cement and hoop iron bond, should ever be adopted, because the paving tile step is quite strong enough, and gives much less trouble in the workmanship, than any very small sort of tiles, such as plain tiles would do. But instead of using tiles 12 inches square and $1\frac{1}{2}$ inch thick, like the common paving tiles of this country, they should be made 15 inches long 12 inches wide and about $1\frac{1}{4}$ or $1\frac{3}{8}$ inch thick only, so that 4 courses might be used for the steps of the principal staircase, and 5 courses for those of a second rate staircase of the same house, the latter of which are always made higher and also usually narrower than the former. And in order to break joint properly, it would be better to make a proportion of half tiles of the same length but only 6 inches wide, than to cut whole tiles in two, for this purpose.

(328) It is to be observed, that we purposely subjected our experimental steps to a more severe trial of strength, than they would be exposed to in a building, by giving them no support any where excepting at the end in-

serted into the wall, whereas in a real staircase of this description, every step has a partial support, throughout its whole length, inasmuch as the front of it rests upon the back of the upper surface of the step immediately below it, as shown in the annexed figures, of which Fig. 1 represents rectangular steps, such as those, which we experimented upon, whilst Fig. 2 represents the form that has been of late more usually adopted in ornamental stone staircases, in which a considerable proportion is wanting of the back of each rectangle of the former construction, both figures being longitudinal sections of a portion of staircase. This must of course make the second arrangement considerably weaker than the first, but long experience has shown that steps of good stone thus formed are strong enough. It would not however be proper to make artificial steps of tiles and cement in this second form, which would render the construction of them troublesome and complex, and would weaken them too much.

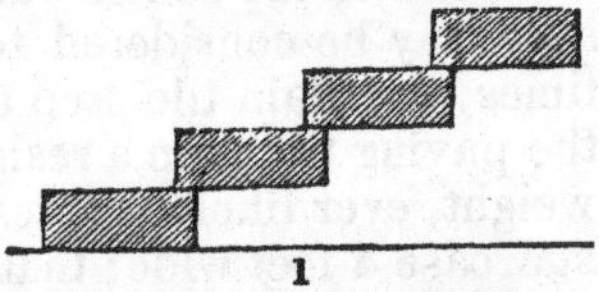

1

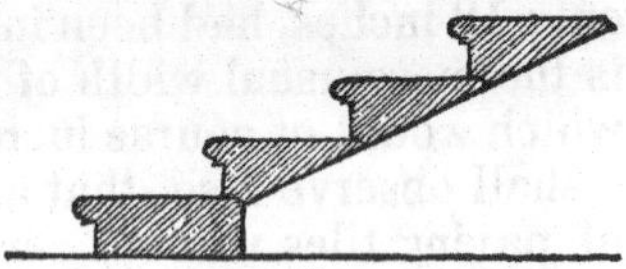

2

(329) Instead of plain tiles, long thin stones such as schists, or coarse slates not good enough for the roofs of buildings, might be used for the same purpose of forming artificial steps, when united by cement and strengthened by hoop iron bond; but in all materials not before tried, it would be proper to make an experimental step beforehand to ascertain the most suitable dimensions of the parts, and the best mode of breaking bond in putting them together. About three months should be allowed for the cement to set, before such steps are let into the walls of a building.

EXPERIMENTS ON THE COMPARATIVE RESISTANCE OF SQUARE PRISMS OF EQUAL SIZE, PARTLY OF SOLID YORKSHIRE STONE, AND PARTLY OF SMALLER PIECES OF THE SAME STONE UNITED BY CEMENT, WITH AND WITHOUT HOOP IRON BOND.

(330) First. We prepared 5 pieces of Yorkshire stone in the form of square prisms, each 24 inches long and 4 inches square, as represented in the first figure annexed, which we broke down by weights suspended from a blunt iron knife edge laid upon the center of the top of each, both ends of the

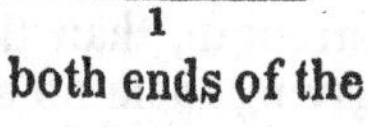

1

experimental prism being supported at the clear distance of 18 inches between the bearings; when their resistances estimated by the breaking weights proved to be 2832, 2758, 3067, 2855 and 2815 lbs, respectively, so that their average resistance was 2865 lbs, the fractures produced being clean and nearly vertical, and also nearly in the center of each stone.

(331) Secondly. We repaired the same prisms by cementing together the fragments of each, by pure or nett cement, as represented in the second figure annexed, and when we thought the cement old enough, to have acquired a respectable degree of resistance, we broke down the whole of these prisms in the same manner as before, by weights of 161, 317, 375, 255 and 284 lbs, respectively, so that their average resistance was only 278 lbs, or scarcely one tenth of that of the solid stone; but this mode of fixing pieces of stone exposed to pressure in the form of beams, is such as would never be adopted in practice, and we only tried it from having the fragments of the prisms before experimented upon in our possession, and not wishing to throw them away, without doing something with them.

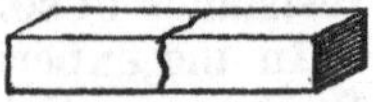

2

(332) Thirdly. We prepared 5 prisms of the same dimensions as the former, each of which was formed by building together five pieces of the same stone, all 4 inches wide and rather less than 2 inches thick, 3 of which were nearly 8 inches long in one course, and 2 of them nearly 12 inches long, in another course, which we united with nett cement, so that there were one horizontal joint and three vertical joints of cement in each prism, as shown in the third figure annexed. On experimenting upon these in the same manner before described, the weights required for breaking them down were 367, 480, 470, 317 and 434 lbs, respectively, their average resistance being therefore 414 lbs. In the first and third trials the prisms were placed with the single vertical joint uppermost, as in the figure, but in the others they were reversed by turning them upside down. Thus the average resistance of the cement joints in these five prisms was only about one seventh part of the resistance of solid stone.

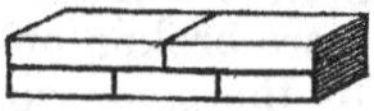

3

(333) Fourthly. We prepared 5 other prisms, built each of 5 pieces of the same stone, cemented with nett cement, precisely like the last in every other respect, but with the addition of one central piece of 1 inch hoop iron, ex-

tending the whole length of the horizontal joint. On experimenting upon these also in the same manner, they broke down under 973, 770, 1016, 625, and 979 lbs, respectively, so that their average resistance was equal to 873 lbs. In the second, fourth and fifth trials, the prisms were placed as in figure 3, with the single vertical joint uppermost: in the others they were reversed. Thus the average resistance of cement joints in these last five experiments, even with the aid of hoop iron, was scarcely equal to one third part of the resistance of solid stone.

In the experiments described in this, and in the two foregoing articles, which were all tried on the 2nd of April, 1838, the cement was 101 or 102 days old, and on examination appeared excellent, though the temperature had been very low, since the prisms, which were kept in a cold room without a fire, had been put together with it. The hoop iron in the five prisms last described was not in an advantageous position, for it agreed with the neutral axis of each, considered as a beam, being at that part which is the least acted upon, and which consequently exerts the least reaction in the breaking of the beam. In all these experiments, the cement joints, and not the stone, gave way; but in some of the last experiments, in which the cement was aided by hoop iron, one or both of the two vertical joints, when undermost, were observed to open, before the upper one appeared to suffer, when only about 2 thirds or 3 fourths of the total breaking weight had been applied; and when the upper vertical joint afterwards gave way, being the part subject to compression, transverse splinters about half an inch thick and 1½ or 2 inches wide were broken off, from one or both of the two upper stones, by the same sort of action, at the time when the prisms finally broke down.

EXPERIMENTS ON YORKSHIRE STONE, AND OTHER MATERIALS, TRIED AS BEAMS.

(334) We next experimented upon the fragments of our three experimental steps, which had all been broken off so close to the wall, that the remaining parts of each appeared to be quite perfect for more than 3½ feet in length, but we thought proper to place the piers which supported their ends in these new experiments, at the clear distance of 27 inches only, iu order that we might compare their resistance with that of three artificial stones, two years old, prepared by Mr. Ranger, for the Works in her Majesty's

Dock-yard at Woolwich, with which I had been favoured by Lieut. Denison, R. E. now acting as the Resident Engineer in that Dock-yard, under Captain Brandreth. These artificial stones, being shorter than the fragments of our steps, would not admit of greater bearings, and I had reason to believe from some former experiments, that a fair trial of comparative resistance cannot take place, except when the length of beams is equal, or nearly so.

(335) First. We broke down the remains of the Yorkshire stone step, by weights applied in the same manner as to our experimental brick beams (309), which we gradually increased until the stone gave way, under a total weight of 13,512 lbs.

(336) Secondly. We broke down the remainder of the artificial step of plain tiles and cement with hoop iron bond, by a total weight of 9497 lbs, the cement being 107 days old. It was broken vertically nearly through the middle, but on one side of this fracture the courses below the undermost hoop iron bonds, had also been broken off by irregular horizontal fractures, owing to the strain upon those hoop irons. The vertical fractures were generally through the tiles, not along the joints.

(337) Thirdly. We broke down the remainder of the artificial step of paving tiles and cement with hoop iron bond, by a total weight of 7596 lbs. the cement being 109 days old. Two vertical fractures took place, both rather irregular, one at the center of the beam, and the other over one of the points of support, and on one side, part of the lower course of tiles had been broken off by the action of the undermost hoop iron bonds, whilst on the other side, part of the top course of tiles had been broken off. There were three vertical joints of cement separated on the breaking down of the beam, all at different points. In other parts the tiles themselves were fractured. This beam was broken into a greater number of pieces than the former, one side of which had fallen nearly perfect.

(338) Fourthly. We broke down in the same manner, the three artificial stones before mentioned, each measuring 3 feet long, 18 inches wide, and 15 inches deep, by weights of 6285, 5141, and 2930 lbs, respectively, so that their average resistance was only 4785 lbs, but rejecting the last as an inferior specimen, I shall take the average of the former two, namely 5713 lbs, which probably would be the true average of this sort of artificial stone, if made of thoroughly slaked lime, which would not be liable to occasional failures,

like Mr. Ranger's system of using unslaked lime, which I have already objected to, for the reasons stated in from Article 203 to 209, inclusive.

COMPARATIVE RESISTANCE OF YORKSHIRE STONE, OF TILES AND CEMENT WITH HOOP IRON BOND, AND OF RANGER'S ARTIFICIAL STONE, DEDUCED FROM THE ABOVE EXPERIMENTS.

(339) By a popular rule which has been laid down, both by Dr. Hutton and Professor Barlow, and which I believe, when length is nearly the same, and other dimensions not very unequal, to be correct enough for practical purposes, the resistance of beams supported at both ends is as the breadth and the square of the depth, directly, and as the length measured by the clear distance between the bearings, inversely.

Hence on reducing the actual resistance of each, as determined by the experiments recorded in the preceding articles, by calculation, to what it ought to have been according to the above rule, had all these experimental beams been of the same dimensions as the Yorkshire stone beam, it will be found, that whilst the resistance of the Yorkshire stone is still measured by 13512 lbs, as before, that of the plain tile beam will be 7889 lbs, that of the paving tile beam will be 9443 lbs, and that of the artificial stone beams will average 1031 lbs.*

Consequently, in these experiments, if we take 8666 lbs, the mean of the resistances of the plain tiles and paving tiles, as a fair proportion for both, because the plain tile beam, though stronger was more shattered, it appears that the resistance of Yorkshire stone as a beam exceeds that of a

* Because the proportional resistances of the Yorkshire stone beam, of the plain tile beam, of the paving tile beam, and of each of the artificial stone beams, the length of all being equal, would be as the numbers 731, 880, 588 and 4050, if they had all been of the same sort of material, these numbers being equal to the breadth, multiplied by the square of the depth of each in inches, according to the above rule.

But their actual resistances, ascertained by the foregoing experiments, and due to the dimensions of each beam, and to its actual strength combined, were as the numbers 13512, 9497, 7596, and 5713 lbs, respectively. In order therefore to obtain their true comparative resistances, due to the strength of the material alone, we must increase the resistance of the paving tile beam from 7596 to 9443 lbs, this being in the proportion which 588 bears to 731, it being smaller than the Yorkshire stone beam, and we must diminish the resistance of the plain tile beam from 9497 to 7889 lbs, and the average resistance of the artificial stone beams from 5713 to 1031 lbs, that is in the proportions which the numbers 880 and 4050 respectively bear to 731, these two last beams being larger than the Yorkshire stone beam, to the standard of which the others are all to be reduced.

beam of equal dimensions composed of tiles and cement with hoop iron bond by about 56 per cent; but that the resistance of Yorkshire stone is about 13 times greater than that of Ranger's artificial stone. This agrees very nearly with our former experiments on small prisms of 2 inches square, and 3 inches between the bearings, by which the resistance of Yorkshire stone was 2887 lbs, (Table VIII), whilst the best of the artificial stone prisms made by us on Ranger's principle, and one year old, was only 209 lbs, (Table VI).

Remarks. Why the comparative resistance of plain tiles considerably exceeded that of paving tiles as a step for a geometrical stair case, I have already attempted to account for; but why the former did not also exceed the latter in the same proportion, instead of being rather inferior to it, when both were tried as beams; and why the resistance of both more nearly approached to that of Yorkshire stone as beams, than as steps, I shall not attempt to explain. Perhaps more numerous experiments might remove the incongruity, which now seems to prevail between these two sets of experiments. It is sufficient, however, for practical purposes, to have proved that tile and cement steps, with hoop iron bond, are quite strong enough for the steps of geometrical staircases, and though this sort of construction is certainly stronger than bricks and cement, aided by the same sort of bond, yet the latter has been fully proved by Mr. Brunel's experimental brick semi-arches and beam, and by Messrs. Francis's experimental brick beam, on a great scale, as well as by my own smaller beams of the same description, to possess a resistance far exceeding any force that could possibly be brought to bear upon brick and cement beams thus bonded, in the practice of Architecture; it being well known, that when great weights such as those of Steam Engines, of shafts bearing heavy machinery, &c, &c, are laid upon masonry or brickwork, they are never placed over an opening, but on a solid pier, deriving its support from the foundations of the building. In this case the materials composing the pier, whether of masonry or of brickwork, whether laid in mortar or cement, are strong enough, provided that the great weights, to which by supposition they are exposed, are not sufficient to crush them. In short they are not exposed to a breaking force like beams, but to a crushing force, and the resistance of bricks against this last estimated power has been estimated at more than

500 lbs, to the superficial inch,* so that a brick pier of only one superficial foot would exert a resistance against such a force, of not less than 864000 lbs. or nearly 400 tons.

FINAL REMARKS UPON THE RESISTANCE OF CEMENT JOINTS. THAT OVER OPENINGS THEY REQUIRE THE AID OF HOOP IRON OR OF SOME SIMILAR BOND: BUT THAT IN ALL OTHER CASES, THOUGH INFERIOR TO STONE, THEY ARE MUCH SUPERIOR TO ANY WEIGHT OR STRAIN THAT CAN POSSIBLY ACT UPON THEM.

(340) Our experiments with the first and second brick beams (310 and 311) sufficiently prove, that cement joints over openings are scarcely strong enough, if deprived of their centering soon, as in the first, and hoop iron bond which multiplied their comparative resistance tenfold, as in the second of those experiments, is so very cheap, and gives so little trouble in the workmanship, that it would be very injudicious to omit it under any circumstances, where it could be useful, especially as it supersedes the more expensive and precarious expedients of timber bond altogether (315).

By analogy, as several of our last experiments proved, that cement joints on a great scale have more resistance, in proportion to stone, than on a small scale, it is probable that the actual resistance of cement joints used for the real purposes of practical Architecture might be more than one seventh part of that of Yorkshire stone, which it appeared to be by our experiments with the small prisms described in Articles 330 and 332. This resistance however is greater than that of sound dry pure chalk which is only one eighth part of that of Yorkshire stone, and of well-burned but inferior bricks which is only one ninth part of that of Yorkshire stone, as was proved by our former experiments (See Table VIII, Article 212).†

* In his Paper on the strength of Materials in the Philosophical Transactions for 1818, Mr. George Rennie states, that he found by experiment, the crushing force for cubes of 1½ inch of chalk to be 1127 lbs, and for brick to be 1265 lbs, which on being reduced, are equivalent to a crushing force of 500 lbs, to the superficial inch for chalk, and of 561 lbs, for bricks.

† The terms well-burned and inferior inadvertently used by me in that Table are not appropriate. The whole of the bricks for our experiments were selected from the brickfields in the neighbourhood, and were well burned and sound to all appearance, such as no Architect could have rejected. And we did not know that any of them were inferior to the others, until several of them had proved so much weaker than our cement joints, that we experimented upon their own resistance by grinding them down to small prisms and breaking them. Those that proved the strongest were entered in the Table as well burned bricks, which ought to have been entered as

(341) Now since cement is stronger than block chalk of the soundest quality, such as we experimented upon, which every one will allow to be capable of bearing the greatest weights, and as it is stronger than at least one-third of the bricks used in the buildings of England; and as there is no instance on record of the bricks in any part of a building having ever been crushed by the greatest weights placed on them, or even having ever been cracked or fractured, excepting by unequal settlements on bad foundations; it follows that cement joints though inferior to stone possess a resistance far superior to any force or weight that can possibly act upon them in the solid parts of walls; and this applies not only to pure or nett cement, but to the joints of cement mortar made with equal measures of cement powder and sand, which from the observations I have made upon old brickwork built with this mortar and broken down afterwards, cannot possess less resistance than the bricks themselves, as the latter were more frequently fractured than the joints.

GENERAL OBSERVATIONS ON THE FOREGOING EXPERIMENTS FOR ASCERTAINING THE RESISTANCE OF STONE, AND OTHER BUILDING MATERIALS.

(342) On considering the results of our experiments on the resistance of small prisms, measuring 2 inches square, at an interval of 3 inches between the bearings, with those made upon larger prisms measuring 4 inches square, at an interval of 18 inches between the bearings, and finally with those made upon larger stone beams, brick and cement beams, or tile and cement beams, measuring from 7 to 12 or 15 inches in depth, and from 12 to 18 inches in width, at intervals varying from 27 to 120 inches between the bearings, we found, after determining the proportional resistance of any two different materials such as Yorkshire stone and concrete, to be as 13 to 1, by the small prisms, that the same proportional resistance also held good or nearly so, on experimenting upon these substances on a much larger scale, as beams, of which I gave an example in Article 339, but on comparing the actual resistances of

superior bricks. Those that proved the weakest were entered as inferior bricks. The whole were well burned, without a single *place brick* or soft under-burned brick amongst them. We suspect the inferiority of some to have been owing to the processes, preparatory to burning, having taken place at too late a period of the season; but the same thing happens every year, so that better bricks than we used cannot be procured.

one and the same material, such as Yorkshire stone, as ascertained by experimenting upon the very small prisms, the intermediate prisms, and the comparatively large beam of that material, I find it quite impossible to apply any general rule, given by former writers, so as to calculate the one from the other, with any degree of accuracy. The popular rule for example, before quoted, entirely fails, as applied to these extreme cases; so that no person who knew that the average resistance of a prism of Yorkshire stone 2 inches square, tried at 3 inches between the bearings was 2887 lbs, could possibly approximate to the actual resistance of the beam of the same stone, 13 inches wide and 7½ inches deep, tried at 27 inches wide between the bearings, which by calculating according to the above rule, he would estimate at 29321 lbs,* whereas its real value by experiment was only 13512 lbs. And in calculating by the same rule the resistance of the small prism, from having that of the beam given, he would arrive at a conclusion no less inaccurate. I do not pretend to be able to deduce any more satisfactory rule myself from those experiments, although I have no doubt that it might be done, if a greater number of experiments with prisms and beams of the same sort of stone were tried, increasing their scantling gradually from 2 inches square to larger dimensions even than those tried by us, and varying also the intervals between the bearings of each several sort, respectively, instead of trying each sized prism or beam at only one and the same interval.

(343) That the difficulties of the subject now under discussion, and which my own experiments were not numerous enough to remove, may be understood, I have inserted the results of the experiments described in the foregoing Articles from 325 to 338, inclusive, in the form of a Table, in which I have included others, although already recorded in former Tables, in order to bring the whole more clearly under notice, in one view.†

* Because as the square of 2 multiplied by 2 and divided by 3, is to the square of 7½ multiplied by 13 and divided by 27, so is 2887 to 29321 nearly.

† Experiment 14 compared with 17, and experiment 16 compared with 18, in this Table appear to prove, that the comparative resistance of beams of the same material and scantling is not as has been supposed, in the inverse ratio of their length measured between the points of support, but in some different ratio; because in estimating by this rule the resistance of the long beam from the actual resistance of the short beam given, it makes it too much, and vice versa on estimating the resistance of the short beam from that of the long beam given, it makes it too little.

TABLE XIX. COMPARATIVE RESISTANCES OF YORKSHIRE STONE, BRICKS AND TILES, AS SMALL PRISMS, AND OF THE SAME SUBSTANCES UNITED BY CEMENT, WITH AND WITHOUT HOOP IRON BOND, AND ALSO OF CONCRETE AND MORTAR, AS PRISMS, BEAMS, &c.

Description of Prism or Beam, with reference to some of the foregoing Tables or Articles.	Dimensions in inches.			Age of Cement or Mortar, in days.	Resistances or breaking weights in lbs.	Whether cracks were observed before breaking down.
	Width.	Depth.	Length between the Bearings.			
Small Prisms.						
1. Yorkshire Stone, Table VIII	2	2	3	..	2887	
2. Superior Bricks ......(VIII)	2	2	3	..	752	
3. Inferior Bricks ......(VIII)	2	2	3	..	329	
4. Plain Tiles ....(XVIII)............	2	2	3	..	1168	
5. Paving Tiles ....(XVIII)............	2	2	3	..	1124	
6. Concrete, the best of Table VI............	2	2	3	457	209	
7. Rosehill lime Mortar, the best of Table VI..	2	2	3	342	210	
8. Blue lias lime Mortar.. (XII)	2	2	3	17	157	
Larger Prisms.						
9. Yorkshire Stone ....(330)	4	4	18	..	2865	
10. Do. in 2 pieces with 1 vertical Cement joint (331)	4	4	18	102	278	
11. Do. in 5 pieces with Cement joints ....(332)	4	4	18	101	414	
12. Do. in 5 pieces with Cement and hoop iron bond .... (333)	4	4	18	100	873	Cracked.
Beams.						
13. Yorkshire Stone ....(335)	13	7½	27	..	13512	
14. Bricks and Cement ....(310)	18¾	12	120	50	498	Cracked.
15. Bricks, Cement and hoop iron bond ..(311)	18¾	12	120	50	4523	Cracked.
16. Halling Lime Mortar and hoop iron bond (313)	18¾	12	120	50	742	Cracked
17. Bricks and Cement ....(316)	18¾	12	36	50	2356	
18. Halling Lime Mortar and hoop iron bond (317)	18¾	12	26½	66	5147	Cracked.
19. Plain Tiles, Cement and hoop iron bond (336)	16¾	7	27	107	9497	Cracked.
20 Paving Tiles, Cement and hoop iron bond (337)	12	7¼	27	109	7596	Cracked.
21. Ranger's Artificial Stones or Concrete..(338)	18	15	27	2 years	5713	
Geometrical Steps, projecting 4 feet.						
22. Yorkshire Stone ....(325)	13	7½	..	..	5514	
23. Bricks, plain Tiles and Cement ......(322)	14½	7¼	..	40	540	
24. Bricks, Cement and hoop iron bond ..(323)	18¾	12	..	68	3560	
25. Plain Tiles, Cement and hoop iron bond (326)	16¾	7	..	105	2396	Cracked.
26. Paving Tiles, Cement and hoop iron bond (327)	12	7¼	..	107	1263	

THAT IN WHARF WALLS HAVING COUNTERFORTS AND BUILT WITH CEMENT MORTAR, THE EXTRAORDINARY ADHESIVENESS OF THE CEMENT INSURES THE STABILITY OF THE PROFILE.

(344) Though the resistance of cement joints, even when three months old, was proved by our last experiments to be considerably less than that of Yorkshire stone, yet our former experiments, and especially that with the two large Bramley fall stones, recorded in Article 216, proved

that in adhesiveness it was equal to one of the best building stones of this country; and the same experiment even proves, that for fixing cantilevers in walls, the resistances of cement, used as plugging to prevent the ends of the cantilevers from moving in the holes, is equal to the cohesiveness of the particles of the same excellent stone, or nearly so; for the only cantilever out of four, that was displaced in the above experiment was that, which after having, like the others, been loaded with more than 7000 lbs, broke off not only the cement by which it was fixed in the hole, but part of the stone itself (217).

Hence in wharf walls, built with cement mortar, in the usual proportion of equal measures of cement and sand, of which the adhesiveness cannot be less than about two thirds of that of solid stone, as may fairly be inferred from the above remarkable experiment (216,) compared with those recorded in the last column of Table XI (243,) it is quite impossible that the wall, if constructed with counterforts, can ever separate from the latter. The whole must necessarily adhere together in one mass, and therefore the wall, even though unequal in itself alone to resist the pressure of earth acting upon the back of it, and tending to force it forward at top out of its proper position, derives full benefit from the counterforts, which the same pressure of earth tends to keep in their proper places, because it acts not only in rear, but also on both sides of each counterfort, and whilst even the former pressure is not entirely unfavourable, the latter is all in favour of stability.* Thus the pressure of earth upon a wall with counterforts is precisely similar to the forces exerted in our experiments upon a couple of bricks or stones, connected by cement or mortar joints. One stone was fixed, and so is each counterfort; whilst the other was pulled away from it, by the weights acting upon it, and so is the wall, which the pressure of earth tends to separate in like manner from its counterforts, which separation is counteracted solely by the adhesiveness of the mortar; and this as I again repeat, if made of cement and sand in the usual proportions, is far superior to the pressure acting upon it.

* According to the former generally received Theory of the Pressure of Earth upon Revetments or Retaining Walls, laid down by Belidor, Dr. Hutton and others, the whole of the pressure of loose earth on the back of the wall was supposed to tend to overset it. Doubting the accuracy of this theory, I tried in 1816, a number of experiments recorded in my Course of Elementary Fortification, which proved it to be completely erroneous; because, as I suspected, part of the above pressure always strengthens the wall.

THAT NEITHER WEAK HYDRAULIC MORTAR, NOR COMMON CHALK LIME MORTAR HAVE ADHESIVENESS ENOUGH, TO PREVENT THE PRESSURE OF EARTH UPON A WHARF WALL, SUBJECT ALSO TO THE ACTION OF THE TIDES, FROM GRADUALLY SEPARATING IT FROM ITS COUNTERFORTS. EXAMPLE OF A FAILURE OF THIS DESCRIPTION AT CHATHAM.

(345) From the favourable reports of puzzolana, I have no doubt that this also will have the power of preventing the injurious effects of the pressure of earth, explained in the preceding article; but experience has proved, that in wharf walls built with weak hydraulic limes, the adhesiveness of the mortar is not sufficient to resist that pressure, as for example in the wharf wall of the old dock at Hull, which separated from its counterforts in the manner that has been alluded to, as was before mentioned in a quotation from Mr. Timperley's description of the works at that port (265 and 266).

A similar failure has recently taken place in the wall of the old Gun Wharf at Chatham, which was faced with Purbeck stone, said to have been laid in trass mortar, with a backing and counterforts of brickwork laid in common chalk lime mortar, and strengthened about 4 or 5 feet from the top by a continued longitudinal bond timber, and transverse land ties also of timber, one passing through each of the counterforts, and at the extreme top by a strong longitudinal capping of timber, connected not only with transverse land ties sunk into the earth in rear, but also with the fender piles in front of the wall.

I have seen an official drawing, of 1789, evidently in reference to a project for the completion of this wharf, a section contained in which very nearly agrees with the present profile of the wall, which is known to have been finished more than 40 years,* and I have ascertained by inquiry, that no appearance of failure was ever noticed until the year 1825, when part of it was observed to have bulged a little forward, but no material change took place until some years afterwards, when a substantial granite coping was laid in front of the wall, as a substitute for the timber capping and land ties, which had become rotten. The weight of this coping, which could have done no harm, had the mortar

* From this drawing one might infer, that a brick wall with counterforts, either finished or perhaps only in progress existed in 1789, which it was proposed to face with stone. But this is conjecture, as I have never seen any document explanatory of the drawing alluded to.

of the brickwork been good, has undoubtedly accelerated the separation of the front of the wall from the counterforts, which action has been gradually in progress, but exerted itself more powerfully, as soon as the bond timber and lower row of land ties also became rotten.* The following figure represents the state of that portion of the old Gun Wharf wall, on being laid open for examination, which has given way in the manner alluded to, the front of it having gradually bulged out irregularly, until it has been forced forward at top, not less than 2 feet out of its original position, whilst a chasm of 15 or 18 inches in width has been formed between the front and back of the wall, the latter having adhered to the counterforts, but not firmly; for all the brickwork of both is shattered, and cracked at top, owing to the weakness of the chalk lime mortar; the whole of which is now in an almost fluid state, communicating no more strength to the brickwork, than an equal quantity of mud would be capable of doing; in consequence of the water from the river having forced its way through at every tide, and kept the whole of the backing in a wet state, in spite of the stone and trass-mortar facing, before described. At the level of the foundation, it is supposed that the brickwork of the wall and counterforts may still remain in close contact as at first.

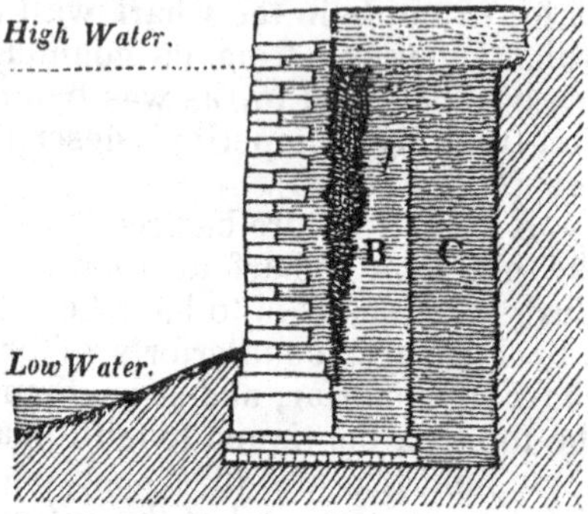

(346) The circumstance of this wall having remained perfect for at least 27 years after it was finished, and of some parts of it still remaining so, may be considered a proof, that the profile was sufficient if better mortar had been used, for the wall, which was about 24 feet high and had offsets or footings at bottom, had an exterior slope of one tenth of its height, and would have been 6½ feet thick at top,

* To guard against this evil, chain cables or strong iron bolts or bars have recently been used by the Engineers of the present day as land ties, for wharf walls. These are particularly necessary in wharfs faced with iron, which has very little stability in itself, and must therefore be aided by long land ties running through the backing of the wharf, and well secured to some immoveable objects in rear.

if the back of it had been carried up vertically, instead of which its thickness was reduced to 4 feet at top, by a step in rear, about 7½ feet below the level of the ground. It had very substantial counterforts (C), measuring rather more than 6 feet square in plan, at central intervals of 18 feet, and terminating about 4 feet below the same level, that is several feet higher than the step at the back of the wall, the lower part of which by being thicker was in itself a sort of counterfort to the upper part of it, in consequence of which the front part of the wall being the thinnest and less capable of resisting the pressure of earth in rear, separated from the back part (B) in rear of the said step, whereas had the whole back of the wall been carried up vertically, the separation resisted by a greater mass of brickwork, would have been less considerable, and would have taken place farther back, entirely behind the back of the wall, and in front of the counterfort.

(347) Our experiments on chalk lime mortar 17 days old, two joints of which measuring 40 superficial inches required weights of 308 lbs, and of 283 lbs, respectively, to tear them asunder, as recorded in Table XIII (277), proved that this sort of mortar has a respectable adhesiveness at first, which appears to be fully confirmed by the fact, that the wall now under consideration remained in a sound state for so many years, which the land ties alone, though a very powerful aid, could not have insured, had the mortar been as bad then as it is now, when its adhesiveness is so very trifling, that any two bricks could be separated by the hand of an infant. The same circumstance seems to occur, in respect to the weak hydraulic limes, for the like failure of the wharf wall of the Old Dock at Hull before alluded to, does not appear to have been immediate, but progressive during a long period of years. This is not precisely what one might have anticipated from any lime having hydraulic properties, such as the Warmsworth lime used in that work, the mortar of which might have been expected gradually to set instead of becoming weaker, in the course of thirty or forty years; and in fact a great part of that mortar did attain a respectable state of induration, but as other portions of it were soft, and the separation of the wall from the counterforts took place in spite of such partial induration, this failure sufficiently proves the propriety of the maxim before laid down by me in consequence, that not even hydraulic limes of moderate energy are to be trusted for the mortar of wharf or dock walls built with counterforts.

(348) To return to the old Gun Wharf wall at Chatham, the great error in the construction of it has undoubtedly been the use of common mortar for the whole of the brick-work, which error however was rather the error of their age, than of the Engineers, who may have given designs for the commencement and completion of this work, the most recent of which was undoubtedly drawn up half a century ago, that is two years before Smeaton published the results of his admirable researches on the properties of water cements, and seven years before the qualities of the Sheppy cement stone had been discovered. Hence the failure, that I have just recorded, is no disparagement either to those Engineers or to their superiors, the then Master-General and Board of Ordnance; notwithstanding which, I would have preferred not bringing it under notice in this treatise, if failures did not afford the most instructive examples in Practical Architecture; and if this did not also prove in the strongest light the necessity, which I have so often before inculcated, of never using chalk lime at all, in any masonry or brickwork exposed to the action of water, even for the backing; for sooner or later it will be sure to fail, as has been proved likewise in various parts of the wharf walls of Chatham Dock-yard, which I shall not enlarge upon, one example being enough.

(349) It has occurred to me on reflecting on the above, that wharf walls of a sufficiently substantial profile without counterforts, have the advantage of not being liable to this sort of separation of the parts, even if the mortar should be of inferior quality: but I am not aware that any of the Civil Engineers of this country have ever dispensed with counterforts, although it has been done in the wharf wall of the Royal Arsenal at Woolwich, and in that of the New Gun Wharf at Chatham, built under the superintendence of the Royal Engineer Department.

(350) *Remarks on Trass. Its efficiency as an Ingredient for Hydraulic Mortar, when judiciously used. That Chemical Analysis proves it to be similar to, and almost identical with Puzzolana.*

I have hitherto said little of trass, not having had an opportunity of examining it personally until lately, when I inspected a brick wharf wall in Woolwich Dock-yard, said to have been built about fifty years ago, with mortar composed of 1 measure of Dorking lime, 2 measures of sharp river sand and ½ measure of trass,* which was then

* These proportions are stated on the authority of the Master Bricklayer of that Dock-yard, communicated to me through the resident Engineer.

being partly pulled down, in consequence of an extension of the Dock-yard, by building a new concrete wall in front of it, which was in progress at the same time. The mortar of this old wall was excellent and every where very hard, except in a small portion immediately under a culvert, and consequently exposed to the action of running water, where it was softish. I have also seen a facing of trass mortar of about 9 inches for the brickwork of a wharf wall in Chatham Dock-yard, which on pulling down the wall proved to be equally good. The circumstance of the trass mortar growing in the joints remarked by Smeaton, and also by me in the facing of the old Gun Wharf wall, was not observable in either of these walls. Mr. Stevenson who used trass for a short time, for want of puzzolana, on commencing the Bell Rock Lighthouse, and also General Treussart in France, both give this substance an excellent character, though they do not consider it quite equal to puzzolana, and they do not notice the defect observed by Smeaton, which from having more experience and information on such subjects, than when I first read his observations on water cements, I now apprehend is not peculiar to trass, but would take place with puzzolana also, if that substance were mixed like the trass mortar of Smeaton's time, with such an injudicious excess of common chalk lime, as twice its own bulk. This proportion of puzzolana would just be sufficient to prevent the lime from dissolving altogether, and being washed entirely away, but not from forming a sort of incrustation gradually growing out of all the joints, like the stalactites from the joints of damp vaults constructed with the mortars even of hydraulic limes (55), but which never occur with cement mortar. In fact the chemical component parts of puzzolana and of trass are so very nearly alike, as will be seen by referring to Article CXXXV in the Appendix, where the analysis of these important substances is given, that the one cannot possess any property not common to the other also.

DESCRIPTION OF THE METHODS ADOPTED IN UNDERSETTING THE FOUNDATIONS OF THE NEW CUSTOM-HOUSE AT LONDON, AND OF A STOREHOUSE IN CHATHAM DOCK-YARD. THAT THE FORMER HAS SUCCEEDED, AND THE LATTER FAILED, TO A CERTAIN DEGREE. PROBABLE CAUSE OF THE DIFFERENCE IN THESE RESULTS.

(351) The New Custom-house of London, a handsome and I believe convenieut building, begun in July 1813 and finished in the latter part of 1817, creditable to the Archi-

tect who designed it in other respects, was built on a piled foundation; badly planned, inasmuch as the woodwork over the heads of the piles was not placed sufficiently low, in reference to the low water of spring tides, to be at all times in a wet state; and still worse executed by the Contractor, who did justice to all other parts of his contract, excepting this, which unfortunately was the part that could the least admit of being trifled with.* Hence in the year 1825, scarcely eight years after the building was finished, the imperfect foundations having in the mean time been gradually giving way, part of the floor of the large apartment called the long room, which was supported on groined arches, fell down, luckily without loss of life, as it occurred at an early hour, before it was crowded with persons on business, although the clerks were seated at their desks, which were arranged along the walls. Gradually increasing cracks announcing settlements, the usual symptoms of an inefficient foundation had previously appeared, and the whole of the walls were now considered in great danger.

Sir Robert Smirke being called in at this period, to remedy the defects in the construction of the former Architect, pulled down the whole of the shattered floor and ruinous vaults, and removed their foundations also, and at the same time he pulled down a part of the adjacent front wall of the Custom-house, facing the Thames, which seemed in great danger; and having cleared out the whole space under this wall as well as under the long room, until he came to a natural bed of sound gravel, he filled it in with a mass of concrete, upon which he rebuilt the portion of the front wall before mentioned, as well as the long room, supporting the

* As the heads of the piles, and the woodwork over them on which the walls were built, were thus liable to be alternately wet and dry, or wet and damp, they must necessarily have decayed in time, even had they been of proper dimensions and driven deep enough into the natural ground, which was not the case; for although the Architect, in a book published by him describing this edifice, which I perused some years ago, declared that the piles were all of proper diameter, and generally driven, if I recollect rightly, to the depth of about 24 feet; it was found afterwards, on taking them out, that they were what may be termed sham piles, the average length of 2378 of those piles being only 10 feet 1 inch, and the length of each varying from 3 feet 6 inches upwards, and many of them being so crooked, especially of the shorter ones, that they resembled commas. Moreover the clumps of 9 piles each, which were intended to support the square piers of the groined vaults under the long room, were not even properly placed for that purpose, inasmuch as the brickwork of each pier had not a fair bearing on the whole of the piles below it, but upon a part of them only. In short these piles were so bad, that until I saw them in the state, which has been described, I could scarcely believe the reports, that I had heard of the very defective nature of this foundation, though by no means exaggerated.

new floor of the latter upon iron columns and girders. In respect to the remaining walls of that extensive building, he decided upon undersetting the whole of them with concrete, and brickwork laid in cement, in a new and very ingenious, and as the result has since proved, in a most judicious manner. Having already noticed this operation in the second note to Article 27, ascribing the acknowledged success of it chiefly to the concrete; and having at the same time mentioned that Mr. Ranger had underset the walls of a Storehouse in Chatham Dock-yard, the foundations of which were giving way with concrete alone, in a very simple and yet efficient manner, as I then believed, and more economical than the former; I should not have brought the subject again under the notice of my readers, if the latter method, which was carried into effect in 1834, had not subsequently failed, which I discovered by mere accident in November 1837,* not having entered that Storehouse in the mean time, as I had no expectation of such failure taking place. Hence it is of importance to point out the circumstances, in which the undersetting or underpinning of the New Custom-house of London differed from that of the Storehouse in Chatham Dock-yard.

In both cases the work was executed in small portions at a time, the brickwork of the original foundation of each of those buildings, being laid open by excavating under it, until a solid substrature was found, which under the Custom-house was a bed of gravel, and under the Storehouse either compact clay or gravel or solid chalk.

(352) *Sir Robert Smirke's Mode of undersetting the Walls of the New Custom House.*

In this operation, which was done in lengths of 10 feet at a time, a space of at least 12 feet in width was cleared out, and substantial transverse timbers were used at intervals of 10 feet to underpin the walls, which were supported in the center only, by strong square posts. The concrete, which consisted of 1 measure of pulverized Dorking quicklime mixed with 7 or 8 measures of Thames ballast, was then thrown in, filling the whole width of the excavation, which when it rose to the proper height was covered with a course of large Yorkshire landing stones, each 6 inches

* An Officer of Chatham Dock-yard having offered to show me some chain cables and buoys, which we had been speaking about, and these happening to lie behind the Storehouse alluded to, I observed whilst looking at them, a crack in the back wall, apparently a new one, which on examination and inquiry it proved to be, and afterwards on entering the building, I found strong proofs of a general sinking or settlement of the walls having taken place.

thick, and averaging 25 superficial feet, the top of which was just 3 feet lower than the bottom of the original brickwork of the walls, which had previously rested upon the woodwork of the piled foundation. This space was filled up with 12 courses of brickwork laid in cement mortar, and wedged up close to the bottom of the original brickwork, with hard materials such as pieces of tiles and slates. This brick and cement work, interposed beneath the original walls and the Yorkshire stones and concrete, was considerably wider than the former, and consequently built with footings or small offsets on each side.

(353) *Messrs. Taylor and Ranger's Mode of undersetting the Storehouse in Chatham Dock-yard.*

In this operation, which was done by small lengths of about 6 feet at a time, but without using any temporary supports, the whole building was underset with concrete alone, composed of 1 measure of Halling quicklime ground and sifted, and 6 measures of Thames ballast, mixed with boiling water according to Mr. Ranger's system, who executed the work himself under the direction of Mr. Taylor then Architect to the Admiralty. The concrete was thrown down in troughs, and rammed in between two sets of boards, placed at such a distance apart, as to suit the bottom of the original brickwork, which was 7 feet thick. These boards were removed when the concrete was finished, so that the whole space excavated, about 14 feet wide, was not filled in with concrete, as at the Custom-house. When the concrete was brought up to within about 1 foot of the bottom of the original brickwork of the wall, it was pressed up against it by a simple and ingenious apparatus of frames and screws, resting on a course of slating, until the whole space under the original foundation of the brickwork, and between that and the solid ground below as aforesaid, was filled with a mass of partly rammed and partly pressed concrete, of which the width at bottom rather exceeded that of the original brickwork, but coincided with it at top.

Whilst Mr. Taylor underset the greater part of the walls in the manner that has been described, he found it necessary to pull down a part of the South-west end of the front wall of the Storehouse, facing the Anchor Wharf, which he rebuilt with cement mortar, upon a concrete foundation, of the same description and dimensions, as that which he introduced under the old walls of the building. When this was done, he caused the numerous cracks in the walls to be stopped and pointed, loose bricks to be taken out and replaced, and several of the flat arches over windows, &c, to be rebuilt; after which every portion of the inside, where those repairs had taken place, was washed with Halling lime, which having a brownish tint, served to distinguish those parts from the remainder of the walls, which had been previously whitewashed with chalk lime. On the outside, the white colour of the new mortar renders all the repaired cracks, or new brickwork put in here and there, no less conspicuous, so that they cannot be mistaken for the original brickwork, of which the joints are of a dark colour. These repairs were done in 1834, and I am sorry to say, that in my recent examination of the same Storehouse, I found that several new cracks had taken place in the walls, two of which are in the brick and cement portion of the front wall, built by Mr. Taylor himself, after he had pulled down the old brickwork, which was in too dangerous a state to underpin. In one of the new flat arches over windows, a brick has dropped down out of its proper place, and in short there are indications not to be doubted, of the whole of the walls having settled, since the concrete foundation was introduced, to what degree it is impossible to say, but I estimate this new settlement as probably amounting to at least 2 inches.

(354) *Remarks on those two different Methods of undersetting the Foundations of the Walls of Buildings.*

On reflecting upon the successful result of the mode of undersetting the Custom-house, contrasted with the failure of that afterwards used in the Storehouse in Chatham Dock-yard, it appears to me that the brickwork laid in cement introduced by Sir Robert Smirke beneath the whole of the original footings of the walls, acted as a great brick and cement beam, which by having the additional bond of the large Yorkshire landing stones, between it and the concrete, has equalized the pressure of the walls of the building upon every part of the concrete, so that at the end of more than 10 years, no crack or other symptom of settlement has yet been discovered.

In the Chatham Storehouse on the contrary, the walls press directly on the concrete, and consequently with unequal force, the solid parts or piers between doors and windows being heavier than where these openings occur, and the result seems to prove that the concrete there used was either compressed upon the solid chalk or ground below, or forced out at the sides, from its width not being much greater than the footings of the brickwork, or both. In fact a person, looking at the inside of the Storehouse now, might suppose, that it was in the same danger as before; but it is not so, because after the concrete shall have been pressed down upon the solid chalk, gravel or clay below, as far as the weight of the building may be capable of forcing it, which effect may possibly have already taken place, no further settlement will insue; and this substance being incorruptible, the same risk is not by any means to be apprehended, as in the decay of its former wooden foundation, which if neglected would certainly have led to the downfal of the building, and which on inquiry I found to have been of such a very curious and extraordinary nature, that it appeared worth while to describe it in detail in the Appendix. (See from Article XXVII to XXX inclusive.)*

THAT THE FACINGS OF CONCRETE OR ARTIFICIAL STONE FOR THE WET DOCKS AND WHARFS OF HER MAJESTY'S DOCK-YARDS AT WOOLWICH AND CHATHAM HAVE FAILED, CHIEFLY IN CONSEQUENCE OF THE ACTION OF FROST, PROVING THAT IT IS UNFIT FOR WALLS EXPOSED TO THE TIDES AND LIABLE TO COLLISIONS.

(355) When Mr. Taylor first proposed to build docks for the largest ships as well as wharfs of concrete or artificial stone on Mr. Ranger's principle, and obtained authority to carry these projects into effect, in Her Majesty's Dock-yards at Woolwich and Chatham, my previous experiments on cements and limes made me consider this construction as a very bold experiment likely to fail, whieh proved to be the case, for after some time it was found necessary to protect the docks alluded to, by facing them with granite (31). More recently the wharf walls in the same Dock-yards, which were exposed alternately to the tides of the Thames or Medway, and to the action of the atmosphere, have given way also, and if not hereafter protected by a facing of granite or of brickwork laid in cement, they will in process

* Part of the foundation of the Storehouse appears to have been so near to the natural bed of solid chalk, that it was quite unnecessary to have used timber in the first instance in preparing this part of the foundation, and I think that it would have been better to have underpinned this part with brickwork laid in cement only, without using any concrete at all. I judge by the drawings which accompany Mr. Taylor's description of the operation, which though not done by scale, induce me to believe that there could not have been a space of more than 3 or 4 feet between the solid chalk and the bottom of the brickwork of the original wall. (See the first volume of the Transactions of the Institute of British Architects.)

of time become a heap of ruins. My first experiments in 1828 and the two following years, induced me to believe that no concrete could possibly resist the action of water in mass; but others tried subsequently to the printing of the first part of this treatise, especially those recorded in from Article 199 to 202, inclusive, made me alter my opinion in favour of concrete moulded with strong hydraulic lime, as artificial stone, and allowed to set in air for a sufficient period, previously to immersion. This last more favourable impression, stated in Article 145, has since been confirmed by an interesting experiment upon one of Mr. Ranger's artificial stones, formed of a mixture of 1 measure of blue lias lime and 6 measures of ballast, tried at Woolwich by Lieutenant Denison, which had been allowed two years to set in air, before he exposed it to the rapid tides of the Thames, which he did for two months by suspending it from a ship's bottom, at the end of which time he took it out and put it into distilled water, when on testing it with the oxalate of ammonia no precipitate took place, and consequently the lime was at this time perfectly proof against the action of running water in mass.* But the late severe winter has developed what appears to be an insuperable objection to the use of concrete for wharf walls in this country, having proved its inability to resist the action of frost, upon those portions of such walls, which were alternately wet and dry. In those parts the frost has destroyed the mortar on the outside, and consequently caused the pebbles or gravel there to peel off in an irregular manner, to the depth of 2 or 3 inches leaving a rough surface, and in some places having produced large holes much deeper than the above average. Such has been the case, in the concrete wharf walls both in Woolwich and in Chatham Dock-yards, the surfaces of which have been more or less degraded by the process described, which in the former Dock-yard had begun to exhibit its injurious effects two years ago, as the concrete wharf wall there had been commenced some time before that in Chatham Yard.† The drift ice, with which both the Thames and Medway were encumbered

* Before he had subjected this artificial stone to the washing process in the current of the Thames, and whilst its surface was probably still coated with that sort of partial efflorescence which always forms at first, on the surface of calcareous mortars and cements (53), Lieut. Denison had previously found on immersing the same block in distilled water, and adding oxalate of ammonia, that a copious precipitate was produced. This does not appear to me to invalidate the inference that may be drawn, from the contrary result obtained after the two months immersion.

† See the Note to Article 145, which was printed in 1836.

during the late severe winter, may also have contributed to injure the surfaces of these walls, as it was observed frequently to shift from one shore to the other according to the wind.

(356) To compare a wharf wall of concrete, with one built with sound brickwork or with stone of good quality, it will be evident, that when the most intense frost attacks the surface of the latter, as it has no action upon the bricks or stones, it can only penetrate a little way into the joints of the mortar, probably not more than an inch in depth, so that when the joints thus injured shall have been pointed with good hydraulic mortar, the wall will be as strong as before. But in a concrete wall, of which mortar not only forms the entire facing, but about one half of the whole mass of it, for the lime and fine sand, equivalent to mortar, are usually equal to the mass of gravel or pebbles in concrete (37); as soon as the frost shall have destroyed any part of this mortar, the loose round pebbles having then lost their bond cannot do otherwise than roll out, and the whole mass will thus be gradually ruined, beyond the possibility of saving it, excepting by the addition of a stone or brick facing.

But even supposing, that a facing of artificial stone two years old, and made of blue lias lime and Thames ballast, should not only be waterproof but also frost-proof, which remains to be proved, the extreme weakness of artificial stone, of which the resistance has been ascertained to be less than one-thirteenth part of that of Yorkshire stone, is an argument not only against using it for wharf walls, which are exposed to the rudest shocks and collisions, but even against using it for the walls of buildings of any importance.

PRECAUTIONS NECESSARY FOR PREVENTING SETTLEMENTS IN THE WALLS OF BUILDINGS ON CONCRETE FOUNDATIONS. THAT CONCRETE BACKING, WHEN USED FOR RETAINING WALLS, SHOULD NOT HAVE COUNTERFORTS.

(357) For the general purposes of Civil Architecture, concrete should therefore, I again repeat, be chiefly confined to foundations; but I conceive that the failure of the new concrete foundation of the Storehouse in Chatham Dockyard has proved, that it is generally, or or least when formed as Mr. Ranger has usually done, with rather a greater proportion of lime than was originally adopted by Sir Robert Smirke, liable to settlements like lime mortar, which in fact forms the principal part of it (353). Hence care must be taken, in commencing the brick footings of a building over

a concrete foundation, not only to use cement mortar and hoop iron bond, in order to do away the necessity of the more expensive expedients of Yorkshire landing stones and chain timbers, but also to construct inverted arches under all the proposed openings for doors and windows, in order to equalize the pressure.

(358) In using concrete for the backing of wharf walls or other retaining walls, care must be taken to connect it well with the stone or brick facing of the wall, but I apprehend, that the wall and its backing should be constructed of a sufficiently substantial profile to dispense with counterforts, because a substance having so little resistance and adhesiveness, as concrete, would admit of the wall in front being forced away from the counterforts, by the pressure of earth acting upon the back of it; as has often occurred to retaining walls and their counterforts, even when built of brickwork. (See Article 266, and afterwards from Article 344 to 347 inclusive.)

(359) In works of Fortification, whilst I have already reprobated the use of concrete for casemates or vaults, yet as the severest frosts seem to destroy those surfaces of concrete only, which are alternately saturated with water and then exposed to the atmosphere, as in the facing of the wharf walls of tide rivers, I see no reason to withdraw the opinion formed by me, previously to the recent failures in her Majesty's Dock-yards at Woolwich and Chatham, that it may be used for retaining Walls not exposed to the action of water, as in the sea wall at Brighton improperly so termed (29), and also for the Revetments of Fortresses in the peculiar situations before mentioned (32), in which it is possible that it might be so much cheaper than regular masonry or brickwork, that although greatly inferior in resistance and consequently liable to be much sooner and more easily breached, whether by battering guns or by mining, this disadvantage would not be a sufficient argument against the use of it in those situations.

A MORE CORRECT ACCOUNT OF MESSRS. FRANCIS'S BRICK AND CEMENT BEAM BEFORE DESCRIBED.

(360) The construction of the experimental brick beam, erected by Messrs. Francis and Sons at their cement works, Vauxhall, having been before described in Article 226, and its downfal in Article 318, I shall here first supply an omission in respect to the construction of that beam, which was built with mortar composed of cement powder

and clean sharp sand in equal parts, and not with nett cement, like Mr. Brunel's and mine, and secondly correct an error in the description, in stating as I supposed, when I first saw it, that it had only 5 pieces of longitudinal hoop iron bond, whereas I have since ascertained, that it had 15 pieces of 1¼ inch hoop bond, namely, 3 in each of the four lowest joints of the brickwork, 2 in the fifth, and 1 in the sixth. I have also ascertained, that the weight, which broke it down, amounted to 50,652 lbs, being rather more than was stated in Article 318.* The circumstance of this beam being built of brickwork laid in cement mortar, and not in nett cement, is of considerable importance, as proving the great strength which hoop iron bond communicates even to the former, which our experiments at Chatham have proved to be much inferior to the latter, both in cohesiveness and in resistance; but as this beam was eighteen months old before it was broken down, it is possible that the comparative inferiority of cement mortar to nett cement, which is very marked during the first two or three months, may gradually diminish as they become older.

(361) *That Cement may be used instead of Lead for certain parts of Roofs.*

In the roofs of low buildings in unprotected situations, where ridges and hips of lead might be exposed to nocturnal depredations, cement cast in moulds to fit the angles in pieces about 2½ feet long, and 1 foot wide, connected at the ends, like timbers scarfed, may be used in preference to ridge tiles and hip tiles. These pieces must be grooved on their under side throughout their whole length, in the manner represented in transverse section in the annexed figure, in order that they may be fixed down upon deal fillets fitted and nailed on to the tops of the ridge boards and hip rafters of the roofs, which fillets must have their upper arrises

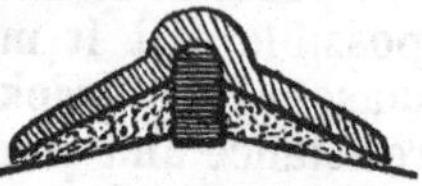

* The corrections in this article are from a paper read to the Institution of Civil Engineers, Feb. 27, 1838, by Mr. Francis describing this beam, of which he had presented me with engravings about a year before, but these being on too small a scale to show any of the pieces of hoop iron bond, I supposed them to be only 5, from my own personal observation, the ends of all the others having been concealed by the brickwork. I was right in stating that it had no deflection, because, though the soffit drooped 1-8th of an inch in the center, it was solely caused by the bending of the planks on which the beam was built. When the cement was 3 months old, it was loaded with 11200 lbs, of scrap iron, which weight in the course of 3 months more was gradually increased to 24000 lbs, when I first saw it (226), which remained about a twelvemonth longer, until the day on which the beam was finally broken down in the course of 2 hours, by gradually increasing the weight to 22 tons, 12 cwt, 1 qr, for which purpose railway bars were used.

or angles rounded off, and are to answer the same purpose with cement, as the wooden ridge and hip rolls used with lead. These pieces must not be bedded on the slates with cement, but with a little common mortar, which can be raked out to remove a broken slate, if necessary. This peculiar use of cement may be considered comparatively, as of inferior importance, but not having been noticed before, it was proper to mention it.

RULES FOR MAKING AN ARTIFICIAL CEMENT, EQUAL IN EFFICIENCY TO THE BEST NATURAL WATER CEMENTS OF ENGLAND.

(362) The mode of effecting this object on a small scale having been sufficiently explained, in the foregoing parts of this treatise, from the attentive perusal of which, the reader will be fully aware of the precautions necessary for insuring success, and of the obstacles to be avoided; it only remains here to make a sort of recapitulation of the substance of what has been said before, in respect to the ingredients, and mode of proceeding, with such modifications of the latter, as may be more appropriate in working on a great scale.

(363) *Ingredients for making the Artificial Cement.*

First, the white or upper chalk of the geologists, which is one of the purest carbonates of lime, and which is always intermixed with thin strata, or with nodules of flint (6). These must be separated, and the chalk either ground dry to an impalpable powder, or by the aid of water to an impalpable paste. The marly or impure chalk usually found near the surface of the ground must not be used.

Secondly, the blue alluvial clay of lakes or rivers, which must also be in a state of minute division, that is extremely fine and free from sand. Hence it is only to be procured in rivers of moderate rapidity such as the Medway.* The brown surface, with which alluvial clay is usually covered must be rejected (84), and care must be taken not to allow it to become stale by exposure to air, which gradually robs it of its blue colour, and at the same time of its virtue as an ingredient for a water cement (85). Where alluvial clay is not to be had, fine pit clay (91 and 92) will answer the same purpose.

* The blue clay from the more rapid current of the Thames is mixed with sand, without removing which it would not be fit for use. But as this cannot be done without extra expense, it is best never to use coarse blue clay at all. Probably the muddy sediment deposited in wet docks, which requires to be occasionally removed, would answer.

(364) *Mode of Preserving the Blue Clay, when it is not convenient to use it immediately.*

In this case, which can seldom occur, but which ought to be provided for, let it be preserved in compact iron vessels of a cubical form, taking care to press it close to the bottom and sides, and covering it with a little water to keep it moist and exclude air; and when required for use, let the brown surface at top be scraped off, and thrown away,* the proportion of which would be very inconsiderable, even if no water were used, because it requires a long time for air to act much deeper than the surface, upon any large mass of clay.

(365) *Proportions of the Ingredients.*

First, by Weight. The best proportion is 100 lbs, of pure chalk perfectly dry, mixed with 137½ lbs, of fresh blue alluvial clay, being equivalent to C4 B5·5, which proved the strongest of all our experimental artificial cement mixtures (187).

Secondly, by Measure. The same proportion stated by measure is nearly equivalent to a mixture of 1 cubic foot of chalk paste, reduced to the stiffish consistency of mortar fit for use, with 1½ cubic foot of fresh blue alluvial clay (166). We have ascertained by experiment, that this consistency will be obtained by mixing 1 lb, of dry chalk powder with 7½ cubic inches of water, which will produce 18 cubic inches of the sort of paste required; so that there will be 96 lbs, of dry chalk to every cubic foot of paste.

(366) *Remarks on the above Proportions, and on the Modes of determining the true Quantity of each Ingredient.*

The above proportions are suitable to the pure chalk of the South of England, and to the blue clay of the Medway, of the latter of which an accurate analysis, with which I was favoured by Mr. Faraday will be given in the Appendix (CXXXIV). If the grey chalk or under chalk of the geologists be substituted for the pure white chalk, or if pit clay be substituted in place of the blue alluvial clay, a different proportion of the ingredients would become necessary to produce a good artificial cement, which proportion can only be determined by experimenting upon various calcined mixtures, in the manner explained in Article 192, by which process, that particular mixture, which is the best, may easily be discovered.

After having decided upon the best proportions of

* This is the principle, according to which we preserved the blue clay for our small experiments in earthenware jars.

the ingredients, it may be remarked, that the first mode of determining the quantity of each, and particularly of the chalk by weight, is the most accurate, and sufficiently convenient for experiments on a small scale; but as chalk is scarcely ever free from moisture, and as the thorough drying of it, without which weighing does not afford a just estimate of quantity (164), would cause much trouble and expense in working on a great scale, the second method by measurement will be the most convenient; and I believe accurate enough. From what we know of the expertness of labourers constantly employed in similar work (9), I conceive that after a little practice they will be able always to mix the chalk paste required for an artificial cement to the average consistency before explained, to which it may be brought by abstaining from measuring it when too wet. If in a very wet state, it must be allowed to dry till it attains a stiffer consistency, before it is measured for the purpose of mixing it with the clay: and should there be any doubt as to the quantity of pure chalk contained in each cubic foot of paste at this period, it may easily be verified, by measuring a small quantity of the paste, then driving off all the moisture by heat, and weighing the chalk powder thus obtained, which if at the before-mentioned proportion of 1 lb, to every 18 cubic inches of paste, will be satisfactory.* The weight of the blue clay when fresh from the River, never varies more than 1 per cent, and from the mean of numerous observations during the last eight months, its average weight proved to be 86½ lbs, per cubic foot.†

* A measure of 4½ by 5 inches in the clear, and 4 inches high, will be large enough for this purpose, and as it will contain 90 cubic inches of paste, the dried powder produced from it ought to weigh 5 lbs. If it should weigh more, it is stiffer than it need be: if less, it is too fluid, and part of the water should be allowed to evaporate. If any other sized measure be used, let its capacity in cubic inches be a multiple of 18, in order to obtain an integral number of lbs.

† There is a much greater difficulty in measuring and weighing any thing very accurately, than people are aware of. Corporal Down, during eight months, that he had the charge of our little artificial cement works, always weighed the blue clay used by him, and found the average of 45 portions fresh from the river to be 86½ lbs, per cubic foot, which agrees more nearly with my first statement of 87 lbs (85), than with the last statement of 90 lbs (164), laid down in former parts of this treatise from much fewer trials. Corporal Down also weighed and measured every portion of chalk powder used, and again measured the same in paste, after having added water, of which he noted the quantity. He found that when more than 7½ cubic inches of water were added to 1 lb, of dry chalk powder, it produced a greater quantity of paste than the standard of 18 cubic inches, and of less specific gravity. By the same series of observations he determined the average weight of 1 cubic foot of dry chalk measured lightly to be 96 lbs, as stated in the preceding article, instead of 100 lbs, as stated in Article

(367) *Mode of grinding the Chalk.* It must first be broken into small pieces, and then either ground in a wash mill, like that of the whitening makers (147); or perhaps a mortar mill, such as was at one time generally employed in great works in this country, and particularly by Mr. Rennie, might answer the same purpose equally well or better; but in this case also water must be used in grinding the chalk; for on trial I found that to grind chalk in its natural state is impracticable, even on a small scale, and to attempt the same on a greater scale would be attended with difficulties, which could not be obviated without adding greatly to the expense.*

(368) *Description of a grinding Mill.* The principal parts of such a mill are two broad vertical iron wheels, carried round by means of a vertical shaft, connected with their common axle, turning upon a pivot in the center of a shallow circular cast iron vessel called *the Pan,* and secured at top to a tie beam of the building, in which the mill is erected. These wheels are purposely placed at unequal distances from the center, one of them being nearer to the pivot, and the other to the outside of the pan. The axle of the two wheels is attached rather loosely to the vertical shaft, so as to admit either of them to rise, in passing over any lump of the materials about to be ground, that may be larger or harder than usual. The wheels may be 4½ or 5 feet in diameter, and from 10 to 15 inches wide at the rim, which grinds the materials, and one of them may be placed at the central distance of 18 inches and the other of 24 inches from the center of the pan. A lever with a yoke for a horse at the extreme end is connected with the vertical shaft, and also with the ends of the axle by some framework over the center of the pan. The distance from the center of motion to the middle of the yoke, or in other words the radius of the horse path, may be 11 feet.

The natural tendency of two vertical cylindrical wheels on the same axis is to move forward in a straight direction, and to pass over without crushing any stone or other obstacle, hard enough to resist their weight. But the forced motion of such wheels round a pivot, especially at unequal

165. He was of opinion, that after a little practice, any careful workman might always mix the paste to the same consistency, and so as to contain the same average quantity of chalk per cubic foot, without either weighing the chalk or measuring the water.

* Because chalk in its natural state is always moist and therefore would cake (174), unless previously broken into very small pieces and thoroughly dried, for which fire would almost always be necessary in this climate (164).

distances, produces a grinding and crushing action upon the materials in the pan below them. As the wheels move, they make ruts of those materials in their course, with ridges on each side, which are continually removed by two iron scrapers, one of which forces the materials thus collected near the outside, and the other those collected near the center, of the pan, into the way of the wheels. These are quite sufficient, in grinding calcined cement, or quicklime, which are both perfectly dry, but when the same sort of mill is employed for mixing mortar, or if it were used as I have just proposed, for grinding chalk in a wet state, it would then be necessary to fix two additional scrapers, in such a position as to embrace the lower part of the back of each wheel, as it revolves, and to throw down the paste or moist stuff, which always sticks to it, in working with water. There is sometimes a hole in one side of the pan, shut by a small iron door, by which the materials when properly ground, or if mortar, when properly mixed, are emptied out of the pan.

The annexed figures sufficiently explain the construction of a mill of this description.

Fig. 1, is the side elevation of the Mill, showing the pan, one wheel and the vertical axis or shaft, about which the wheels revolve, with the horse lever and its yoke, and part of the wooden and iron struts and braces, by which it is strengthened. One of the additional scrapers grasping the back of the wheel as before mentioned, is also represented.

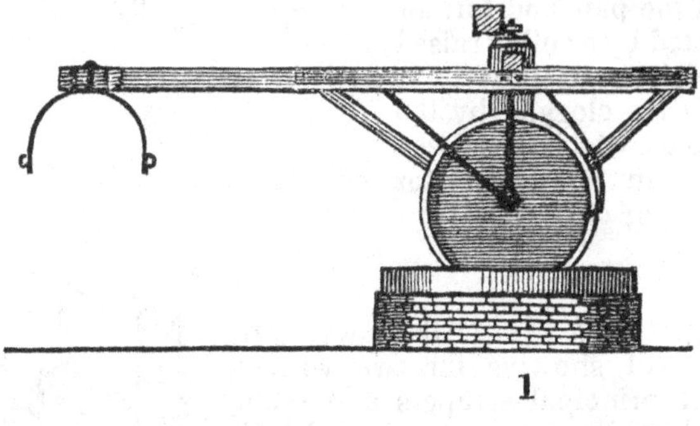

1

Fig. 2, is the plan of the upper part of the mill, showing a sort of wooden frame work, in the form of a square and its diagonals, by which the horse lever is braced.

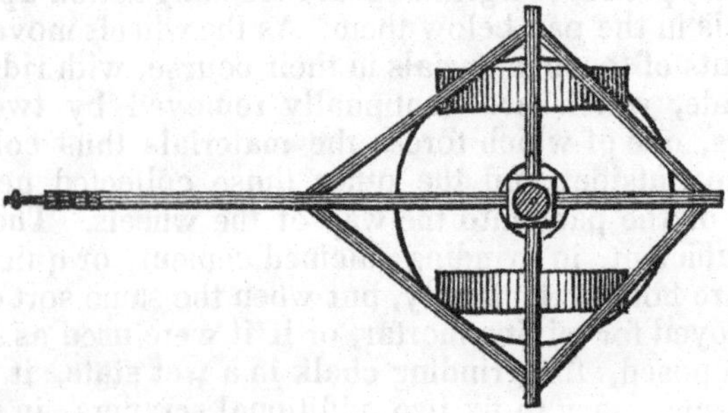

2

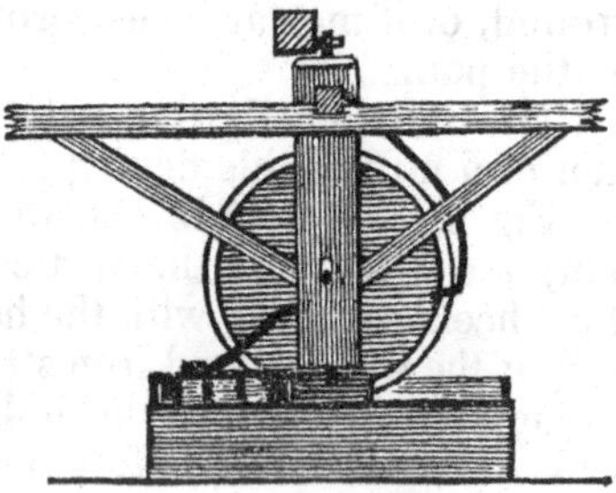

3

Fig. 3, is a section of the lower part of the mill, corresponding with the elevation, showing the two principal iron scrapers before-mentioned, acting within the pan, and a pair of vertical teeth, between them, which are carried by a rod, attached to the shaft. The pivot of the shaft works in a socket made in a central circular step, raised above the bottom of the pan, and further protected by a collar raised still higher, to prevent it from being clogged by the materials, which has been omitted in the figure for the sake of clearness.

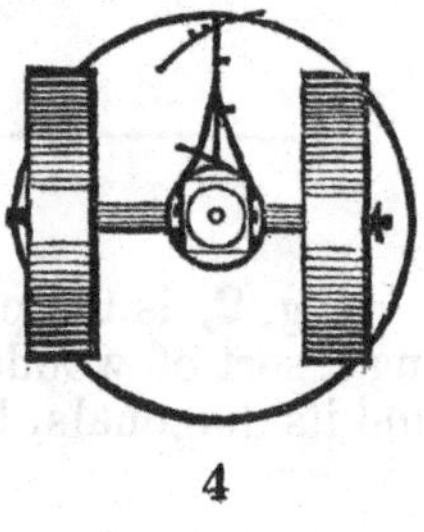

4

Fig. 4, is a plan of the lower part of the Mill, showing the two wheels, and the principal scrapers and teeth, which have been mentioned, and their connection with the shaft.

Fig. 5, is the end elevation of the mill, in which are shown the positions of the two additional scrapers before alluded to, which serve to clean the backs of the wheels, when the materials in the pan are to be mixed or ground by means of water.*

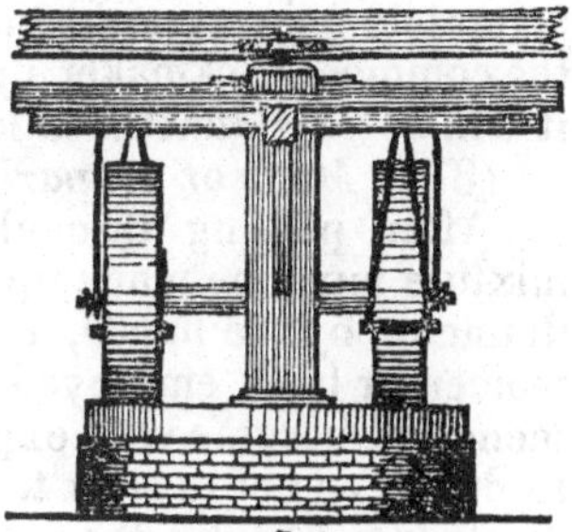

5

(369) *Mode of mixing the Chalk and Blue Clay.* The former, when ground with water in one of the mills just described, would probably be in too fluid a state for immediate use. The superfluous water must therefore be partly drained off, and partly evaporated, by allowing it to dry under cover from rain, until the chalk paste is brought to the proper consistency before stated (365). If too dry, let water be added to restore it to that consistency. It must then be mixed with the blue clay, by means of a couple of small measures, the capacity of which must be as 1 to 1½, the former for measuring the chalk paste, the latter for the clay. Let the contents of these measures be thrown alternately over each other into a pugmill, until it is quite full; and then let the mill be set in motion, and as the mixture contained in the first millful might not perhaps be sufficiently incorporated, let it pass through the mill a second time; which will not be necessary afterwards, when fresh chalk paste and fresh blue clay shall be continually added, which entering by the top will be intimately mixed before they get to the bottom. In respect to the precise size of the measures for the chalk and clay, the smaller they are the better, provided only that they are not made so inconveniently little, as to cause unnecessary trouble and loss of time, in the measurement. Probably some cheap and simple apparatus might be contrived, to facilitate the process of measurement and the partial mixing alluded to, but I am persuaded, that the final and thorough and intimate incorporation of the ingredients, upon which the goodness of the cement will chiefly depend, cannot be effected either by manual labour, or by any other means with which I am

* The above description agrees nearly with the construction of a mill recently erected in Chatham Dock-yard, for mixing mortar, which with the exception of having the additional scrapers for cleaning the wheels, is similar to one, that had been previously used by Mr. Ranger for pulverizing the quicklime, for his artificial stone and concrete in the same Dock-yard.

acquainted, in so perfect and satisfactory a manner, as by the common brickmaker's pugmill, a machine which I think it will be very difficult to improve upon.

(370) *Mode of preparing the Raw Cement for the Kiln.*

After passing through the pugmill, the raw cement mixture must be made up into balls of about 2½ inches in diameter by the hands, in which after a little practice the women or boys employed, for it is scarcely the work of men, will become very expert. These balls must be allowed to dry so as not to stick together, when in contact, nor to be easily crushed by the superincumbent weight to which they will be exposed in the kiln, for which I conceive that 48 hours exposure to air under cover from rain will be enough, as half that period usually sufficed in preparing the raw cement balls for our little experimental kiln at Chatham. Smaller balls than the above would be liable to spoil by exposure to air (86), and much larger ones, though less liable to injury from this cause, would not be so convenient for burning.*

(371) *Of Kilns for burning Cement. Description of that in Sheerness Dock-yard.*

These differ in nothing from the common inverted conical-frustum-shaped lime kilns, and the size may vary without inconvenience according to circumstances; but when not built partly underground, on the side of a cliff, as is sometimes the case, they are usually built in the external form of small cylindrical brick towers, with strong iron hoops, and sometimes with strong vertical bars also, to prevent the fire from splitting the brickwork.

For example, the cement kiln in Her Majesty's Dock-yard at Sheerness, designed by Mr. Rennie, and which is considered of a very convenient construction, is a mass of brickwork measuring 17 feet in external diameter, and 21½ feet in extreme height. The hollow inverted conical frustum measures 8 feet in clear diameter at top, and about 5½ feet at bottom. A nine inch ring of brickwork incloses this space, surrounded by brickwork in mass. At the bottom of it there is a sort of small solid dome 2 feet 3 inches high, to throw off the calcined cement, and let it fall down through the ash-holes of four openings or *Eyes,* as they are techni-

* This method, which, if the number of alternate measures of chalk and clay be noted, records also the quantity of raw cement prepared, is much simpler and better than that first adopted by us in 1829, and persevered in for several years, as described in Article 108. It is also much better than Mr. Frost's method, which is not only injurious to the quality, but causes a great waste of the raw cement mixture. (See the Appendix, Article XV.)

cally termed, at the bottom of the kiln, formed in recesses, which are arched over, and increase by splays and gathering courses from more moderate dimensions near the inside, to 6 feet 3 inches by 7 feet 6 inches in extreme width and height, at the outside. These ash-holes are 2 feet 6 inches wide, and 18 inches high to the crown of the flat arch which covers them, and over each of them, at an interval of 15 inches higher, there are fire holes 12 inches square within the same recesses, having iron bars at top to support the brickwork above them.

The three following figures represent this construction, in which I have omitted nothing but a wooden fence nearly 6 feet high inclosing the whole top of the kiln, instead of which a railing, or a low thin parapet wall of brickwork, to prevent the labourers working there from falling over, is more common in other kilns.

CEMENT KILN IN H. M. DOCK-YARD, SHEERNESS.

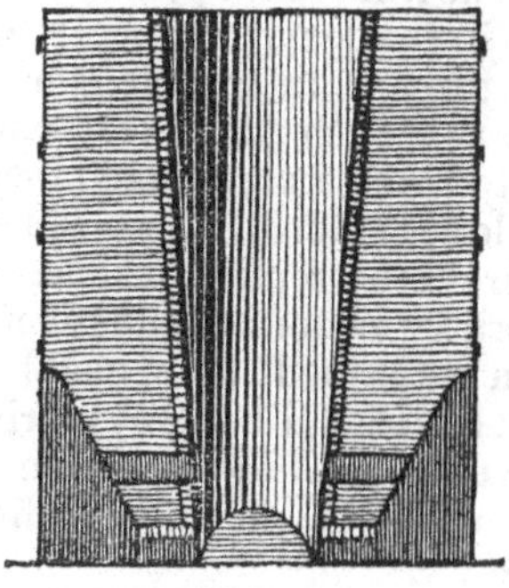

1. *Section.*

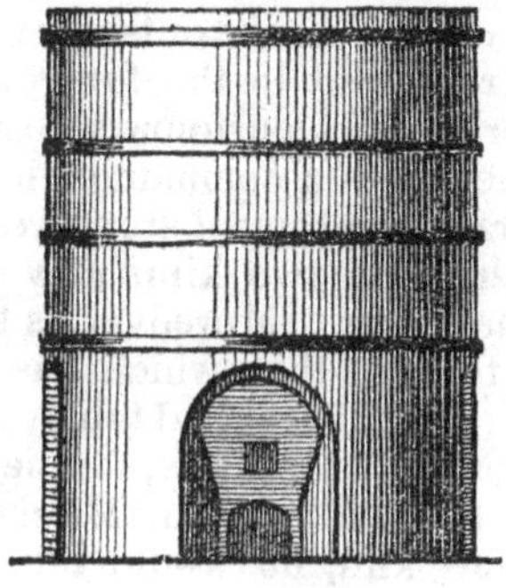

2. *Elevation.*

There are four wrought-iron hoops, inclosing the brickwork, as represented in Fig. 2, each 3 inches wide, and 3 eighths of an inch thick, which are in several pieces connected together at the joints, by strong vertical iron bolts like those of a hinge. The plan, Fig. 3, is supposed to be a horizontal section, taken on the level of the eyes or ash-holes, in which the hollow parts are left blank and the solid shaded.

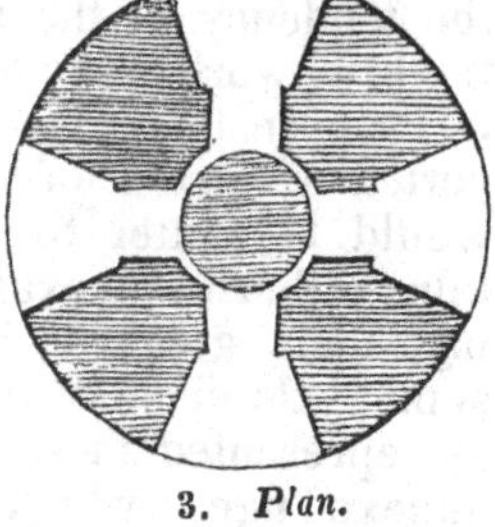

3. *Plan.*

(372) Mr. James Mitchell, the resident Engineer now in charge of the Works in Sheerness Dock-yard, informs me that this kiln can contain nearly 30 tons of broken cement stone, measuring on an average 26 cubic feet to the ton, together with the whole of the fewel necessary for burning it, which varies according to the management of the workmen employed. The bottom of the kiln is first filled with wood and shavings, after which the coals and cement stone are laid in alternate layers, the former being broken so small, that they occupy very little more space, than is necessary for filling up the interstices between the strata of the latter, each of which is usually 1 foot in height or thickness. Three days after the kiln is lighted, the workmen may begin to draw the calcined cement, whilst by laying on more coals and raw cement stone at the top, so as to keep it continually burning, they may afterwards draw the kiln once in every 24 hours. Every ton of cement stone is said to produce 21 bushels of cement powder. The workmen who throw in the coals and cement stone from the top, ascend by a ladder from the outside, which it was deemed unnecessary to represent in the foregoing figures. In private cement works, it is customary to have two such kilns near each other, with a communication at top, and a crane fixed there to raise the cement stone from below, and drop it into whichever of the two kilns may require to be replenished at the time being. Moreover as the brickwork of all kilns is not so thick as that, which has been described, it is usual to increase the width at top by a sort of circular balcony carried all round externally, for the convenience of the workmen.

Remarks. Mr. Mitchell informs me, that in building a new kiln, he would introduce chains in the middle of the brickwork, in addition to the external hoops, to resist better the tendency of the fire to split the brickwork. My own opinion is, that instead of a regular inverted frustrum of a cone, it would be better to build such kilns rather of an oval form, having their greatest diameter at some little distance from the top, as represented in section in the annexed figure, which is supposed

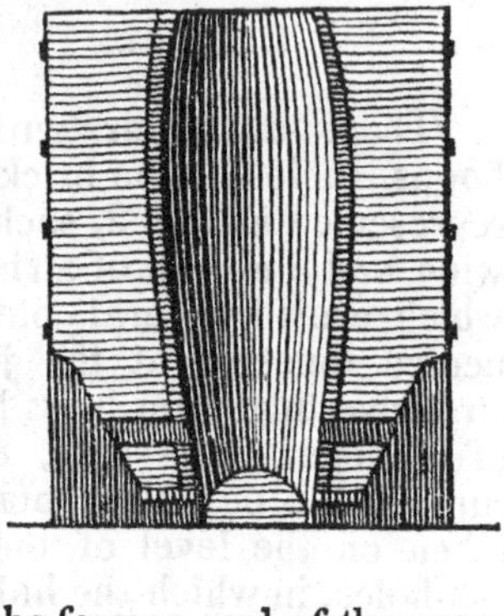

to be a kiln of the same height as the former, and of the same diameter at bottom, and from thence increasing to 8 feet, which is its greatest internal diameter, at about two thirds of the way up, and again diminishing to 6 feet in clear diameter at the top.

(373) *First and usual Mode of grinding and sifting calcined Cement.*

There are two modes of grinding and sifting the natural cements, which would be equally applicable to the artificial cements. The first differs in no respect from the mode of grinding corn and sifting the flour, the same sort of hoppers, mill stones and bolting apparatus being used for both ; and to the best of my information, it has been adopted by all the private cement manufacturers of this country.

(374) *Second Mode of grinding and sifting Cement.*

This was I believe first adopted in Sheerness Dock-yard, by order of Mr. Rennie, where the cement has been ground for many years, by the same sort of mill, which was always used by that eminent Civil Engineer for preparing the mortar for the facing, if not for the whole, of the masonry of works of any importance.

I before described the usual construction of these mills, when they are worked by a horse, (368), but in the cement mill of Sheerness, and in the mortar mill of Ramsgate pier, both worked by steam, and both constructed by direction of Mr. Rennie, a different mode of communicating the motion has been adopted, by making the pans constantly revolve with the materials, which they bring under the action not always of iron wheels, but sometimes of cylindrical mill stones, causing these wheels or stones to turn also, but in the same vertical position nearly, because the axles are immovable, with the exception of a slight vertical play for the purpose before explained; and in this construction each wheel or stone has its own axis.

The cement, ground in Sheerness Dock-yard by the mill that has been described, was thrown by a labourer into a sieve, containing 17 wires to the inch, which was shaken by the steam engine, and the sifted powder was then packed in casks or sacks for use, whilst the coarse parts left in the sieve were thrown back again into the cement mill, to be ground better. Two tons of cement powder were ground daily by this process.

(375) *A new Mode of grinding Cement, proposed to be adopted in Sheerness Dock-yard.*

This ingenious arrangement is now in progress under the direction of Mr. Mitchell, who obligingly lent me his drawings for inspection.

He proposes still to use the same mill of Mr. Rennie's construction, which before did the whole of the work, instead of which it will now be required only to grind the calcined cement roughly or imperfectly, which will be continually lifted out of the pan by an apparatus of small buckets attached to an endless belt, moving in a trough round a couple of small wheels, on the principle of the dredging machine or chain pump, which are to discharge this coarsely ground cement into a couple of hoppers, each fixed over a pair of common horizontal millstones, by which he thinks that it will be ground so fine as to require no sifting, and from thence it is to be led down by troughs into casks placed below for receiving it. These are to be continually shaken in order to compress the cement and fill them well, after which their heads are to be properly put in by the cooper, to render them air tight. The whole of these operations are to be performed by the same 14 horse power steam engine, by which the cement was ground before, which also works some lathes in a turner's shop above the cement mill. Mr. Mitchell is of opinion, that 3 tons of cement may be ground daily by this apparatus.

REMARKS. From my own observation of the action of the common grinding mill described in Article 368, and especially from its efficiency for pulverizing quicklime for Mr. Ranger's artificial stone, I am of opinion, that this sort of mill is not inferior to the flour mill for grinding cement; and therefore it becomes a question of economy, as to the prime cost and repairs of each, which a person proposing to set up a manufactory of artificial cement ought to investigate fully, before he decides which to use; and in either case, notwithstanding Mr. Mitchell's authority, I would recommend, whichever mill may be chosen in preference, to have a very fine sifting apparatus attached to it, with the means of throwing back all the coarse parts to be reground, until an impalpable powder shall be obtained. If he should prefer common grinding mills, which may be worked either by horses or by steam, the manufacturer would naturally consider well the probable extent of his business, before he put himself to the expense of erecting an engine.

(376) *Rules for making an Artificial Cement, when hard Lime Stone only can be procured.*

Supposing that no chalk can be procured, but only hard lime stone, which must be burned and slaked before it is mixed with the clay, as it would be too expensive to grind, and supposing further that it is of the same quality as the

Plymouth marble, or nearly a pure carbonate of lime, the same proportion of chalk to clay by weight, which made the best cement mixture, will also fix the proportion of lime stone to clay; but instead of weighing the former in its natural state, it will be most convenient to weigh it as quicklime after it comes out of the kiln, in which state 40 lbs, of lime to 100 lbs, of blue clay,* or 39 lbs, of lime to 1 cubic foot of this clay will be the proper proportion. Let therefore the lime fresh from the kiln be weighed in portions of 39 lbs, and mixed with sufficient water, to form lumps of slaked lime paste of a thinnish consistency; and in about 24 hours afterwards, when they will be thoroughly slaked, let each of these lumps be mixed with 1 cubic foot of the blue clay and the whole incorporated with the pugmill, after which the process of making the mixture into balls, and drying, burning and grinding will be the same as in working, with chalk paste and clay.

(377) *Probable Proportion of Fewel, for burning Artificial Cement on a great scale.*

It appears to me that 1 measure of coals will be required for burning 8 measures of the raw cement balls prepared as before directed (370), that is whilst they are still moist, except at the surface; because though Mr. White the present proprietor of Frost's artificial cement works informs me, that 1 measure of coals has been found sufficient for burning 10 measures of that cement, it is in consequence of its having been always thoroughly dried before it was put into the kiln, which I consider inadmissible, as being highly injurious to the quality of artificial cement.

(378) *Probable comparative Expense of Artificial Cement.*

I shall conclude by observing, in reply to a question that I have often been asked, but which I have never had the means of answering accurately, as all our artificial cement at Chatham was made by manual labour, that I con-

* Because the best proportion of chalk or of natural lime stone to the blue clay by weight was before stated in Article 365, as 100 to 137½, and the produce of 100 lbs, of pure carbonate of lime is only about 55½ lbs, of quicklime. But as 55½ to 137½, so are 40 to 100 nearly, which last proportion has been chosen as more simple and more easily remembered than the former. Again as 40 to 100, so are 39 to 97½, the last number being the weight in lbs, of 1 cubic foot of fresh blue clay.

This proportion of 1 to 2½ differs from that of 1 to 2 before stated in Article 96, which was adopted to suit our first artificial cement mixture of 1830, but which the after experience of 1836 induced us to reject, as not yielding the best cement, that could be made from the same ingredients. (See Article 146 and those immediately following it.)

sider it probable, if an artificial cement manufactory of chalk and blue alluvial clay were set up on a great scale, in a convenient situation, and in a judicious manner, according to the rules which have just been suggested, that it might be sold at least 25 per cent cheaper, than the natural cements, with equal profit to the maker.

THE END.

APPENDIX

TO THE

OBSERVATIONS ON LIMES, &c.

(I) *Preliminary Remarks.*

With the exception of some quotations or extracts from former Authors, especially from Mr. Timperley,* the foregoing Treatise consists either of original experiments, or of facts for which I can vouch from personal observation. I now propose briefly to notice the labours of former writers or of practical men bearing on the same subjects, according to information obtained from books or from inquiry; and shall put into one connected view, first what has been done in this country, and secondly the proceedings of some eminent men on the Continent, who have experimented upon natural and artificial hydraulic limes.

(II) *Mr. George Semple's Proceedings.*

It appears from a Treatise on building in Water, published by Mr. Semple at Dublin in 1776, that he had been employed there in rebuilding Essex Bridge, which he finished early in 1755, after taking down the old bridge. The lime used by him was Roach lime, which had been proved by experience to have the property of setting under water, but in the second part of his Treatise he falls into an error in declaring, that 'although some sorts of lime 'set much sooner and harder under water than others, yet that any 'good lime, properly mixed and tempered with sharp clean sand, 'will bind and cement as effectually under water as above,' and after making this assertion he defines 'good lime to be that which

* Mr. Timperley's Account of the Harbour and Docks of Kingston-upon-Hull, from which I have quoted largely in from Article 263 to 273 inclusive, was honoured by the Council of the Institution of Civil Engineers with one of the Telford Premiums, awarded by them, from the interest of a handsome legacy, which will probably amount to more than £3500, bequeathed to that Society by their late excellent and distinguished President, to be expended annually in honorary rewards for the best Papers on Professional subjects. I did not know this circumstance, when the above Articles before mentioned were sent to press, as it was not mentioned in the first volume of the Transactions of that Institution, in which Mr. Timperley's communication was published.

'is made of the hardest closest grained and consequently heaviest 'lime stone that can be found,' whilst he condemns all chalk lime as 'being bad, in consequence of its being produced from the softest 'stone in nature.' This opinion, the fallacy of which has since been proved by modern Chemistry, was generally received in Europe at that period, and had even been held as an axiom in Practical Architecture since the time of the Romans, and therefore we must not consider it a disparagement to Mr. Semple.

The first part of his Treatise relates chiefly to the construction of new Essex Bridge, where having observed that a part of the gravelly bed of the River there had been indurated by lime thrown down upon it, when the old bridge had been built, he turned his attention to concrete, and having studied the works of Palladio and Alberti, who explained the system often used by the ancient Romans, of building walls in coffers or cases of small materials grouted; he proposes afterwards, in his second part, the same system for building the foundations of bridges over rivers, stone piers in the sea, &c, by means of concrete, laid in what he calls coffers, but which should more properly be termed Caissons, because he generally proposes to make them with wooden bottoms.*

Mr. Semple appears to have been an intelligent zealous man, with very little previous education, who laments the want of information and examples on the subject of bridge-building under which he laboured, when the rebuilding of Essex Bridge was first proposed to him. In fact, it appears that at that time, there was scarcely a book on the subject in the English language, in consequence of which he requested a friend who was travelling on the Continent, to procure for him, at any expense, all the books, drafts or plans, that could afford him any kind of instruction for laying the foundation of a bridge about 25 feet under high water in a rapid river, and to send them with all expedition.

This commission his friend executed with zeal, and as Mr. Semple expresses himself, 'it happened at that juncture that Colonel 'Belidor had compleated his 4th volume of Hydraulic Architecture, 'which was sent me with the other three volumes; and also a per- 'spective view of the men at work in a coffer dam, at the bridge at 'that time rebuilding at Orleans. The language I was a stranger 'to, but on turning over the plates, I quickly perceived his con- 'struction of coffer dams as we now call our inclosure. My 'drooping spirits then instantly revived, and I immediately went on 'my work with vigour, and entertained the most sanguine hopes of 'success.'

(III) *Of the Treatise entitled Experiments and Observations*

* In Article 27, I ascribed the recent extensive adoption of the use of concrete in this country for foundations, &c, first by Sir Robert Smirke and afterwards by others, to a similar observation of the effects of lime on the gravelly bed of the Thames, in building Waterloo Bridge, which is perfectly correct, though Mr. Semple's recommendation, which had no influence whatever, was of prior date, and the occasional use of concrete is probably prior to the Christian æra.

made with the view of improving the Art of composing and applying Calcareous Cements, and of preparing Quicklime, &c, &c, published in London in 1780, *by Bry. Higgins*, M.D.

This gentleman states, that he was induced to investigate this subject from its connection with Chemistry, on which he had delivered public lectures in the Metropolis. Unfortunately the only limes that he experimented upon were the chalk lime then generally used in London, and stone lime, but as the latter was not the blue lias, but Plymouth lime stone, which has no hydraulic properties, although coming under the definition of a marble from its hardness and capability of receiving a fine polish; it is not to be wondered at, that he should have pronounced it to be no better than chalk lime, by which he means the common white chalk lime, and no other.

After the experiments of several years, Dr. Higgins took out a patent for what he termed *a cheap and durable Water Cement or Stucco, for building repairing and plastering Walls, and for other Purposes*, which consisted of mortar made with 1 part by weight of stone or chalk lime, 7 parts by weight of coarse and fine sand, usually 4 of the former and 3 of the latter, and 1 part by weight of fine bone ash, mixed with lime water carefully prepared for the purpose. The lime to be well burned, slaked by putting it on a fine sieve, plunging it under water and drawing it out repeatedly, and rejecting every part that would not pass through the sieve, as being underburned, which he considered highly detrimental. The sand was to be washed in running water to drive off every particle of clay or dirt, and to be sharp or angular, not round. In making his mortar, he first mixed the two sorts of sand, and spread them six inches thick on a board, and then wetted them with lime water, adding next what he calls his purified lime in several portions, mixing and beating this and the sand together, according to the custom of that day, and finally he added and mixed up the bone ash also, which was to be sifted much finer than that used for cupels. This substance was to prevent cracking, and to promote plasticity and induration by attracting carbonic acid gas from the atmosphere, and he stated that it also prevented dampness in inside work. In all cases, even in wetting the materials or the stucco itself, he invariably prescribed the use of lime water. As he attached great importance to fresh slaked lime, he recommended that all the lime that could not be used immediately, should be preserved in air-tight casks.

He divided his mortar into several sorts, 1st, *Water Cement coarse grained*, 2dly, *Water Cement fine grained*, for giving a fine surface to the former when required, 3dly, *Water Cement of the coarsest sort*, cheaper than the former, in which he used shingle or coarser sand in addition to his usual coarse and fine sand, 4thly, his *Cement for Water Fences*, by which I presume he means wharf walls, reservoirs, &c, in which he omits 2 thirds of the bone ash, and substitutes pulverized Tarras or Trass in lieu of it, and 5thly, he proposes to make *Artificial Stones*, by alternate layers of his own cement and of flints, hard stone or brick, put together in moulds, and allowed to dry well before they are exposed to rain.

(IV) He mentions that several houses, chiefly in London, had been stuccod with his cement, but as my own observations on chalk lime, and the sort of stone lime used by him, which is no better, induce me to believe them both equally unfit, not only for hydraulic purposes, but even for the outside stuccoing of buildings; I presume that his inventions, which consisted in attempting to improve bad limes by expensive arrangements must have died a natural death, being not only much inferior to the Sheppy and other English cements afterwards discovered, but even to the mortars and stuccos of blue lias lime.

Smeaton's Researches, which were of much prior date, unfortunately were not published until many years afterwards. If Dr. Higgins had been aware of the more enlightened and comprehensive views of that eminent Civil Engineer, his own zeal and industry better directed might have led to more useful results.

(V) *Experiments to ascertain a complete Composition for Water Cements, with their Results, as contained in Book third, Chapter fourth, of the Narrative of the Building, &c, of the Edystone Lighthouse, &c, by John Smeaton, Civil Engineer, first published in London in* 1791.

Smeaton's observations upon the limes and cements then known are most valuable, and do the highest credit to his sagacity, since at the time that he was intrusted with the execution of that arduous and memorable work, he seems to have been almost new to the subject, and yet in a short time arrived at juster views than any former writer.

After having compared together the various limes of England, with a view to determine that most suitable for the mortar of a work of such importance, and so much exposed to the violence of the sea, as the Edystone Lighthouse; and having formed an accurate opinion of their comparative goodness, from his own direct observation, by a very judicious practical mode of testing their water-setting powers, he consulted a friend well versed in Chemistry, in which science he candidly owns his own incompetency; and having been instructed by him in that simple sort of chemical analysis, which enabled him to distinguish the pure carbonates of lime from those containing other ingredients also, he discovered that every stone producing a lime capable of setting under water, had a proportion of clay in its composition. The perusal of this important fact first induced me to attempt to form an artificial cement with a mixture of chalk and clay, as I have acknowledged in Article 1 of the foregoing Treatise; and I consider it probable, that it has also served as a hint to others, who have entered into the same pursuit, whether in this country or on the continent, where Smeaton's labours were well known; although they have not all acknowledged the obligation, as I felt it my duty to do.

In announcing this important discovery, by which he overset the prejudices of probably more than 2000 years, maintained by all former writers on the subject, from Vitruvius to Mr. George Semple inclusive, Smeaton's words are as follows. 'Having now found,' that is from the experiments before alluded to, 'a species of materials, and 'a method of compounding them, very competent to our purpose; 'and having plainly seen, that there was a great difference in the 'effect, arising from the different nature of lime burnt from different 'kinds of lime stone; and that its acquisition of hardness under 'water did not depend upon the hardness of the stone; inasmuch as 'chalk lime appeared to be as good, as that burnt from Plymouth 'marble; and that Aberthaw lime was greatly superior to either, 'for the purpose of aquatic buildings, though scarcely so hard as 'Plymouth marble; I was very desirous to get some light into some 'of the sensible qualities, that might probably occasion the difference, 'or at least become a mark of distinction. I therefore applied to my 'friend Mr. Cookworthy, whom I had found at all times ready to 'afford me his assistance, wherever his knowledge could be of use to 'me. He taught me how to analyze limestones: and though my 'chemical friends will be at no loss upon this subject; yet as it is 'very possible, that some of my readers may be no more acquainted 'with chemistry than myself, for the sake of these I will describe 'the process, as being useful for all those who are concerned in 'building to know.' (See Article 192 in the Chapter before quoted.)

After explaining the simple sort of analysis alluded to, which it is unnecessary for me to repeat here, he describes his own further experiments on various limes, with that clearness and precision, by which his writings are distinguished, after which he lays down for the first time in Europe, in opposition to the false doctrine of ages before alluded to, the true principles, upon which the value of any lime for the purposes of Hydraulic Architecture depends, in stating which I shall transcribe his own words.

'Perhaps nothing will better show, that the qualities of lime for 'water mortar do not depend on hardness or colour, than a compari-'son of the white lias of Somersetshire, (which though approaching 'to a flinty hardness, has yet a chalky appearance) with what is 'called near Lewis in Sussex, the *Clunch Lime*; a kind of lime in 'great repute there for water works, and indeed deservedly so. This 'is no other than a species of chalk, not found, like the lias, in thin 'strata, but in thick masses, as chalk generally is: it is considerably 'harder than common chalk, but yet of the lowest degree of what 'may be denominated a stony hardness; it is heavier than common 'chalk, and not near so white, inclining toward a yellowish ash 'colour. This stone, when analyzed, is found to contain $\frac{3}{16}$th parts of 'its weight of yellowish clay, with a small quantity of sand, seemingly 'of the crystal kind, not quite transparent, but intermixed with red 'spots. Hence the fitness of lime for water building seems neither 'to depend upon the hardness of the stone, the thickness of the 'stratum, nor the bed or matrix in which it is found, nor merely on 'the quantity of clay it contains. But in burning and falling down 'into a powder of a buff coloured tinge, and in containing a considerable quantity of clay, I have found all the water limes to agree. Of 'this kind I esteem the lime from Darking in Surrey to be; which is 'brought to London under the idea of its being burnt from a stone, 'and in consequence of that, of its being stronger than the chalk 'lime in common use there; though in fact it is a chalk, and not 'much harder than common chalk: it contains $\frac{1}{17}$th part of light 'coloured clay of a yellowish tinge.

'There is in Lancashire a lime famous for water building, called 'Sutton lime; I have lately had an opportunity by favour of John 'Gilbert, Esq. to get a specimen of the stone in its natural state. I 'had long since seen it in the Duke of Bridgewater's works, both 'in the burnt stone and slaked, made up for use, and in the water. 'I observed that it agreed with the lias in being of a buff cast. The 'stone itself is of a deep brown colour, and the piece I have is from 'a stratum about three inches thick, with a white clayey coat on each 'side. The goodness of the quality as water lime, does not therefore consist in the colour before it is burnt; for we have already 'seen blue, whitish, and now brown, to be all good for that purpose; 'but they all agree in the colour or hue, after they are burnt 'and quenched: and having analyzed the Sutton limestone, I find 'it to contain not only near $\frac{3}{16}$th parts of the original weight of the 'stone of brown or red clay, but also $\frac{1}{42}$dth of fine brown sand; so 'that in reality I have seen no lime yet, proved to be good for water 'building, but what, on examination of the stone, contained clay: 'and though I am very far from laying this down as an absolute 'criterion, yet I have never found any limestone containing clay in 'a considerable quantity, but what was good for water building; and 'limes of this kind all agree in one more property, that of being of 'a dead frosted surface on breaking, without much appearance of 'shining particles.'

(VI) As I conceived that every Engineer or Architect would wish to study the very interesting work from which I have made the foregoing extracts, it was not my intention to have quoted so largely from Smeaton's admirable chapter on water cements, if I had not found, that his meaning has been misunderstood, and that he has been charged both by M. Vicat in France,* and more recently by Mr. George Godwin junior in this country, with having fallen into the same error as Dr. Higgins, who from his writings was evidently unacquainted with any of the excellent water limes of England, and ignorant of the properties of hydraulic limes generally, which Smeaton had investigated with so much ability and success.

Mr. Godwin's words, in treating of the limes most proper for concrete, are these. 'In ordinary cases, stone lime should always

* L'Ingénieur Anglais Smeaton, et le Docteur Higgins prétendent au contraire que la craie et le marbre, qui offrent a-peu-près les deux extrêmes en dureté parmi les espéœs calcaires, donnent de la chaux qui procure au mortier une egale bonté. (See the author's Recherches expérimentales sur les chaux de construction, &c, section première, chap. II, published in 1818).

'be employed—that from Dorking when attainable. Both Dr. 'Higgins and Mr. Smeaton contend that chalk lime, when well 'burned and used with care, is fully equal to stone lime. Experi- 'ments, however, do not appear to bear out their views, probably 'from the difficulty of fulfilling the chief condition; and I should 'advise, notwithstanding the authority of such names, that chalk 'lime be never used for the purpose of concretion.'

Having placed this quotation from Mr. Godwin's Essay on the nature and properties of concrete in juxtaposition with my former quotations from Smeaton, the reader will clearly see, that Mr. Godwin has imputed an error to Smeaton of his own creating. In respect to Dr. Higgins, I do not consider, that even he was so far wrong as Mr. Godwin alleges, in stating that chalk lime was as good as the sort of stone lime experimented upon by him, namely the Plymouth marble; but he was in error in evidently supposing that all stone limes were alike. And Mr. Godwin himself appears to have fallen into the same error, which is aggravated by his having adopted another mistake, which was exposed nearly half a century ago by Smeaton, that of the workmen of London, who have improperly given the name of *Stone* to what is in reality a *Chalk*, I mean the argillaceous chalk of Dorking, which misnomer, like the equally inappropriate term *Roman Cement*, only serves to mislead the public, and to cause confusion of ideas. The term *Stone Lime* should therefore be abolished, because even if it did not involve the glaring absurdity of including chalks as well as stones, there are no two sorts of hard calcareous stone that yield lime of the same quality.

(VII) Having ascertained that the blue lias was the strongest of our English water limes, Smeaton experimented upon Trass, then called Dutch Tarras, and on Puzzolana, a mixture of which two substances with lime afforded the only hydraulic mortars then known in Europe, as the still more powerful water cements of England had not yet been discovered. Observing that Trass mortar grew out from the joints, owing I believe to the very injudicious mode of using it which prevailed in this country, as described by him, which was to reverse the proper proportions of the ingredients by mixing 1 measure of Trass with 2 of common lime, Smeaton rejected that substance altogether,* and adopted Puzzolana in preference, which he met with by mere accident, as there seems every reason to believe, that it had never been used before in England; and it may be remarked as a very curious circumstance, that he was ignorant of the properties of Puzzolana, until he read of it in Belidor's works, to which distinguished Frenchman Mr. Semple also acknowledges that he had himself, about the same period, become indebted for his success in

* Smeaton with his usual sagacity discovered by experiment, that equal measures of Lime and Trass made a better water setting mixture, than the injudicious excess of the former ingredient; but he seems to have imagined that the defect of growing out from the joints, so as to project beyond the original face of the wall, was inherent in Trass under all circumstances.

the rebuilding of Essex Bridge, as was before mentioned.* Such was the very imperfect state of the Art of Civil Engineering in this country about the middle of the last century, which has since made such rapid progress towards perfection.

Smeaton observes that his Edystone mortar, composed of equal measures of puzzolana and of blue lias lime in the state of slaked lime powder, which he considers the strongest water cement, that it was possible to make with any of the ingredients, then known and used for that purpose, set well under water, but not quickly enough to resist the violent wash of the sea; to guard against which he sometimes caulked the outer part of the joints with the same mortar mixed with oakum, and sometimes he protected them by an external coating of gypsum or plaster of Paris. He observes that his Edystone mortar set so slowly in joints exposed to air, that for some time it seemed almost stationary. On visiting the Edystone in 1787, nearly 30 years after the Lighthouse was finished, he found that the mortar was sensibly corroded in most parts of the upper works, but not on an average above the 6th part of an inch deep. In the parts wet every tide, it had not failed at all.

(VIII) Smeaton also experimented upon several substances, having like puzzolana the property of improving limes, and communicating to them the property of setting under water, such as the scales from a Smith's forge, and minion or iron stone calcined, or having undergone the first process in making iron: and he states that whenever he saw a reddish or brownish stone, he generally tried it, and in one instance he found a piece of coarse deep brown sand stone, which on being burned, pulverized and mixed with lime, produced a mortar equal to that from forge scales.

Having mentioned the principal hydraulic mortars recommended by him, in from Article 247 to 250 inclusive, I shall not repeat them here, and for the further proceedings of this eminent Civil Engineer, whilst investigating the properties of water limes and mortars, I beg to refer the reader to his own interesting account of the Edystone Lighthouse, and, for still more detailed instructions respecting the proper mode of using puzzolana, to the 3rd volume of his Tracts.

(IX) *An account of the Bell Rock Lighthouse, including the details of the Erection and peculiar Construction of that Edifice, to which is prefixed a historical View of the Institution and Progress of the Northern Lighthouses, &c, &c, by Robert Stevenson, Civil Engineer, F.R.S.E. &c, &c. Edinburgh*, 1824.

In building this Lighthouse, nearly opposite to the mouth of the river Tay, which was commenced in 1807 and finished in 1810, Mr. Stevenson acknowledges that he took Smeaton's system for what he calls his text book, but he was by no means a servile copyist, there being great ingenuity and originality in many of his arrangements, and equal energy and perseverance in all his proceedings.

* Belidor was also the first, who entertained just notions on the subject of Military Mining, in opposition to an absurd Theory, which prevailed in France and in all Europe, until he subverted it by his celebrated experiments on surcharged mines, by him styled *Globes of Compression.*

His mortar, which is the only part of his interesting book, that refers to our present subject, differed from Smeaton's in his having added 1 measure of sand to 1 measure of puzzolana, and 1 of Aberthaw lime, and also in having measured the latter in the state of quicklime powder, which from the circumstances that were explained in Articles 292 and 293 of the foregoing Treatise, render their respective proportions of lime, though nominally the same, very different from each other. But I have described his mortar, and remarked upon it so particularly in other articles, that it is unnecessary to enlarge further upon it here. The addition of the sand may be considered an improvement upon Smeaton's system, so far as economy is of importance, though in adhesiveness, resistance and water setting powers, I consider from my own experiments that Stevenson's mixture, which was nearly equivalent to blue lias 1 P1 S1 of Tables XII and XIII (277), is rather inferior to Smeatons, which was nearly equivalent to blue lias 1 P2, of the same Tables.

Mr. Stevenson pointed his external joints with English cement, to secure them against the action of the waves, which he observes washed and wasted a great deal of his slow setting mortar in the lower parts of his work, which were covered by every tide at high water.

(x) *An Historical Description of the Suspension Bridge constructed over the Menai Strait in North Wales, &c, &c, from Designs and under the Direction of Thomas Telford, F.R.S.L. and E. President of the Institution of Civil Engineers, &c, &c, by William Alexander Provis, the resident Engineer, &c, &c.*

The workmen began to quarry their stone at the Penmon quarries in 1818, but the first stone of this justly celebrated work was not laid till the 10th of August 1819. The bridge was practicable for carriages on the 30th of January 1826, when the Holyhead Mail first passed over it, but some additional bracing was executed soon afterwards, to prevent for the future the undulation or vibration of the superstructure, which took place during a very violent gale of wind.

In the specification it was directed that 'all the masonry from 'the foundations to an offset 3 feet above high water was to be 'built in well worked mortar of Aberthaw lime and sand,* and 'the rest of the masonry in mortar made of lime from the fragments 'of the Penmon stone, mixed with a proper proportion of sand.'

After thus describing the mortar to be used, Mr. Provis adds, 'though known to almost every Practical Builder in England, it 'may not be improper here to observe, that mortar made of Aber'thaw lime mixed with sand sets extremely well. If masonry 'built with it were covered immediately with water, a thin crystal'lization would be formed on the surface of the mortar in two or 'three hours, and in as many days it would become perfectly har'dened and attached to the stones.'

* Mr. Provis writes Aberddaw according to the Welsh orthography of this name. A portion of the masonry of most of the piers is under water every tide.

'Where the masonry, after being built, is to be subjected to the 'action of a strong current of water, it is better to point the joints 'with British* (sometimes called Roman) cement, which setting 'immediately prevents the mortar being washed out of them. The 'advantages of Aberthaw lime mortar compared with cement are 'its taking a better hold of the stones, and being much cheaper.'

The former objection to the use of cement, which I believe is still general amongst practical men, will no doubt be removed by our experiments at Chatham (216), which have proved its adhesiveness to Bramley-fall stone to be equal to the tenacity of that excellent stone itself.

(XI) *Information respecting the Mortar used in erecting the Landing Wharf for Steam Packets at Hobbs' point Milford Haven, near Pembroke, communicated by Captain (now Major) Savage, R.E. who superintended that Work.*†

The foundation of the extreme end of this wharf was laid at 32 feet below the low water mark, and 57 feet below the high water of spring tides, and no less than 6200 cubic feet of the masonry were built by the diving bell. The mortar used was composed of 1 measure of Aberthaw lime in lumps from the kiln, mixed with 2 measures of sand. Captain Savage, who had before used the blue lias lime of Lyme Regis with equal success on the Cobb or Wharf of that town, as stated in Article 26, informed me that this mortar set so well under water, that about a cubic foot of it, which had been left on the wall without being used, whilst the masons went to work on other parts, was afterwards found to be nearly as hard as stone on their returning to the same spot, which was only accessible by the diving bell; notwithstanding which Captain Savage generally protected the external joints by cement.

(XII) *Of the Specification of Mr. Parker's first Patent, entitled a Patent for making a certain Cement or Terras, to be used in aquatic or other Buildings and for Stucco Work.*

The specification was dated the 27th of July 1796, but the the patent itself was granted to James Parker of Northfleet, in the county of Kent, gentleman, on the 28th day of the preceding month. When he first discovered the extraordinary properties of the natural English cements, Mr. Parker designated them in the above manner, and in one part of his specification he uses the expression *a Water Cement.*‡ He describes his cement, as being obtained from certain stones, or argillaceous productions or nodules of clay, containing cal-

* English Cement is the most proper name, none being found in Scotland.

† This landing wharf was commenced from a design of Col. Fanshawe. R.E. The first stone was laid in January 1830, and the work was finished in 4 years. Some alterations were made in the wharf whilst in progress, and the buildings attached to it were chiefly designed as well as executed, by Captain Savage himself.

‡ This has been changed into the unintelligible phrase *a Water & Cement*, by mistake of the Scribe, who has also changed Mr. Parker's nodules into *Noddles* of clay, throughout the whole of the original Patent, which I was permitted to peruse on paying a fee at the proper office near Chancery lane.

careous veins and usually a drop of water in the center. He states that these must be burned in a kiln like common lime, but with a stronger heat, and that they become warm and soften, but do not slake afterwards, when water is thrown upon them. He states that the calcined nodules must be ground to fine powder by mechanical means, and that when mixed with a moderate quantity of water, namely about 2 measures of water to 5 of powder, a cement is formed which he declares to be much stronger for the purposes above specified, and one that sets sooner and harder, than any mortar or cement prepared by artificial means. He provides in his specification for occasionally burning and mixing his cement with lime and other stones, clays or calcined earths, and includes in his patent all stones or concretions of clay having the same properties.

I always thought it probable that Mr. Parker must have been skilled in Chemistry, until I perused his patent, the peculiar wording of which has convinced me to the contrary, as he has reversed the proper definition of the water cements of England by calling them argillaceous productions containing calcareous veins, &c, whereas the calcareous matter is much more prominent in their composition than the clay. I am now of opinion, that he must have taken the hint from Smeaton, whose book was only published a few years before, because the rude sort of analysis pointed out by that author, would render it possible for any person, though not a scientific Chemist, to ascertain that the Sheppy cement stone was a mixture of lime and clay; and if Mr. Parker followed Smeaton's other rule to burn and pulverize every reddish or brownish stone, by which of course must have been meant stones, that did not slake after calcination, and mix them with lime, he would have found that the Sheppy cement stone improved common lime, and from thence it was only one step more, to try it without adding lime to it.

Although I am perhaps only indulging in a speculation, which may be imaginary, as Mr. Parker may possibly have discovered the Sheppy cement, without being indebted for it to Smeaton's book; yet I consider it most certain, that after the publication of that book, it was scarcely possible that the valuable properties of the water-cement-stones of England, undiscovered for so many ages, could have remained much longer unknown to the world.

It was not until some years afterwards that Mr. Parker designated the valuable water cement, of which he first discovered the use, by the improper title of Roman Cement, which serves only to mislead the public, who with the exception of professional men, generally believe it to be a rediscovered secret of the ancient Romans.

A curious instance of this sort of mistake occurred in a Report made by Mr. Benjamin H. Schlick a Danish Architect, to the Academy of Fine Arts of the French Institute, on the 25th day of November 1826, upon the subject of the Thames Tunnel, who amongst the causes of the success of that wonderful undertaking, gravely enumerates 3dly, 'the advantage of the cement of the

'Romans,' upon which he observes, that 'the modern arts in re-'discovering the Roman cement, have achieved an important 'victory.'* Instead of this absurd title, which no more applies to the cements of England, than Chinese porcelain would to Wedgwood's ware, and even less so, for the Romans had no knowledge of this sort of cement at all, it ought to be designated as Sheppy, as Harwich, or as Yorkshire cement, or if compounded of two of those, let it be called English cement, or Parker's cement, from the name of the ingenious discoverer of its admirable properties; and in concluding this subject, let it be fully understood that my impression of his having taken a hint from Mr. Smeaton is no disparagement to the merit of Mr. Parker's discovery, which is of immense importance, for as I observed before, it enables the Architect to execute with ease many constructions, which, without such powerful cement, would have been absolutely impracticable (235). In respect to Smeaton, though he discovered a great principle, and thus threw daylight into an important practical subject, which before his time was involved in utter darkness, and thereby cleared the way for others; yet he himself neither discovered any new hydraulic lime nor cement, but worked with limes previously known and even appreciated in their respective localities, though their value was not understood beyond those narrow bounds, and even there the cause of their superiority was entirely unknown.

(XIII) *Of the Patent granted to Maurice St. Leger of St. Giles' Camberwell, Gent. for a method of making Lime, dated May* 1818.

I believe that several patents for making artificial cement may have been taken out in this country, none of which have come into general use except that known by the name of Mastich. Of all that I have seen, the only one that involves nearly the same process afterwards adopted by me, is that of Mr. St. Leger, before quoted, of which the following copy was sent me a few years ago by a friend, who was acquainted with my proceedings, and was struck with the resemblance, and who informed me that he had extracted it from the new monthly Magazine of the 1st of August 1819. It is as follows.

'I the said Maurice St. Leger do hereby declare, that the na-'ture of the said invention, and the manner in which the same is to 'be performed is as follows, viz. I take chalk, stone, or any other 'substance from which lime can be obtained, which I pulverize, to 'which I add common clay or any other substance containing 'alumine and silex, which I increase or diminish according to the 'required strength of the lime. I mix them together, and add water 'to them until they become a paste of the consistence of common 'mortar. I then make the said paste or substance into lumps. 'These lumps after being thoroughly dried by natural or artificial 'heat, I put into a kiln, and expose to the action of fire in the

* The words are these 'Les arts modernes, en retrouvant le ciment Ro-'main ont fait une importante conquete.'

'usual way of making lime. The degree of heat must depend upon 'the size and quality of the lumps, but I find the lumps have been 'sufficiently exposed to the fire, when they can be broken by the 'hands. Instead of chalk, stone, or such other substance as is above 'mentioned, ordinary lime, slaked or pulverized, may be substituted; 'but in that case the compound does not require to be so much ex-'posed to the action of the fire. The quantity of clay, or other 'substance containing alumine and silex, to be added to the chalk 'or such other substance as aforesaid, or to the lime slaked and 'pulverized, must depend as well upon the quality of the chalk, 'stone, or other substance, or lime, as upon the quality of the clay 'or other substance containing alumine and silex. But I find in 'general, that from one to two and twenty measures or given quan-'ties of clay, or other substance containing alumine and silex, to 'every one hundred measures or given quantities of chalk, stone, or 'other substance, or of lime as above mentioned, is the proper 'quantity or proportion.'

A French gentleman of the same name, having established a manufactory for making artificial hydraulic lime according to M. Vicat's system, at Meudon near Paris, about the same period, I have not the smallest doubt of his being the identical individual, who took out the above patent in England in 1818, upon which I shall remark, that the proportions, are such as may produce a hydraulic lime, which was evidently the intention of the Patentee, but not a water cement: and as the excellent natural English cements had then come into general use in London, to which Mr. St. Leger's lime must have been infinitely inferior; and as the Dorking and Halling limes, and even the blue lias must have been much cheaper, I cannot discover that his patent ever came into operation in this country: and if any person had attempted to make artificial cement from it, he must have failed, not only from the inaccuracy of the proportions, but from the want of precision in respect to the nature of the clay; as will be evident from the failure of my own first attempts in 1826, though working with clay of a quality comprehended in Mr. St. Leger's specification. (See Article 1.) The process adopted at his manufactory at Meudon will afterwards be described.

Of Mr. Frost's Artificial Cement Works, in Swanscomb Parish, Kent, now the property of Mr. John Bazley White & Son.

(XIV) I am informed that Mr. Frost obtained possession of the ground on which he established these works, in 1825. I visited them in December 1828, at the time when I had succeeded in my first experiments on a small scale. At that time I found that Mr. Frost had made two sorts of artificial cement, viz.

First, his British Marble, or Silicate of Lime. He gave this name to a cement composed of a calcined mixture of pulverized flints and chalk, which was perfectly white and might be used either as a stucco, or for an artificial stone. The flints after having been burned in a kiln, were quenched in water whilst red hot, and ground in a circular basin by four chertz stones dragged round on a pavement of the same material covered with about a foot of water, by the revolutions of a frame moving round on a vertical axis, to which they were attached; and thus the calcined flints were reduced to a very thin paste. When in this state a proportion of fluid chalk ground in another mill was let in, and the two ingredients mixed together. After being thoroughly incorporated, the wet mixture was allowed to run off into a back or reservoir on a lower level, and after attaining a tougher consistency, it was cut down in lumps and dried in sheds, and finally burned in a common lime kiln; and then ground and preserved in casks. Mr. Frost informed me, that one measure of this artificial cement might be mixed either with one measure or with one measure and a half of sand. From my own experiments, I found afterwards that this cement was not well adapted for hydraulic purposes, and as a stucco it was far too expensive, and consequently all his attempts to introduce it into the London market proved ineffectual. Two specimens of it applied as stucco are still to be seen on the same premises, one of which has a very fine, the other a rough surface, and both are extremely hard.

Secondly, Mr. Frost's Artificial Cement of Chalk and Clay.

(xv) In this his second sort of artificial cement, Mr. Frost acknowledged that he had profited by the experience of M. Vicat, having undertaken a journey to France for the express purpose of consulting that gentleman, and inspecting his operations. But instead of working with slaked lime to be burned a second time, according to M. Vicat's favourite system, Mr. Frost used chalk ground in two mills, each having 4 iron wheels and 2 harrows, with sand trap or pit near it for getting rid of the sand, on the same principle, as those of the whitening makers (147), but of much superior workmanship and requiring greater power to work them, than the common whitening makers' mills in this neighbourhood, which are of the rudest construction and worked by one horse only* The fluid chalk ground in these mills, and the clay, which was the fine brown pit clay of the neighbourhood of Upnor, were then mixed together in a washmill precisely similar to that used by the brickmakers, in which they prepare the same ingredients, but in very different proportions, for making *Marl Bricks* or *Malms*, as they are vulgarly termed. When sufficiently comminuted and incorporated together, the fluid mixture of chalk and clay was by the opening of a small sluice allowed to flow into backs or reservoirs, also similar to those of the brickmakers, where it usually remained some months,

* The whitening makers alluded to are labouring men with small capital, none of whom use more than one wheel in their respective mills, which is not always of stone or iron, being sometimes built of blocks of old ship timber bound round the rim with pieces of hoop iron. The horses which turn these wheels are not very powerful ones, and usually move round in so small a circle, that they derive much less advantage from leverage, than is allowed in better constructed grinding mills.

and acquired the consistency of a stiff paste; in cutting into which, regular strata, generally of tints differing a little from each other, were observable, each of which was the produce of one day's contents of the washmill. In this state the raw cement was cut out of the back in lumps which were laid on open shelves, in drying sheds, similar to those of the whitening makers. When sufficiently dry, these lumps were broken into smaller pieces of a convenient size, and burned in kilns of the common inverted cone-like form, in the same manner as lime, that is with alternate layers of fewel and cement. Owing to the lumps of raw cement breaking irregularly, and also generally splitting with the strata before described, there was a good deal of rubbish, composed of small fragments and dust, which was thrown away, as being unfit for the kiln. When sufficiently burned, the calcined cement was ground by 2 pairs of the same sort of millstones which are used for corn, but without the bolting or sifting apparatus of the corn mill, after which the powder was preserved in casks or sacks as usual. The flint mill, chalk mills, wash mill and grinding mills, that have been described, were all worked by a steam engine said to possess a 30 horse power, which Mr. Frost erected for the purpose, and he also built a wharf on the Thames, with which he communicated by a railway upon a raised road constructed by him across the intervening marsh, for the use of his works, which are close to the London Road, on the north side of it, and to the eastward of Ingress Park, with a lane between.

(XVI) From a motive of delicacy, as being engaged in the same pursuit, I did not ask Mr. Frost his proportions, but I communicated with him for several years, and always gave him full information of my own proceedings, which I hoped might be of use to him. I found him at one time in some perplexity, as he said that the mortar of the wharf built with his artificial cement according to M. Vicat's rule, had proved defective, and he lamented, that whilst all his experiments on a small scale succeeded, the very same process repeated on a larger scale often failed, which he ascribed to his kilns being damp, or not well proportioned, and I believe that he altered them in consequence.

My own opinion on this subject is, that successful experiments on a small scale will never fail, if repeated on a great scale, unless there be some difference in the mode of proceeding, which the experimenter may not always detect. For example in my own first experiments on a small scale, in making artificial cement, the whole process, from the mixing of the ingredients to the burning of the mixture and the proving of the calcined cement, only occupied a few hours, and I always obtained good cement; but when I tried the same process on a great scale, the same mixture entirely failed, first from my having washed it, and afterwards owing to my having allowed it to remain several weeks exposed to air before it was burned, which spoiled it in both cases.

Mr. Frost must no doubt have been equally expeditious with his experiments on a small scale, whilst in working on a great scale, he always occupied several months between the grinding of his ingredients and the burning of the mixture, as has been stated.

On finding that the blue clay of the Medway answered so well with me, he sent barges to dredge for it below Chatham Dock-yard by permission, and on trial adhered to it, at least for some time afterwards. He also tried my proportion of chalk and clay of 1830, and I believe adopted it as nearly as he could approach to it by a different mode of measurement, for he measured the chalk in lumps and the clay in mass, both by cart loads. I frequently recommended him to abstain from washing his clay and drying his mixtures, in the mode previously adopted by him, as my own experiments appeared to me to have proved, that washing and drying either spoiled or materially injured the peculiar raw cement mixture, used by me and afterwards adopted by him (see from Articles 108 to 111 inclusive); and he informed me that in one experiment in which he had acted on my system, mixing his ingredients by the pugmill and burning the mixture immediately, he had obtained better cement than by his his usual process, which however he did not finally alter. Mr. Frost informed me that his cement was first used in the erection of Hungerford Market, both for the foundations and as stucco. At one time he added ochre to the mixture of chalk and clay, in order to improve his artificial cement, which I believe he did on the same principle that had induced me to use the protoxide of iron, in experimenting on a small scale (88).

(XVII) When Messrs. Francis, White & Francis, cement manufacturers, were in treaty with Mr. Frost to purchase his artificial cement works in 1833, Mr. White the second partner referred to me for my opinion on the subject. I assured him that my experiments, of which I shewed him the results, had convinced me that artificial cement might be made of chalk and blue clay, perfectly efficient for all hydraulic purposes, as well as for stucco; and I stated that by doing away the washing and drying, which I considered prejudicial, the qnality of the cement might be much improved. Afterwards Messrs. Francis and White purchased Mr. Frost's business, on his retiring to America, which they carried on, until by a separation of the firm, it was made over entirely to Mr. White and Son, on the 1st of January 1837. This gentleman wrote to me, that having repeatedly tried Frost's cement, against the best natural cement, by setting out bricks from a wall with both, he considers the cohesive strength of the former to be in the proportion of about two thirds only of that of the latter, which he does not think a disparagement, 'that which is artificial,' to quote his own words, 'not being so perfect, as what nature gives us, yet' he observes that 'Mr. Frost's cement is so good, that it may be used with perfect 'safety for all building purposes, where it is not required to harden 'in water. The latter it will not do so effectually as the Roman 'Cement.'

In reply to such observations, I have assured Mr. White, that if he will lay aside Frost's system of washing and drying borrowed from M. Vicat, as well as his proportions, in which without knowing them precisely, I am sure that the chalk predominates too much over the clay:* in short if he will adopt the process and proportions recommended by me in the foregoing Treatise, that he cannot fail to produce an artificial cement, equal in efficiency for all the purposes of hydraulic as well as of common Architecture, to the best water cements of nature. Messrs. Francis inform me that they also propose to make artificial cement on Frost's principle, since the dissolution of their partnership, but neither they nor Mr. White appeared to attach any importance to the manufacture of this sort of cement, for they invariably recommended their customers to purchase their natural cement in preference, though more expensive, and never disguised their opinion of the inferiority of Frost's cement. But as I think it probable, that the making of artificial cement will sooner or later become general in this country, if they should not think it worth their while to adopt a better system, some other manufacturers will supplant them in the peculiar branch of their business now alluded to, by producing an excellent artificial cement, instead of the very middling article made by them, and known in the market by the name of Frost's cement, of which they themselves, as I again repeat, do not scruple to admit the inferiority.

On cementitious Architecture, as applicable to the Construction of Bridges, by John White Architect, &c. &c. published in London in 1832.

(XVIII) Mr. White's attention appears to have been drawn to the subject of cements, in consequence of having amongst other competitors given in a design for building a new bridge over the Thames, in lieu of old London Bridge, which must have been submitted to the Bridge Committee early in 1823, and afterwards to a Committee of the House of Commons.

The first mode of construction suggested by Mr. White is the cementitious, which he defines to be that consisting of small materials of almost every form and quality, so well united as in fact to approach to a perfectly solid congruent body. He proposes to use either Parker's Sheppy cement or Atkinson's Yorkshire cement, or puzzolana for this purpose; the two first where immediate adhesion is required, the last where time for setting can be allowed. He observes that ' none of these cements will shrink, they attach well to each ' other, to stone and to brick, so that if the stone or brickwork is

* I do not know Mr. Bazley White the cement manufacturer's present proportions, nor whether he now uses the brown pit clay of Upnor or the blue clay of the Medway, nor whether he continues to employ ochre as one of the ingredients; but the colour of his cement certainly approaches much more nearly to that of the natural cements than mine, which is composed of chalk and alluvial clay only. Mr. White junior informed me, that the machinery on the premises being all adapted to Mr. Frost's system, was an objection to any change, even for the better.

'well set and fitted in, an almost homogeneous structure will be 'obtained. Admitting this to be the case, the form of the edifice 'itself can only be so far important, as that it should afford sufficient 'substance to be united, capable of carrying its own weight, and 'the adventitious weight, which public traffic may impose upon it, 'the shape and size of the united materials not being of much im-'portance, provided they do not approach to regular squares, or that 'rounded bodies are not introduced in too large proportions. Per-'fectly vitrified bricks may be equally employed with stone, but not 'to form the external face, which is supposed to have a moderately 'tooled plain worked surface, and the arch stones following the 'curve as secants to the circumference.'

With regard to the stone, he observes that 'the cementitious or 'adhesive bridge may be of best Kentish rag, best Dartmoor granite, 'best Aberdeen granite, selected Portland stone or brick mixed, all 'in small sizes. The only essential point is that there should be 'an uniform external facing of best rag or granite, well united with 'the internal materials.'

He also proposed a bonded bridge of large stones cut in the usual manner, that has hitherto been generally adopted for important bridges, all nicely fitted with very thin joints, and those of the arches, which were segments of a circle, all radiating properly, and he suggests that the stones might be tooled by a steam engine.

He estimated his cementitious bridge at £357,285, and his bonded bridge at £470,750.

I had not seen this pamphlet, and consequently was not aware of the above circumstances, when I adverted to Mr. White's experiments on artificial puzzolana, in Article 259, which were first published in 1826, in volume XX of the Quarterly Journal of Science, No. XXXIX, as an article, the 8th of that number, on calcareous cements, in which Mr. White's object appears to have been to ascertain the comparative adhesiveness and resistance of cement, common mortar, and puzzolana. He states that puzzolana, whether foreign or British, for by the latter name he terms his own artificial puzzolana,* if reduced to a very fine powder, lost considerably its power of adhesion, though it was more plastic. He therefore inferred that the greater the variety of dimension of the particles, the greater would be the entanglement of the asperities, and consequently the greater the adhesion. Of the mortar made, it also appeared that the finer the lime could be ground, the more perfect would be the combination, and the harder the mortar obtained; because the hard particles of puzzolana being in a state of actual contact, no compression was likely to take place; and which in fact, upon the subsequent investigations, proved to be the case.

* I have seen a Paper printed in 1824 containing directions for using Patent English Pozzolane, but I do not know whether this patent was taken out by Mr. White, who writes Pozzolano, in the Pamphlets published under his own name. I believe that the proper Italian orthography of the word is *Puzzuolana*, implying *Terra*, from the Latin *Terra Puteolana*, which I have simplified by omitting one superfluous letter.

After this he observes, in respect to artificial puzzolana, 'that the 'real difficulty which existed was the obtaining a perfect knowledge 'of the best state, and the best mode of indurating properly the clay 'itself, for if the burning of the clay was such as to cause vitrification, an imperfect mortar was made. Perfect glass, scoria and 'pumice stone produced very inferior mortar; perfect puzzolana 'appeared to be made when a chalky clay was so indurated by 'fire, as to put on the appearance of an incipient vitrification only.'

(XIX) In his first experiment, tried on the 9th of August 1824, a pier 3 feet 4 inches wide, 1 foot 10½ inches thick, and 5 feet 8 inches high, built one month before of hard sound London burned stock bricks, and mortar of 1 part ground lime, 1½ part sand, and 2 parts rough puzzolana, was elevated by applying screw jacks on each side, course by course, beginning at 15 courses or 3 feet 9 inches from the bottom. This elevation from its foundation took place, without any separation of the courses, until the screw jacks were placed 2 courses or 6 inches from the top, when these two courses separated from the remainder of the pier, which was then thrown down, when it again separated at eight courses from the foundation, showing that the middle was not dry.

In his second experiment, tried on the 21st of April 1825, a pier 35 courses of brickwork high, being 6 feet wide, 3 feet thick, and 8 feet 10 inches in height, built about 9½ months before, of the same sort of bricks, with a cement composed of 3 parts of artificial puzzolana ground and sifted, having no particles bigger than one 8th of an inch, and one part of ground Dorking lime unslaked and fresh mixed, was first elevated from its base by a strong chain grappling it at nine courses of brickwork from the top; again in the same way at six courses from the top; and at last the whole pier was suspended by a set of lewises let into the middle of the top of the pier, about 15 inches deep. The weight of the pier was about 9 tons.*

In preparing for his next experiments, he had adopted a suggestion of Mr. Brunel and Mr. (now Sir Robert) Smirke, who were present on both the former occasions, that it would be useful to try the comparative adhesive properties of these cements, by building other piers of such dimensions as might enable them to be laid horizontally, and have weights placed on them in the middle; and accordingly 9 piers were constructed on the 21st and 23d April, in a vertical position, and of the following dimensions, viz: 6 feet high and about 14 inches square.

The first pier (A), consisted of bricks united by pure fresh Roman cement, by which I believe the author meant Sheppy cement, and was accidentally broken in laying down, at a part where the cement had partially set before adhesion had taken place. A fragment of this pier converted into a beam 3 feet 6 inches in length, was laid on two supports, as shown in the following figure, at the

* Mr. Brunel, and Sir Robert Smirke were present at both of these experiments. Mr. Matthew Wyatt and Mr. Smith, &c, were also present at the first, and Sir Thomas Baring, Admiral Sir Edward Codrington, Colonel Lowther, Dr. Chumside and nearly 200 spectators at the second.

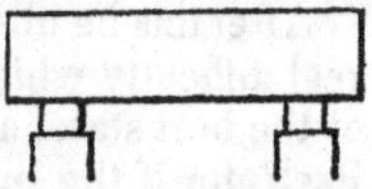

clear distance of 2 feet 6 inches apart, and being carefully loaded over its center with weights added by half a hundred at a time, it supported 1232 lbs, and broke down under 1288.

His second pier (B) was built with mortar composed of 3 parts of puzzolana and 1 part stone lime, reduced to putty as common mortar. When this pier was similarly placed on the supports it broke in the middle, but one of the fragments being similarly placed and loaded bore 448 lbs, and broke under 504 lbs.

The third pier (C) was built with the same puzzolana mortar as the former, but broke in turning it round.

The fourth pier (D), which was built with 3 parts of pure sharp sand and 1 part of stone lime, broke into three pieces on attempting to lay it on the supports.

The fifth and sixth piers (E and F), which were composed of 3 parts of sharp washed sand and 1 part of chalk lime, crumbled to pieces on attempting to place them.

The seventh pier (G), which was built with Atkinson's Yorkshire cement and sand in equal parts, supported 560 lbs. laid over the center of it, when supported at the clear interval of 4 feet between its bearings; and broke under 616 lbs, near to one of the supports, in a place where it was found that the cement had evidently not adhered equally to every brick.

The longest fragment, when tried in the same manner, but with the clear distance between its bearings reduced to 3 feet, supported 2240 lbs, for an hour, and broke under this weight in consequence of a shock caused by the breaking of the next pier.

The eighth pier (H), which was built with 4 parts of puzzolana and 1 part of air slaked stone lime, broke in laying down. On trying a second experiment with it, as on the preceding pier, it bore 616 lbs, and broke in two pieces.

The whole of these experiments were tried on the 12th of May of the same year, and therefore the cement and mortars used were 19 or 21 days old at that period.

The large pier before tried in the second experiment was next examined by breaking into it with iron wedges and sledge hammers, and splitting it in two places transversely. It resisted long, but it was found that many bricks had not taken the mortar. It resisted still more when they split it longitudinally. It was then also split vertically. Mr. White states that the mortar, which was composed of 3 parts of artificial puzzolana and 1 of Dorking quick-lime powder, and which was at that time ten months old, was not yet completely dry, and that it would require another summer to give it all the toughness and tenacity it can acquire.

Eight days afterwards, the fragments of the piers broken down in the former experiments, were subjected to a vertical pressure, at Mr. Bramah's at Pimlico, by his hydrostatic press.

First trial. The pier (A), built with Sheppy cement, section

196 superficial inches, cracked with 48960 lbs, and was entirely destroyed by 92160 lbs.

Second trial. The pier (B), built with 3 parts of puzzolana and 1 part of lime, section 180 superficial inches, being overlaid with sand at top escaped compression, so that the result was uncertain.

Third trial. A smaller pier, 9 inches square, built of 3 parts of puzzolana and 1 part of ground lime, section 81 superficial inches, was compressed a very little with 18720 lbs, and entirely destroyed with 24480 lbs.

Fourth trial. The pier (G), built with Atkinson's Yorkshire cement and sand in equal measures, section 196 superficial inches, was cracked on two sides with 37440 lbs, and entirely destroyed with 80640 lbs.

Fifth trial. The pier (H), built with 4 parts of puzzolana to 1 of lime, section 81 superficial inches, cracked with 28,800 lbs, fracture increased with 48960 lbs, entirely destroyed with 51840 lbs.

Sixth trial. The pier (C), built with 3 parts of puzzolana to 1 of Dorking lime, section 81 superficial inches, was fractured on one side with 31680 lbs; fracture increased with 43200 lbs: fracture again increased with 48960 lbs. This experiment was not pursued till the pier was entirely destroyed.

Seventh trial. The pier (D), built of 3 parts sand to 1 of stone lime putty, section also 81 superficial inches, was compressed one eighth of an inch, and cracked with 40320 lbs; and was fractured in five places and entirely destroyed with 46,080 lbs.

Eighth trial. On a pier of Portland stone, 14 inches by 12 inches and 2 feet 7 inches high, section 168 superficial inches. A sudden fracture was produced, which divided the stone into two pieces in the center of the widest side with 173½ tons or 388640 lbs. The upper end of the stone was bedded in puzzolana, which was compressed into a cake of five eighths, one fifteenth thick in a wet state, which cake remained quite solid after the fracture of the stone.*

(XX) *Inferences drawn by Mr. White, from the above Experiments.*

First, that an important adhesion of brickwork had taken place by the use of puzzolana, sand and lime, in the short period of 30 days.

Secondly, That from the use of puzzolana and lime in the proportions specified, almost all the advantages required from a good building cement were obtained.

Thirdly, That Lord Mulgrave's or Atkinson's (Yorkshire) cement had, in the short space of 23 days produced an induration, which was sufficient to maintain almost any weight brickwork was capable of for openings in buildings. The effect probably would have been the same in Parker's, had the material not set before the

* The experiments of the two last days were made in the presence of Mr. Brunel, Sir Robert Smirke, Major-General Sir Alexander Bryce, and other gentlemen, amongst whom was the Earl of Southampton on the first occasion. I think Mr. White is right in stating the witnesses to his proceedings, which is always satisfactory.

bricks were fixed in it. Further, that puzzolana had not in that period produced an equal adhesion, and that common mortar had produced hardly any; and it appears from the splitting of the large piers thrown down on the 21st of April, that an increasing induration took place. This was evident from the nearly equal fracture of the bricks and cement.

Finally, the incompressibility of mortar being one of its material qualities, it results that Parker's, Mulgrave's, and puzzolana are so far equally useful, that brickwork composed with them will bear on each superficial foot, before the bricks will crack, about 23 tons, or 51520 lbs, that 50 tons, or 112000 lbs. will totally crush such brickwork; and that Portland stone of the best quality will not split with less than 173½ tons, or 388640 lbs, and that a bedding or joint of puzzolana mortar is not destructible with that weight.

As the results of the experiments before described by Mr. White, to ascertain the weights required for cracking and for breaking his various mortars, appeared to me too discordant to admit of the inferences he has drawn, as to the Sheppy cement, the Yorkshire cement, and his own puzzolana mortar, being all equally useful, and all capable of being cracked, and afterwards entirely destroyed by the same average weights, I took the trouble to analyze the results of his crushing experiments with Bramah's press, and found that the mean weight, which cracked or compressed the two cement piers was only 31730 lbs, that for the puzzolana mortar piers 46933 lbs, and that for the mortar pier in which sand and Dorking lime were used 71680 lbs, per superficial foot; and that the crushing or totally destroying weight was for the cement piers 63477 lbs, for the puzzolana mortar piers, supposing that pier (C) would have required as great a weight as (H), 75913 lbs, and for the mortar pier 81920 lbs, whilst the breaking weight for Portland stone was 333120 lbs. per superficial foot. Thus from his own experiments, and contrary to the inferences drawn by him, it appears that the use of cement produced the weakest, and Dorking lime mortar far the strongest of his brick piers, whilst his own puzzolana was of intermediate strength between these two, that is so far as their resistance to a cracking or crushing weight are concerned. As these results are contrary to all the experience of the building profession, it being well known that brickwork laid in cement is stronger than that laid in the best mortar, I am afraid that Mr. White's experimental piers must have suffered from careless workmanship, many of the cement joints being imperfect according to his own account, and those of his large experimental puzzolana pier being also imperfect.

Hence his experiments only prove the great inferiority of the resistance of good bricks to that of Portland stone, and the great inferiority of chalk lime mortar to the other mortars and cements experimented upon by him; but I cannot discover that they afford any proof of his own artificial puzzolana mortar being stronger than that made of Dorking lime and sand: on the contrary, the only apparent superiority of one of his puzzolana mortar piers (B) was that in falling it broke into two pieces instead of three, which may have been by accident; but in comparative resistance it was far inferior to the Dorking or stone-lime-mortar pier. His other puzzolana mortar pier (C) was an entire failure. In fact I discovered myself after much more numerous experiments, that it was useless to attempt to arrive at the comparative strength of cements, by uniting them with materials of so very little resistance as common bricks, and this remark would apply to Mr. White's experiments also, even if the workmanship of his brickwork had been perfect: but as that was not the case, it is useless to attempt to deduce any just conclusions from the results of experiments tried with materials so badly put together.

I am sorry to be compelled to criticise Mr. White's proceedings, but it is essential to arrive at truth in a subject of so much practical importance, and as I think that he has drawn erroneous inferences from his own experiments, it would be false delicacy not to notice them. In one respect his labours have been of great use, if as I suspect they drew Mr. Brunel's attention to the same subject, for I think it probable that the unsatisfactory resistance of Mr. White's brick and cement piers, experimented upon as beams, must have led Mr. Brunel to increase the resistance of beams of this description by hoop iron bond, an improvement of the great importance of which I was not convinced, until I tried the same construction with and without it.

Extracts from and Remarks upon a Prize Essay on the Nature and Properties of Concrete, and its application to Construction, up to the present Period, by George Godwin junior, Associate of the Institute of British Architects, published as the first Paper of the first Volume of the Transactions of that Society, 1836.

(XXI) In this Essay, after stating that a sort of concrete was used by several ancient nations, especially by the Romans, which they styled *Cæmentum,* and afterwards by the Anglo-Saxons and Normans in this country, as well as more recently by the Italians who termed it *Smalto,* and by the French under the name of *Beton* as described by Belidor,* Mr. Godwin remarks, that Mr. George Semple of Dublin, who published in 1776 a Treatise on building in water, appears to be nearly the earliest modern English writer, that treated of the construction of artificial foundations by concretion.

Upon these statements it may be remarked, that the *Cæmentum* of the ancient Romans, and the *Beton* of the French are not precisely what we now understand by the term concrete, nor were they applied to the same purpose. They were composed of regular mortar mixed with pebbles or small broken stones, and alternating with thorough-bond courses of wall tiles or of flat stones in the

* To whom Mr. Godwin might have added the Moors, for I have observed works of small masonry or of rubble masonry grouted, in Spain, Minorca and Sicily, which are supposed to have been built by them.

Roman works, at moderate intervals; and in the French *Beton* as described by Belidor, alternating also with courses of rubble stone at smaller intervals. In short they were masonry of small materials, and were used, not for foundations properly so speaking, but as walls : for when the Romans built in this manner, they prepared their foundations with much larger stones, and never commenced the small masonry resembling our present concrete, until they had got above the surface of the ground, that is in building on dry land*: but in working under water, after driving two parallel rows of piles as a casing, both the Romans and after their example the French used this sort of small masonry or *Beton*, without reference to the nature of the foundation at all, and solely in order to enable them to construct their wall without being under the necessity of making coffer dams, and of pumping out the water, which could not be dispensed with in regular masonry. And though Belidor and others have used the term foundation for this sort of work, it is by no means appropriate, being applied to walls of uniform profile, which might be raised as high as 18 or 20 feet or more, on both sides, until they reached the surface of the water, after which they were finished with regular masonry, carried so much higher as the nature of the work might require.

(XXII) In respect to Mr. George Semple, I agree with Mr. Godwin that he was the first modern English writer, and what is more I consider him the first writer of any age, who proposed the sort of concrete which recently has been so much used in this country, for he recommends grouting a mixture of gravel and sand, or a mixture of small broken stones and sand, with quicklime; but he so far differs from the judicious principle now generally adopted, that instead of confining his concrete to foundations or backing, which are either buried under ground, or protected in front by regular masonry, he proposes to build the piers of bridges or other walls with it under water, in tide rivers and on the coast, at the depth of from 16 to no less than 36 feet at low water, above which level he finishes with regular masonry, and in addition to the depths above mentioned, he supposes a rise of from 10 to 20 and even 24 feet more, at high water of spring tides, so that in one example the extreme depth at that time of tide is 60 feet. He proposes to build these great concrete walls in caissons of timber, having strong bottoms and sides, and as they were all to be perpendicular, and as he says that any sort of lime made from hard lime stone would answer the purpose, which we now know to be an error; Mr. Semple's concrete walls, unless built with some very strong hydraulic lime such as the blue lias, which with him would depend upon accident, would never set at all, but would remain a soft pulp containing

* See Archæologia, vol. IV, paper VII, entitled 'Remarks on the Antiquity and the different Modes of Brick and Stone Buildings in England, by Mr. James Essex of Cambridge,' read before the Society of Antiquaries, the 8th of December 1774.

a mass of pebbles and sand, so feebly united, that nothing but the sides of the caisson would hold them together, in the event of a flood inland, or of a storm on the coast.

Hence it must be acknowledged that Mr. Semple's ideas are extremely crude; not so his practice; for in the construction of new Essex Bridge Dublin, he used regular masonry founded upon a piled foundation, after having cleared out the water by proper coffer dams. In short, excepting in his predilection for semicircular arches to the exclusion of all others, he then followed the most approved practice of bridge-building, and it was not till more than twenty years afterwards, when he published his Treatise on building under water, that he indulged in those rash speculations, which he never carried into execution. Nor was his system of building under water with concrete ever copied since, until a few years ago, when Mr. Taylor, at that time Architect to the Admiralty, employed it for the first time, but under much less difficult circumstances, than those contemplated by Mr. Semple, in the new wharf wall of the Royal Dock-yard at Woolwich, which was constructed by Mr. Ranger, under his direction, but not until concrete had previously been generally used and with perfect success, for the foundations of very important buildings, and for the backing of wharf walls.

(XXIII) To return to Mr. Godwin's Essay, he next remarks, that when Smeaton examined the ruins 'of Corf Castle, Purbeck, 'at a time when he was actively engaged collecting facts relative to 'the strength of materials, he "found that its solidity, as usual in 'ancient edifices, did not consist in having been built with large 'hewn stones throughout; for the filling in of the walls was of rough 'rubble and fragments from the quarries, the interstices being en- 'tirely filled up with mortar that had undoubtedly once been fluid, 'and in that state poured in; and from the nature of the component 'matter, as well as time, the whole mass had become strongly 'cemented together." The mortar was composed of lime, with a 'considerable admixture of sharp sand and pebbles.'

'After this,' Mr. Godwin adds, that 'when he (Smeaton) was 'about to construct the first lock on the River Calder, induced by 'the consideration of this circumstance, he introduced what he calls 'a rubble backing, composed of 4 parts barrow lime, 2 of coarse 'sand, 2 of fine sand, 2 of ground minion (siftings of the iron stone 'after calcination) and 8 of pebbles, the largest not exceeding in 'size a horse bean. This was in 1760.'

Smeaton himself describes the back part of the walls of the River Calder works differently, as having in general been done with rubble or rough stone from the quarries, where the interstices were large and open, and required a good deal of matter to fill them, and for this reason he mixed pebbles with his mortar, in the proportions before described in the last quotation from Mr. Godwin, and he calls this mixture *Pebble Mortar*. Hence Mr. Godwin's description of Smeaton's rubble backing is imperfect and incorrect, for by omitting the stones entirely and noticing the mortar only, he leaves it to be inferred that the backing of those works was a concrete, whereas, like the walls of Corf Castle, which Smeaton professedly copied, it was in reality a rubble masonry.*

(XXIV) Mr. Godwin next observes 'when the generic term 'concrete,—derived it is hardly necessary to say from the Latin word '*Concresco*, to grow together, to congeal, clot or thicken,—was first 'applied as a special appellation to the particular composition under 'consideration, can hardly be pointed out, the transition was so 'easy; but as this must have resulted from the general and frequent 'use of this concretion, the term perhaps can only date from that 'period, when its use *became* general and frequent, probably not 'longer than 15 or 20 years ago. It is true that in 1800, it was 'used by Mr. Ralph Walker at the East India Docks, but at that 'time it was so little understood, that Mr. Rennie the elder, who 'was engaged there, laughed at the idea, and when he found that 'Mr. Walker was in earnest, declared he would have nothing to do 'with the execution of it. Mr. Walker however persevered, and 'notwithstanding the natural soil was exceedingly unequal and bad, 'in no one case did it disappoint his expectations.'

Having always understood, that concrete was first used in this country for the foundations of the Penitentiary at Millbank in 1817, and never having heard of its being employed at the East India Docks, as stated by Mr. Godwin on the authority of Mr. Macintosh, I was induced to inquire into the fact, and I find that Mr. Godwin has been misinformed, and that my own previous impression was correct. Mr. Walker the present distinguished President of the Institution of Civil Engineers, informed me, that as the Nephew and Pupil of the late Mr. Ralph Walker, he knows that concrete was not employed for the foundations of the London Docks, instead of which well washed gravel only was used for those foundations, and I am not surprised that Mr. Rennie should have doubted the efficacy of this sort of foundation, for though a natural bed of gravel makes an excellent foundation, it is usually indurated by some cementing matter, being seldom or never found in the loose state, in which Mr. Ralph Walker used it. On cutting into one of his Dock walls afterwards, it was found that a portion of the lime, dripping down from the wall, had accidentally converted a small part of the gravel beneath into concrete towards the front of the wall, but that in rear, the gravel was still quite loose. This information was confirmed by Mr. Henry Martin now in charge of the Works and Buildings of the East India Docks, who thus expresses himself in a letter upon this subject, now in my possession, dated the 8th of December 1837. 'In answer to your inquiry respecting the matter 'contained in Colonel Pasley's letter of the 2d instant, I have to

* I believe that Smeaton used common lime for the backing mortar, and barrow lime only for the face mortar, of the River Calder Works, and in the latter he did not use pebbles, these being added merely to fill up the large interstices of the rubble masonry.

'observe, that seeing some time since the paper referred to, I took 'some pains to ascertain the accuracy of the statement relating to 'Mr. Ralph Walker, and searched the minute books, specifications, 'contracts and other papers in the Dock Company's possession, 'but found no mention made of the circumstance. I then enquired 'of some artificers, who assisted to build the premises. They re- 'membered a part of the foundation of boundary wall being formed 'of gravel washed in and punned,* but thought no lime was 'used. Upon excavating in several places, I found this to be the 'case, I am therefore inclined to think, that Mr. R. Walker did not 'use concrete at this place, and am more convinced, when I know 'that there is not a single building put up by him at this place, that 'have their foundations not timbered.'

After having stated, but as I have shown erroneously, that concrete was used at the East India Docks in 1800, Mr. Godwin observes, 'From that time to the year 1815, we find but a few 'isolated instances of its adoption.† In this year however, if I am 'rightly informed, it was used with success at the Penitentiary, 'Millbank. The soil was unstable in the greatest degree, and two 'towers erected with some attention to this fact entirely failed; they 'were taken down, having split in several directions, a body of con- 'crete was introduced, upon which others were built, and they stand 'well to this time.'

(XXV) This statement also is not quite accurate. No concrete was used in England until the year 1817, by Sir Robert Smirke, who was called in and appointed Architect to the Penitentiary, in consequence of the failure of the two first pentagons, previously built by another Architect, who used the same sort of foundation adopted by Mr. Ralph Walker at the East India Docks, namely loose wet gravel, which giving way as stated by Mr. Godwin, this gentleman was superseded by the then Lords of His Majesty's Treasury. I have ascertained on inquiry, that the chief cause of the failure of the gravel was an error in judgment on the part of the Architect, in not extending it to a sufficient distance beyond the external walls of those pentagons, which stood on the very edge of it, so that these walls only gave way, for the internal walls resting on the center of the gravel foundation, which was laid over the whole area, afterwards covered by each of those pentagonal buildings, showed no symptoms of failure. The foundations of another part of the Penitentiary, built previously to the employment of Sir Robert Smirke, also failed. This was the Porter's Lodge or entrance, which was founded upon piles that became rotten in a few years

* A technical expression by which they meant rammed.

† The only isolated instance mentioned by Mr. Godwin is in a note, in which he states that 'a concrete of Kentish rag stone and liquid mortar 'was placed between the heads of the piles on which Waterloo Bridge was 'built, in 1811.' This does not appear to me to come under the proper definition of concrete, certainly not of a concrete foundation.

afterwards, as I had an opportunity of observing, when they were removed and a concrete foundation introduced.

Thus the merit of first successfully and judiciously applying concrete, on a great scale and with perfect success, to the foundations of an extensive and important public building, on the very worst description of soil, for the Penitentiary was built on soft alluvial mud mixed with peat, is due to Sir Robert Smirke, who neither took the hint from the Roman Cæmentum or French Beton, nor from Mr. George Semple, but from a fact mentioned to him in conversation by the late Mr. Rennie, of part of the gravelly bed of the River Thames having been converted into a sort of concrete, by the accidental sinking of a barge full of lime during the period occupied by the erection of that bridge; which was a much more convincing proof of the probable advantages of this hitherto untried expedient, than the similar observation made by Mr. Semple on taking up the foundations of old Essex Bridge at Dublin in 1753, when he observes 'that the bed of sharp gravel on which the piers had been 'built was actually petrified seemingly into a close solid stone, by 'the small quantities of the petrifying qualities of the lime, that had 'sunk down into it in about 70 years, but principally since erecting 'the effigy (a statue of George the 1st) in 1722.' I have before mentioned that this circumstance induced Mr. Semple to recommend the extensive use of concrete, but in a less judicious manner, than was afterwards adopted by Sir Robert Smirke, without reference to the contents of Mr. Semple's Treatise, which never led to any practical results, as it might perhaps have done, had his projects for the application of it been less extravagant, some of them being not merely injudicious, but chimerical.

Having myself watched the progress of Practical Architecture in the south of England for many years past with attention, I was particularly struck by the great importance of the recent improvement of using a mass of concrete for foundations, in soft soil, in which the precarious and expensive expedient of piling would formerly have been deemed indispensable, and I never heard it ascribed to any other person than to Sir Robert Smirke, until I read Mr. Godwin's elaborate Essay on the subject of concrete, in which the name of that eminent Architect is not even mentioned, though in his very first attempt he used it at the Penitentiary in a manner, that has since been generally copied, both in the proportions of the ingredients and in the mode of using them; and which I do not think has been improved upon by those who have deviated from his example. The only thing we do not owe to him is the term *Concrete*, which is certainly more applicable than the phrases, *Grouting* and *Grouted Foundations*, which he himself at first made use of.

I shall conclude by remarking, that all my own Observations on Concrete, as contained in the first sheets of the foregoing treatise, were actually printed before Mr. Godwin's Essay made its appearance; and though his opinions in some points will be found to differ from mine, I abstain from entering into any discussion relating to them; but in respect to several of his statements as to matters of fact, on which the well merited reputation of such men as Smeaton and Sir Robert Smirke may in some measure depend,* I thought it desirable to point out those inaccuracies, into which Mr. Godwin has been led, from too hasty a perusal of the writings of the former, and from want of correct information as to the practical operations of the latter, and of some of his cotemporaries.

(XXVI) In Appendix A, to Mr. Godwin's Essay, extracts are given from the specification of Mr. Abraham, containing his rules for making the concrete foundations of the new Bridewell, Westminster, which appear extremely judicious; and in one precaution perfectly original. I allude to his having required the contractors to sink an engine well 6 feet in diameter near the middle of the ground, or where directed by the Architect, for the purpose of draining the water from the foundations, whilst the same are being put in, and keeping the water from the concrete until the same is perfectly consolidated. By means of this well, the water was to be drained from the sites of the buildings and boundary wall, to the depth of at least 5 feet below the bottom of the concrete under the boundary wall, or 16 feet 6 inches below the highest tide level. From my own experiments, I am persuaded that this precaution could not fail to insure the object in view, of consolidating the concrete more quickly and effectually, than if it had been left in a wet state: but at the same time I should not pronounce it to be a thing absolutely necessary, because even the loose gravel first used for foundations by Mr. Ralph Walker, if confined at the sides, which was not the case in the foundations of the two first pentagons of the Penitentiary, is incompressible when thoroughly wetted; which fact no doubt Mr. Walker must have known, and which has recently been confirmed by my own experiments; and therefore such gravel must sink equally. Hence a moderate proportion of lime slaked with water, just sufficient to fill the interstices of clean gravel and sand, and to change this incompressible mass from a loose into a solid state, cannot fail to form an efficient foundation, whether it set quickly or slowly, whether into a moderately hard, or into a very hard, substance.

* In a lithographed Treatise on Practical Architecture, written for the use of the Junior Officers of the Corps of Royal Engineers in 1826, of which I sent copies not only to the Commanding Engineers at all the Stations of the Corps, but also to several Architects and Builders, and to my excellent friend the late Mr. Telford, requesting them to favour me with their opinion and remarks, I ascribed the first introduction of the use of concrete to Sir Robert Smirke. If I had been wrong, Mr. Telford would have corrected me, as he always watched the progress of professional improvement. In his own writings, all of subsequent date to the East India Docks, but of prior date to the building of the Penitentiary, namely the articles Bridge, Civil Architecture, and Navigation Inland, in the Edinburgh Encyclopædia, he does not even mention concrete at all, though he afterwards used it himself at the St. Catherine's Docks; and I am not aware that it ever was mentioned by any recent author on Civil Engineering or Architecture, prior to 1817.

Of the Foundation of the North Storehouse on the Anchor Wharf, Chatham Dock-yard, which was underpinned with Concrete by Mr. Ranger in 1834, under the direction of Mr. Taylor.

(XXVII) This operation has been described by Mr. Taylor in the first volume of the Institute of British Architects, of which he is a fellow, and at the time it was going on, I inspected the work occasionally, without any great attention, but sufficient to satisfy me of the correctness of his description of the mode of proceeding.

Two extensive storehouses, built in the same alinement, occupy almost the whole length of the Anchor Wharf in Chatham Dock-yard, and were nearly similar to each other in most respects, when first built, excepting in their foundations, one of which has remained perfect, whilst the timber of the other having decayed, a complete derangement of the walls and floors was the consequence, after which Mr. Taylor caused part of the front wall near the S. W. angle* to be pulled down and rebuilt on a new foundation of concrete, and at the same time the whole of the rest of the foundations were underpinned with concrete, as before-mentioned. This storehouse is the most northerly of the two, and is described by Mr. Taylor as being 540 feet long, 50 feet wide, and 5 stories high, which is the case in the center and near the ends, but in other parts it is only 4 stories high. Towards the North end of this storehouse, there are vaults under the ground floor, and at this end the foundation rested on planks and sleepers supported by piles. In other parts it rested on transverse planks, with sleepers below them of very large scantling, laid alternately in longitudinal and transverse courses, and Mr. Taylor particularly mentions, that in one part there were no less than nine feet of timber heaped up in this preposterous manner, and I suspect that excepting where piles had been driven, this timber foundation was bedded on the chalk, the surface of which is said to have been in a shelving or sloping form, and consequently required unequal thicknesses of timber in different parts, to prepare it for the brickwork. In the section, which Mr. Taylor has given to illustrate his mode of underpinning the foundation, four courses of large timber and one course of transverse planks are shown below the brick footings of the wall, which at that part appear only to have been about 4 or 5 feet distant from the solid chalk. Since the recent settlements in the new concrete foundation, which gave this building a sort of interest with me, that I did not attach to it before, I inquired into what circumstances could have caused such an injudicious construction to be adopted, and having ascertained, that one of these two storehouses was built by Messrs. Nicholson and Co. of Rochester, I called on Mr. Nicholson, who informed me, that the first storehouse, of which the foundation is still perfect, was

* Which in Mr. Taylor's paper is called the N.W. angle by mistake, for the Anchor Wharf lies more nearly North and South. than East and West. The two storehouses are of the same height, and apparently of the same elevation, though the South Storehouse is much longer than the other, and somewhat narrower, but the difference does not strike the eye of a person in the Dock-yard.

built by his Father and Mr. Martyr of Greenwich as Contractors,* under the directions of an Architect, who very judiciously required the foundations of all the walls to be carried down to the solid chalk below, not however without using some transverse sleepers and longitudinal chain bond timbers, according to the custom of that day, but at such intervals apart, that their decay could not injure the stability of the walls.

(XXVIII) The construction of the second storehouse, which is said to have been finished in 1798, and which was intended to be nearly a fac simile of the former, was intrusted to the superintendence of the Master-Shipwright of the Dock-yard, who instead of going down to the solid chalk, built his brick footings upon those extraordinary masses of woodwork, that have been described, which being composed of very large old well seasoned oak ship timber, he considered proof against all future decay. The result however proved his anticipation to be erroneous, for when Mr. Taylor caused these timbers to be removed, the whole of them were rotten, except the under part of the piles, which having been continually wet, were in a sound state, and therefore he allowed them to remain.

(XXIX) Another error of judgment, or at least what appears to me to be such, though by no means unprecedented, was committed in the construction ot this storehouse.† The tiers of story posts, which assisted in supporting the girders, were not all placed vertically over each other. For in the ground floor, there were 3 story posts to each girder, whereas there were only 2 to each girder of all the upper floors. Hence three and in some parts four tiers of story posts of the upper floors, had a false bearing, resting on the middle of the unsupported parts of the girders over the ground floor, in consequence of which, as these girders were giving way, it became necessary to introduce a great number of additional posts, to prop them up.

I am informed that the lower story posts rested on blocks of stone, and these may probably have had a bearing on the solid chalk, for the story posts have not settled like the walls, some of the girders which must originally have been placed on a horizontal level, or with a very moderate camber, being now 7 inches lower at the ends than in the middle, so that they have a strangely curved appearance, which has given to the floors in some parts a greater convexity than is usual in the decks of a ship; but at the south end in particular, owing to the settling of the end wall, the drooping of the floor is very extraordinary, and its level altogether irregular. Moreover whilst the wall plates have settled along with the walls in which they are imbedded, they have parted company with the ends of many of the girders, which remain unsupported, excepting by the story

* They finished this storehouse in 1783, before Mr. Baker entered into partnership with them, after which they erected another large building in the same Dock-yard, of which the foundations are also perfect.

† The story posts of the adjacent storehouse are disposed in the same irregular manner, but neither they nor the girders have given way in consequence.

posts before mentioned, so that whilst some of the ends of the girders have descended so far, that they still remain in contact with the wall plates, upon which they all originally rested, others stand at various heights from 1 sixteenth of an inch to at least 3½ inches above them, and these separations are quite irregular, the ends of the girders not having like the wall plates preserved any general uniformity of level.

(xxx) It is a great pity that the vacant spaces alluded to, under the ends of the girders, had not been all filled in immediately after the walls were underpinned, or that some record had not been kept of their state at that period, which would have enabled us to judge, what proportion of these separations took place before, and how much after that operation. At present, this is involved in great uncertainty, and but for the new cracks in the walls, and partial displacement of the new arch bricks, which can neither be mistaken nor disputed, as being proofs of recent settlement, it might be supposed that no such process had taken place at all, since the concrete was put in; for in the absence of precise details, Mr. Taylor's description of the separations of the girders and wall plates at the period, when he first resolved to underpin the walls, might apply equally well to their present state.

The opinion stated by me, that the whole of the walls have settled at least 2 inches, since that period, is deduced from observing that in some parts of the walls, between those girders which have been abandoned by the wall plates, spaces of rather more than 2 inches in height of red brickwork have become visible, immediately under the floors supported by the ends of those girders. This proves beyond the shadow of doubt, that in those particular parts, the floor supported by those hanging girders has retained its original position, or nearly so, whilst the walls have sunk down since they were last whitewashed; for the men employed in that operation ran their brushes into every part that was accessible; and whitewashed the whole inside of the walls from floor to floor, which was done in 1833, partly by bricklayers of the Dock-yard, paid by the day, and partly by convicts, just one year before the underpinning of the walls was finished. Hence I consider it proved, that a settlement of rather more than 2 inches has taken place since that whitewashing, of which settlement it seems reasonable to suppose the greater portion, if not the whole, to be of subsequent date to the underpinning, because the Master-bricklayer and his workmen, who repaired all the defective parts in the inside of the building, and washed them with Halling lime immediately after that operation, assure me that they do not recollect having seen any of those patches of red brickwork, at that period, which they have observed since in going round the storehouse.

I fear that many of my readers will consider the mention of these details as unnecessarily prolix, but the question, as to the merits of Messrs. Taylor and Ranger's mode of underpinning defective foundations with concrete alone, is of great practical importance to all Architects and Engineers, and to the public Service generally; and if I had merely expressed my opinion, that their mode of proceeding was an imperfect one, the assertion that it had failed, might be met by a counter-assertion to the contrary, unless I had also stated the grounds upon which I had formed that opinion, and shown that I had taken every pains to investigate the subject, which I did by personal reference to the Foreman and Labourers in charge of the store, as well as to the Master Bricklayer, and to several of the men who had been employed under him, in the pointing and special washing of all the repaired parts with Halling lime. The whole of these persons are unanimous in their opinion as to the new cracks which have appeared since this special whitewashing, but in respect to the general settlement as proved by the red brickwork coming to light, they cannot vouch for the precise period, at which this red brickwork first made its appearance.

(XXXI) After describing the alarming state of the building in 1834, when it was generally considered so bad, that nothing short of pulling it down and rebuilding it would do, Mr. Taylor states, that in order to avoid the immense cost of this arrangement, which was estimated at £70,000, exclusive of the additional expense of removing the stores, &c, he recommended that it should be underpinned, and that his first impression was to adopt the system so successfully used at the Custom-house under the direction of Sir Robert Smirke by Messrs. Baker; and that after having received directions to obtain their terms for executing it according to this system, and having opened the ground in several places to discover the nature of the foundations and their state, in order to form the necessary calculation of the cost, Mr. Baker gave in a tender for these works of £8800. He farther states, that as it had occurred to him, that a large portion of the expense consisting of 3 or 4 feet of brickwork in cement, according to Sir Robert Smirke's system, might be saved, by pressing the concrete, by means of strong screws, up to the under side of the old wall, he suggested this expedient to Mr. Baker, who declined to do it, as he did not consider that the building could be made secure in that manner.

The sum demanded by Mr. Baker for underpinning the wall according to the same system adopted at the Custom-house appearing to the Admiralty Board very large, Mr. Taylor adds that he was directed to obtain terms from some one who would undertake to adopt the method of screwing up; and consequently Mr. Ranger was applied to, who he says executed the work in much less time than had been proposed by Mr. Baker, and for a sum not exceeding one half of his estimate.

The event has proved that Mr. Baker was correct in his objections to this mode of underpinning the storehouse, which I heard him express at the time; but it did not strike me, that his apprehensions were likely to be so soon verified.

(XXXII). *Of the Metallic Cement.*

Messrs. Benson and Co. proprietors of copper works, have recently proposed a substance to be used like puzzolana for improving lime, namely their copper slag, which is usually in a coarse state like sand, and which was formerly always thrown away; until they used it in erecting a small building attached to their works, which they did merely in consequence of their common sand being expended, and in consideration of this building being of little importance. The mixture of this ingredient with lime produced a mortar of such very superior quality, that they were induced to offer their copper slag for sale in the London market, at the premises of Messrs. Howard & Co, Finsbury Wharf, City Road, under the title of Metallic Cement, which the workmen of the Metropolis have abbreviated into Metallic.

Mr. Troughton their agent gave me some specimens of mortar and of concrete, each made of Aberthaw lime and metallic cement mixed with sand and with Thames ballast, which were extremely hard, together with some of the raw material in the form of rather coarse sand, which it was my intention to have experimented upon; and I am sorry that I did not try it in sufficient time for the present Treatise; for as a substance only useful in combination with lime, it may fairly claim a place in a work on calcareous cements. I shall however report on its character from the best authority.

Sir Robert Smirke, who to my knowledge has always been anxious to try every promising expedient held out as an improvement in Practical Architecture, which he has usually done in the first instance on a small scale and at his own expense, before he would recommend it for any public or private building, of which he was the Architect, first used the metallic cement with lime and sand, in stuccoing some of the offices at his country seat at Stanmore, in 1832 or 1833, and afterwards he permitted it to be used as a less expensive material, instead of mastic, in the fronts of some of the houses of the new approaches to London Bridge, at the request of the persons about to build there, who were required, in so doing, to adhere to the elevations and specification prescribed by him, as the Architect of those new Streets. The first trials of it, both at his own house and in some houses in King William Street, in which the work was executed by Mr. Bellman, a plasterer of reputation, gave him great satisfaction, as it proved considerably harder, than the stucco composed of the same lime, which was usually blue lias mixed with sand alone, would have been. Messrs. Baker have also informed me, that the stucco composed of 1 measure of Rosehill or of blue lias lime, mixed with one measure of metallic cement and 1 measure of clean sharp sand, which they have used at various times, has also evinced the same sort of superiority. But Sir Robert Smirke added, that more recently he thought it prudent to cease from recommending or permitting the use of the metallic cement, in consequence of the stucco of several buildings done with it having failed, especially at and near Fenning's Wharf, where it cracked or peeled off, in an unseemly manner, in about two years after it had been applied. On inquiry into the circumstances, he found that it had been generally complained of, by the operative plasterers, paid by taskwork, in consequence of its requiring much more labour to apply it properly, than is necessary either in laying on lime stucco without this ingredient, or cement stucco. In order to remove this inconvenience, and in short to please the workmen, a proportion of plaster had been mixed with the copper slag, which mixture, though it had effected the object in view, deprived the stucco of that strength and durability, which the copper slag appears to possess when mixed with lime and sand alone. Mr. Troughton had previously communicated to me the results of several experiments, each tried with 1 measure of copper slag mixed with from 1 to 3 measures of blue lias lime, of Merstham (equivalent to Dorking) lime, and of chalk lime, both with and without sand, and all applied to brickwork both as stucco and as mortar, of which the mixture of blue lias and of copper slag in equal measures was found to be the strongest; and he assured me that the mixtures of the blue lias and Merstham limes with copper slag evinced much greater adhesiveness, than I consider the best mortars of the same limes to possess; but they all appear to have set much more slowly than cement, for in sticking out bricks from a wall by them, though the results, as reported by him, were such as would not be unsatisfactory even in cement of a fair marketable quality, it was necessary to support the bricks by planks, for what period he did not mention, but I suspect for several months; and so indispensable was this sort of aid, that he admitted that those cemented by the most promising of his experimental mixtures broke down, in consequence of the plank being deranged.*

Upon the whole I have no doubt, that by substituting a proportion of copper slag, instead of part of the sand usually employed in making mortar or stucco, with the blue lias, Dorking or Merstham limes, that superior hardness and adhesiveness will be obtained; but at a greater expense, partly because this materiel is dearer than the best sand, and partly because, as Sir Robert Smirke informed me, it requires more labour in the manipulation to execute the work properly; but no experiments have yet been brought forward to prove, that its efficiency for the purposes of Hydraulic Architecture is equal to that of puzzolana, which communicates water-setting properties to common chalk lime.

(XXXIII) *Of Hamelin's Patent Mastic Cement.*

This very ingenious composition, which forms an excellent stucco, shall be noticed here, though invented by a Frenchman, in

* Two bricks were fixed to a wall, by joints of Sheppy or Harwich cement, to which 7 others were attached by the mixtures of blue lias or of Merstham lime with copper slag, and all supported by a strong plank as aforesaid, until the time of trial, after which each of these little piers projecting horizontally was loaded with bricks piled up vertically upon the extreme end of it, until it broke down. When thus loaded, the brickwork projecting from the wall had the appearance of the capital letter L.

consequence of its having been frequently used in England, as for example, for the fronts of the United Service Club Pall Mall, and of many other buildings. I first saw the preparation of it at Messrs. Francis and Co.'s Nine Elms Vauxhall, who informed me that it consisted of litharge or red lead, mixed with calcined and pulverized flints, and with the dust of Caen, Bath or Portland stone, forming a dry mixture, which they sold under the name of Mastic Powder.

In preparing it for use, it must be mixed with old raw linseed oil, well incorporated together, turned over with a shovel and trodden or beaten down, in a trough or box, until the powder shall have entirely and equally imbibed the oil, and become in appearance like moistened sand, and of homogeneous colour, free from those spots, which are observable, before the ingredients are sufficiently mixed. The walls to be stuccod with this composition must be thoroughly cleansed with a birch broom, and brushed over with boiled linseed oil, after which the mastic is applied by a plasterer's trowel with considerable pressure, before the boiled oil is absorbed. The work is thus brought to a straight surface with the rule, and finished with the hand float. In casting ornaments, it must be pressed into the moulds much harder than is done either with plaster or cement. The mean thickness of this stucco ought to be three eights of an inch.

It is more expensive than cement, both in the prime cost and workmanship, for unless the latter be executed by first rate plasterers, and with the greatest care, the mastic which is applied in a state resembling wet sand, will not adhere to the wall. On the first introduction of mastic in this country, a reference was made to Sir Humphrey Davy, whether the oil might not avaporate in time; who as I am informed did not think it probable that it would; and certainly, if protected by a coat of paint from time to time, a failure of this sort can scarcely be expected. The remarkable property which linseed oil appears to possess of hardening stone, has been proved by many of our models at the Royal Engineer Establishment, Chatham, made of chalk, and afterwards saturated with boiling linseed oil poured over them, which has rendered the surface of this soft substance as hard as Portland stone. The stone frieze of the Athenæum club-house, Pall Mall, which has been decorated with figures in alt-relief copied from the Elgin marbles, was treated in the same manner, by advice of Sir Humphrey Davy, who was consulted on the occasion by Mr. Decimus Burton, the Architect of that fine building.

(XXXIV) *Of the Resinous or Bituminous Cements, which have been proposed, or used in this Country.*

First, Fitz Lowitz's Cement. This cement which has been proposed but never used, at least to my knowledge, has been sufficiently described in the Note to Article 42.

Secondly, Lord Stanhope's Cement. This composition, formed of Stockholm tar, pulverized chalk and sand, as described in the Note to Article 229, was used not only for the flat roofs of Buckingham Palace, but also for the roofs of several houses in Regent street, by order of the late Mr. Nash, Architect for that street. Having heard that these roofs had been subject to continual leaks, I made a personal inspection of that of the Palace, and found on the contrary that it was in a satisfactory state, and Mr. Hogg clerk of works informed me, that it had required no repairs at all, the only inconvenience being the fluid state, which the composition assumes in a hot sun, which renders it unfit to walk upon, in summer, notwithstanding the slates.*

(XXXV) I shall now proceed to notice what has been done on the Continent, as far as has come to my knowledge; and although I cannot pretend to have read many foreign books on the subject of calcareous mortars and cements, and there may be numerous incidental notices of importance on this subject, to be found in periodical or other works, with which I am unacquainted; yet the two French writers, whose names I have had occasion to mention more than once in the foregoing Treatise, and whose opinions I shall chiefly dwell upon, have so fully described the labours of their predecessors or cotemporaries in the same subject of inquiry, that trusting to their information, where my own reading fails me, I hope thet nothing of importance will be found wanting, in the following abstract of the proceedings of our continental neighbours; to render which more intelligible, I have reduced the weights and measures used by them to our own standards, which I have done by means of Sanders's Tables, a little work which every one who studies the French works, either on Architecture, Mechanics, or Civil or Military Engineering, ought always to have before him.†

* A composition somewhat resembling Lord Stanhope's cement, but applied to a different purpose, has recently been proposed by Mr. Cassel as a bituminous road surface, for which he has taken out a patent. It consists of mineral pitch boiled with a quantity of prepared gravel, sand and road scrapings, in certain proportions, laid on hot and ironed smooth by very hot irons, to vary in thickness from 3 inches at the center to ½ inch at the sides. It was applied about three years ago, to 100 yards of the turnpike road opposite to Trelleck place Pimlico, and has been highly recommended by Mr. Alexander Gordon. A similar invention of prior date has been used in France, which will afterwards be noticed.

† This most useful little book entitled 'A Series of Tables in which the 'Weights and Measures of France are reduced to the English Standards,' by the late Christopher Knight Sanders, Lieutenant in the Royal Engineer Corps, was published by Baldwin, Cradock & Joy, in 1825. It was undertaken at my request by that zealous and amiable young Officer when under my command at Chatham, whose premature death was universally regretted by all who knew him. There is one deficiency, which arose from his having reduced the French measures of capacity into the gallons, hogsheads, and bushels of our former legal English wine measure and corn measure, the Imperial measures which have one and the same gallon both for wet and dry goods, not having been established by law at the time, when he compiled his Tables. This could easily be rectified in a new Edition of these Tables. I was induced to urge Lieutenant Sanders to undertake this work, not only from its great utility, but from having seen a similar set of Tables, published when the British Army was in France, in which, on recalculating a few of the reductions stated, I found so many errors, that it appeared better to calculate new Tables than to correct and republish the work alluded to, which must have been done in haste, as there can be no doubt of the competency of the individuals concerned in it, who are highly distinguished in their respective professions.

Of Mr. Loriot's mode of making a Cement or Mortar.

(XXXVI) This was described in a pamphlet, published in Paris in 1744, entitled 'A Memoir on a discovery in the Art of Building made by Monsieur Loriot, Mechanic and Pensionary to the King,' in which is published by order of His Majesty, the method of composing a cement or mortar proper for an infinity of works, as well in building as in decoration.

In the 31st page of his pamphlet, he declares that the admixture of powdered quicklime in every sort of mortar or cement, made with slaked lime, is the most effectual method of giving it every desirable perfection, and that is the chief discovery which he announces. In the next page he gives the following prescription. 'Take one part of brickdust finely sifted, two parts of fine river 'sand screened, and as much old slaked lime as may be sufficient 'to form mortar with water, in the usual method, but so wet withal 'as to serve for the slaking of as much powdered quicklime as 'amounts to one fourth of the whole quantity of brickdust and sand. 'When the materials are well mixed, employ the composition 'quickly, as the smallest delay may render the application of it 'imperfect or impossible.'*

In another part of his pamphlet he explains, that in naming one fourth of quicklime powder, over and above the total quantity of sand and brickdust, he has stated the mean proportion which he considers proper for lime of middling quality fresh from the kiln. If the same lime had been burned some time before it was used, more would be required, but if it were lime of a superior quality made of hard stone, which absorbs a good deal of water, then he considers that a smaller proportion would suffice.

He also notices several little varieties, such as a mixture of coal dust for outside and of plaster for inside work, and the use of calcined clays ground to powder, instead of brickdust, &c, and he asserts that the cement or mortar thus formed by lime, sets much quicker than any other sort then known, and that being perfectly waterproof, it is the best composition for the lining of cisterns, and the coating of casemates or subrerraneous vaults.

Both Dr. Higgins and M. Rondelet, a recent French writer of reputation, quote but disapprove of M. Loriot's method, which he asserted to be a rediscovered secret of the ancient Romans. M. Rondelet's objections being from personal observation are the most conclusive. He observes that the quicklime added to moist mortar, absorbs the water contained in the first mixture so rapidly, that it makes it set almost as quick as plaster of Paris: and that when employed for works subject to the action of water, it appears to produce the most favourable effect, and to be even superior to puzzolana, by setting much more quickly. But as the quantity of lime thus employed is almost double of what experience has established

* I quoted the above from Dr. Higgins, before I saw M. Loriot's pamphlet. Having sinse perused it, I find the Doctor's translation so good, that I have not altered it.

as the best for making good mortar, it follows that M. Loriot's mortar loses after a certain time the superiority which it appears to possess at the period of using it, whilst common mortar on the contrary acquires in process of time a consistency and a hardness which are continually increasing, and which in the end become equal to that of hard stones and well burned bricks.

Having examined stucco made about 15 months before at the Observatory at Paris, under M. Loriot's direction, M. Rondelet found that it had a very hard and smooth but very thin surface, and that on removing this kind of skin, the under part was much weaker and softer than good mortar made with lime and tile dust.* He also observes, that the quantity of quicklime added to Loriot's mortar makes it too dry for masonry, and especially for walls above the ground and of moderate thickness, and that it does not combine so well with stones or bricks as common mortar, besides which he adds that it is very expensive, requiring double the quantity of lime used in common mortar, about one half of which has to be reduced to quicklime powder by a costly and troublesome process.

(XXXVII) M. Rondelet describes also a memoir of M. Morveau, who proposed some modification of M. Loriot's process, by letting lime slake in the air and then reburning it, which I consider it unnecessary to notice in detail, nor shall I dwell upon an after observation of M. Rondelet himself, who says that 'owing to its 'property of producing an immediate effect, M. Loriot's method 'may be employed with success, in an infinity of circumstances 'where it is necessary that mortar should set quickly,' for I consider that the use of twice the quantity of lime, commonly employed in mortar is to be reprobated; and though the addition of tile dust may improve common lime, I cannot understand in what way adding quicklime powder to stale mortar, both being made of the same lime, can possibly convert a common into a hydraulic lime. In short I have no faith in such niceties.

Belidor's Observations on Limes and Mortars.

(XXXVIII) The state of practical knowledge and the opinions entertained in France, on the subject of limes, and of common as well as hydraulic mortars, may be collected from Belidor's 'Science of Engineers in the conduct of Works of Fortification and of Civil Architecture,' published by him in 1739, and from the fourth volume of his 'Hydraulic Architecture,' published in 1753.

In book 3d chapter 3d of the former work, he treats of lime and the mode of slaking it, in which after premising, that lime is as it were the soul of masonry, which it ought to unite as firmly as if composed of a single stone, he states that in order to obtain good lime, very hard, heavy and white lime stones ought to be used; so that no lime is so good as that which may be made from white marble. He further observes, that limestone fresh quarried is better

* The expression in the original is 'le bon mortier de ciment,' (see Article 257).

than that which has been kept in heaps; and that the stone of moist and shaded quarries is better than that of dry ones. He adds that the Boulogne limestone which is of a yellowish colour is very good, but that the other limes of Picardy and Artois are but middling, being made from a white soft stone differing very little from chalk, which he says produces the worst of all limes.

In respect to the slaking of lime, he observes that after lime is drawn from the kiln, one must take care to see that the workmen use the proper quantity of water necessary for that purpose, because too little water burns lime, aud too much drowns it.

After this he quotes a maxim of Philibert Delorme a former French writer on Architecture, who declares that lime is good, when well burned, white and fat, and he recommends the mode of slaking proposed by that author, as the best. This is done by putting the lime into a basin or pit, covering it regularly with one or two feet of sand, and throwing water over the heap, taking care afterwards to add more sand, if it should burst this covering in slaking, in order that the smoke may not escape. By this process, it is converted into a mass of grease, which being opened at the end of two or three years will resemble cream cheese, and will be so fat and glutinous, that on breaking into it, it will be difficult to draw the tool out again, and in this state it will form an excellent mortar for plastering and stucco.*

Belidor describes the lime of Metz as being excellent, though it does not slake into a putty; for he says that when treated in the manner that has been described, so hard a substance is produced at the end of a year, that it requires iron wedges to break it. He states that this lime slaked under the proper proportion of sand, and made into mortar soon afterwards, becomes so strong, that when mixed with coarse gravel from the river alone, without either stones or bricks, it forms a sort of masonry used for most of the cellars in the city of Metz, which will resist steel pointed tools, after the mortar has set.

Here Belidor describes all the properties of a good hydraulic lime, and a sort of concrete for artificial stone made from it, of which the strength appears to have been exaggerated, but according to the ancient prejudice then prevailing, he attributes the excellence of this lime to its being produced from a very hard stone.

He then concludes by remarking, that lime in the state of dust, by which of course he means stale lime slaked by exposure to air, was good for nothing, because its salt having changed its nature and virtue, it had no longer the property of uniting with masonry.

(XXXIX) In his next chapter, he describes the valuable properties of puzzolana nearly in the same terms as Vitruvius had done

* With the exception of keeping lime 2 or 3 years, which is never done, this is the common method of slaking lime used from time immemorial by the masons and bricklayers both of France and of this country, when not required to follow some other peculiar mode, pointed out by the Architect or Engineer, under whose direction they are employed.

1600 years before, but observes that it had been little used in France, and he also describes Dutch Tarras or Trass, but erroneonsly, as an earth found in the environs of Cologne, and prepared for use by burning as well as grinding, whereas it is usually found as a stone, and if in an earthy state requires no burning.* He considers Trass when unadulterated, in which state he says it was difficult to procure it, to be equally proof against the action of water, the vicissitudes of the seasons, and the alterations of wetness and dryness.

He mentions the ancient and well-known expedient of pounded tiles as a substitute for puzzolana or trass, and states that stones *(cailloux)* collected in the fields, or pebbles *(galets)* from rivers, when calcined and pulverized will answer the same purpose.

(XL) In treating of mortar in the chapter following, having previously acknowledged, that it would be impossible now to keep lime slaked for 2 or 3 years in pits, according to the practice of the ancient Romans, which he considers to be the best for all purposes; he proposes to make two pits near each other, a small one, in which the lime is to be put and worked up continually, whilst water is thrown upon it to slake it, and when in a fluid state it must be allowed to run over into the second being a larger pit, from whence after having attained the consistency of white cream cheese, it must be taken out and mixed with 2 measures or with 1½ measure of sand to 1 measure of lime putty, according as the lime increases more or less in bulk in slaking; but if the lime be very fat, and produced from a good stone, he says that 3 measures of sand to 1 measure of lime may be used. He prefers river sand to pit sand, and justly requires that both the sand and the water used for mortar should be clean.

He describes *Tournay Cinders*, as a mixture of small pieces of lime and coals caked together, which fall down under the grates of the kilns in burning an excellent hard blue limestone found near Tournay, which cinders having very strong water-setting properties, are collected and sold.

He states that puzzolana and trass, as well as tile dust, and Tournay cinders, should be mixed with lime, in the same manner, and if I understand him rightly in the same proportions as sand, for masonry exposed to the action of water; but when the work is of intermediate importance between this and the masonry of common buildings, he says that a mixture of sand and puzzolana, &c, in equal parts, may be used with the lime instead of all puzzolana, &c, or all sand.

(XLI) Thus it appears, that towards the middle of the last century, the ancient error, that hard lime stones yield the best lime, prevailed in France, which as I before remarked, Smeaton, who

* Smeaton, who is accuracy itself, and more recently M. Sganzin in France, describe it as a stone. Belidor's error was afterwards copied by M. Rondelet. If Trass were ever burned at all, which I doubt, it must have been only to facilitate the grinding, as is done in the case of flints.

was himself in some points the pupil of Belidor, soon after detected. I need also scarcely point out the absurdity of supposing, as stated by Belidor, that pebbles from the fields or from the beds of rivers, taken indiscriminately and calcined and afterwards pulverized, would form an excellent artificial puzzolana, without reference to their nature, whether siliceous, argillaceous, calcareous or granitic, from which in itself one might infer the very imperfect state of chemical knowledge at that period, if some curious statements by Belidor of the opinions of the Chemists and natural Philosophers of his day on the subject of gypsum,* lime and mortars, did not prove that the proper chemical analysis of minerals was then unknown.

Belidor states, that the chemists had ascertained that the hardness of bodies is occasioned by the salts distributed throughout their whole mass, which bind their particles together, and that when lime stone is burned, the greater part of the sulphureous and volatile salts, which belong to it, are driven off by the violence of the fire, and thus it becomes porous, &c, and that afterwards on mixing it with sand, a fermentation is caused by the sulphureous particles still remaining in the lime extracting the volatile salts which are also found in sand, which therefore combine with the lime and fill the pores of it, and by beating up and mixing the lime and sand well together in making mortar, this process is expedited and promoted; and thus the lime of the mortar recovers by degrees the original hardness which it possessed as a stone before it was burned. In another part he observes, that when the building stone of any district is a lime stone, the best masonry is obtained by using lime of the same sort of stone for the mortar, because their volatile salts will agree better, than those of different sorts of lime stone and lime.

Belidor being a respectable mathematician, but apparently unacquainted with chemistry or geology which were then in their infancy, and he never having been personally employed in any branch of Engineering or of Hydraulic Architecture, though he has illustrated the state of both in his time with considerable ability, and with extraordinary industry, it is no disparagement to his talents, that he should have stated the received opinions of his age, when erroneous, without objection, and even with applause, at a period when Lord Bacon's rule of rejecting all hypothesis, and questioning nature by continual experiments, had not yet been fully adopted by men of study.

(XLII) *Belidor's Rules for plastering the roofs of Casemates, to render them water-tight.*

* Gypsum or Plaster of Paris being used in that capital for the general purposes of building, is amply treated of, by all the French writers on Practical Architecture, but I have omitted Belidor's observations upon it, in consequence of its being confined to inside ornamental work only in this country. His account of the reason assigned by the Chemists, why plaster allowed to become stale is unfit for use, owing to a supposed change in the state of the salts contained in it, is very curious, and would in the present day be considered perfectly absurd.

He states in the 11th chapter of the same book, that the cement for this purpose is usually made with Tournay cinders, beaten and prepared once in every 4 or 5 days, for six weeks, but only mixed with water the first time; or it may be made of 1 measure of good quicklime and 2 measures of trass, beaten and prepared in the same way, or instead of trass, the same proportion of puzzolana or of well burned old tiles may be used; but in all cases the substance used must be ground to powder by a handmill, and passed through a baker's sieve, and must be mixed with the lime several times, without using water, except at first.

The masonry over a range of casemates, usually finished with ridges and valleys, like common roofs, when it is to be coated with the cement above described, must have 5 or 6 months allowed for drying and for settling. The joints must be raked out, and the surface well cleaned by watering it, after which the cement is laid on fresh mixed for an inch and a half in thickness, which is spread equally and beaten both longitudinally and transversely with little bats only 2 inches wide to press it better into the joints, and afterwards it must be floated with smooth irons rounded at the ends, to render the first coat smooth, until it begins to set. For a certain time the surface must be mopped every day, with a cloth mop as large as the head, dipped into a basket of thin cement; then the iron is passed over it, after which it is covered with straw till next day to prevent the heat from blistering it. This process of mopping, floating and covering with straw, must be repeated until no more cracks are perceptible on the surface, after which it must be only mopped for 5 or 6 days more.

Upon the coat of cement, 4 or 5 inches of coarse sand or gravel must be applied, over which must be laid 18 inches of earth well beaten, and thus they must continue, if I understand the author rightly, adding layer over layer of those substances, until the proper level for the terreplein or terrace over the casemated arches shall be obtained. He says that this mode was adopted for covering the tower-bastions of New Brisack.

He also mentions the famous orangery of Versailles, the haunches of the arch of which were covered over with dry rubble stones with lime dust between the joints, to the height of 18 inches, over which were laid alternate layers of lime dust 4 inches thick, and of well-washed pebbles 12 inches thick, until they reached the top of the arch, after which a bed of mortar was laid upon the last layer of pebbles. He says that this construction answered so well, that though the top of the orangery is only a terrace, no injury ever happened to the arch.

He mentions another method also as having been practised, which is to put a bed of dry rubble stones, having their joints filled with lime dust over the arch as before, then to lay on 4 inches of lime dust, over which a layer of well-beaten clay 12 inches thick, then a layer of pebbles also about 12 inches thick mixed with lime dust, and over this a bed of mortar, 3 or 4 inches thick, to receive the earth.

These modes of finishing casemates or vaults, the first of which only appears to be waterproof, are very curious, the first from its repeated and tedious mixings, beatings, moppings, &c, the others from the use of lime powder. I do not think that any Architect or Engineer of the present day would copy them.

(XLIII) *Belidor's Description of Foundations formed in deep Water, by throwing in dry Stones.*

In the fourth volume of his Hydraulic Architecture, book 3, chapter 10, section 1, he describes a mode of building moles in deep water with dry stones, thrown in so as to obtain a slope on each side of double the height, similar in short in principle to the celebrated Plymouth Breakwater, executed 60 years afterwards by the late Mr. Rennie, but far inferior to that gigantic work in boldness of conception, and in the style of execution, especially in respect to the size of the stones.

As Belidor's Breakwater, which he calls *Fondations à pierres perdues*, literally foundations with stones lost or thrown in at random, do not come within our present subject, I pass them over, and proceed to the contents of the next section of the same chapter.

Belidor's Description of Beton Mortar, and of the mode of building with it, in deep water.

(XLIV) In preparing to build a wall of this sort, Belidor proposes to drive two rows of sheeting piles, to correspond with the proposed profile of the wall, one row vertical for the rear, and another at a slope of one fifth for the front of the wall, to be driven in portions of about 60 or 70 toises, from 138 to 150 English feet at a time, as he does not propose to pump any part dry. But in addition to these piles, he drives some stronger ones outside of, but connected with the former, and carrying longitudinal caps or wales, upon which he places machines, first for dredging the intermediate space, that the foundation of the beton masonry may be made lower than the ordinary level of the bottom; and afterwards for letting down the materials composing that mixture into their proper places. The beton is merely hydraulic mortar formed with lime, mixed with puzzolana, trass, tile dust, or Tournay cinders, in the usual proportions approved for masonry subjected to the action of water, which is afterwards mixed with as many broken stones as it will bear, that is not more than to admit of every stone being well bedded in this sort of mortar, so that the stones shall not be in direct contact with each other.

Belidor observes, that the rules for making and mixing beton most approved by M. Milet de Monville, whom he mentions as having tried various proportions, and having had great experience in this sort of work, enlightened by sound principles, were as follows.

Having prepared a level well rammed spot, one must take 12 parts of puzzolana, trass, or Tournay cinders, with which a circular inclosure must be formed of 5 or 6 feet in diameter, upon which must be spread equally and regularly 6 parts of clean sharp sand. Within this circle must be laid 9 parts of quicklime pounded by an iron rammer, to make it slake more easily, which may be done with salt water, if for a maritime work, and it must be stirred about from time to time with proper tools, until it is reduced to lime putty, after which the sand and puzzolana mnst be thrown in, and well incorporated with the lime. The whole being well mixed, 13 parts of broken stone, and 3 of pounded forge-cinders or slag; or if the latter cannot be obtained, 16 parts of broken stone or pebbles, none exceeding the size of an egg, must be added to the mortar before prepared, and the whole well moved about, turned over and stirred by men with spades for an hour, after which it must be formed into heaps, and allowed to set, until it becomes so hard as to require a pickaxe to break it, which state it may attain in 24 hours in hot summer weather, but in 2 or 3 days in winter.

(XLV) The beton is let down into the space between the two rows of piles in a cubical box, which Belidor says may contain 27 cubic feet of French measure and ought not to exceed 64, as it would then become too unwieldy. When this box arrives to within 3 feet of the spot where its contents are to be deposited, the bottom of the box which is fixed by a couple of gudgeons at one side is thrown open by a rope, and the mixture is discharged accordingly, falling perpendicularly downwards out of the box.

He remarks, that although the beton is let down dry, it falls asunder and spreads itself on reaching the bottom. The position of the machine is changed continually, until the beton discharged from it covers the whole space within the piles for 10 or 12 inches in depth, after which a course of rubble stones of moderate size, none exceeding a quarter of a cubic foot, are spread over it. These stones being carefully arranged along side of each other sink into the beton which has become soft; but which afterwards sets again, so that after 3 or 4 months, the masonry becomes indissoluble, and improves still farther by age. This bed of stones having been covered with a new layer of beton, the workmen must commence a second course of stones, and thus they proceed until their work arrives to within 6 or 7 feet of the surface, after which troughs and baskets may be used instead of the box, taking care to prevent the mixture from being injured by throwing it down from a height rashly, in which case the heavier parts would fall to the bottom before the others, and the whole essence of the lime would be turned into milk.

(XLVI) Belidor observes, that he saw this method put in practice at Toulon, in forming one of the jetties or piers in the new basin; and remarks it was a practice of which the ancients had left us the example, and states, that though hitherto confined to the coasts of the Mediterranean, there was no reason, why it should not answer equally well on the ocean or in rivers, to found a wharf, or the piers of a bridge, in places which are never dry, and he considers it preferable to foundations executed in caissons.

He gives a statement of the materials for making 16 cubic toises, or 3456 cubic feet of French measure of beton with their weight, which is as follows.

			Lbs.
	942 feet cube of puzzolana, at	90 lbs. per ft.	84780
	471 ditto of sand, at	115 ditto....	54165
	1020 ditto of broken stones, at ...	80 ditto....	112200
	235 ditto of forge cinders or slag, at	80 ditto....	18800
	706 ditto of pounded quicklime, at	76 ditto....	53596
	618 ditto of stone, at	160 ditto....	98880
Total ..	3992 cubic feet of materials, weighing in all..		Lbs. 422421

He observes that 30 men were employed for 12 days of 8 hours each to prepare the beton, and make it into heaps, and 24 days for the rest of the work, which being done by galley slaves chained together in pairs, he observes that hired labourers would have executed it much sooner, though not so cheap.

(XLVII) *Of the sorts of Mortar, used by M. Perronet for the Bridge of Neuilly, at Paris in* 1768.

This gentleman who was Architect to the King of France, first Engineer of Bridges and Roads, Knight of the Order of St. Louis, &c, in a very interesting work published in 1788, has given a description of the construction of the Bridges of Neuilly, Maxence, and others executed according to his designs and under his personal direction, of which he has explained the minutest details by elegant plates. In his specification for the mason's work of the Bridge of Neuilly, which was one of his finest works, planned in 1766, but of which the masonry was commenced and the first pier finished in 1768, he requires that two sorts of mortar shall be used, one which he calls white mortar, to be composed of 1 measure of Vernon lime mixed with 2 measures of sharp clean river sand from the Seine; the other for the foundations and parts exposed to the action of water, was to be composed of equal measures of lime, and of an artificial puzzolana, which he calls cement according to the custom of the builders of France at that time, which was to be made by grinding tiles from the tile kilns of Neuilly in a mill, and sifting the tile dust thus obtained before it was used. He particularly prohibits the contractor from using cement obtained by grinding bricks. If the Neuilly tiles were made of as fine a clay, as is generally used by the tile makers in the neighbourhood of Chatham, namely the brown clay of Upnor, I have no doubt that M. Perronet's artificial puzzolana must have formed a very serviceable hydraulic mortar. He being one of the most eminent Civil Engineers of France, the above may be considered a correct example of the most approved practice of the French at that period.

(XLVIII) *M. de la Faye's method of making Mortar, said to have been published in* 1777.

M. de la Faye alleges that he discovered his method of

making mortar, by the attentive perusal and examination of some obscure passages respecting the nature of lime, in Vitruvius and in the Treatise of St. Augustin on the City of God, the latter of which is probably metaphorical; and thus, he also holds out that his own mode of proceeding, is a rediscovered secret of the ancient Romans. His rules are as follows.

You must procure lime of hard stone, fresh from the kiln, and cover it well to preserve it from damp air or rain. Lay it on a clean wooden platform, in a dry place under cover, and provide some dry casks and a large vat three fourths full of clean river water, not of a mineral quality. Two labourers must be employed, one to break the lime into pieces of about the size of an egg. The other must fill a flat basket with this broken lime without heaping it, after which he will plunge it under the water, and keep it there until all the surface of the water begins to boil. He will then draw out the basket, let the water drain off for a moment, and pour the lime into a cask. He will proceed in the same manner, until he has wetted all his lime, and filled the casks with it to within two or three inches of the top. Then the lime will throw out considerable heat, rejecting most of the water with which it has been drenched, and emitting fumes and opening its pores, it will fall down into a powder, and finally lose its heat, passing into the state of slaked lime powder, which Vitruvius calls *calx extincta.* This must be done in a place where air circulates freely, in order that the workmen may stand so as not to be incommoded by the hot vapours of the lime, during the process of slaking. As soon as this is over, the casks must be covered with large pieces of cloth or bags of straw. Well burned lime fresh from the kiln will heat and fall down in powder quickly; but if long burned and therefore somewhat stale, or if underburned, it will heat slowly and will be very badly divided.

(XLIX) Such is M. de la Faye's mode of slaking lime, which the French have since termed *Slaking by Immersion,* to distinguish it from the *Common mode of Slaking* with excess of water, and from *Spontaneous Slaking*, by exposure to air. Lime slaked by the common method forms a thin paste; whilst by the other two, it is reduced to a powder.*

In making mortar with pit sand rough to the touch, such as the Romans called *Fossilitium*, or with sea sand or river sand, M. de la Faye recommends three measures of sand to one of lime, but in using earthy sand, white yellow or red, and fine or soft to the touch, he recommends two measures of sand only to one of his lime: and in mixing his mortar, he recommends water enough to make what he calls a fat mortar, unless the sand be obtained in so very wet a state from the sea or from a river, that extra water may be dispensed with.

* The lime used by Smeaton in building the Edystone Lighthouse was slaked in 1755, but as a matter of convenience only (V), according to the method afterwards proposed by M. de la Faye, as an important improvement upon the common mode of slaking.

For mortar that is to be used as cement, he recommends mixing two thirds of sand and one third of tile dust, and then to incorporate two measures of this mixture with one measure of lime.

M. Rondelet, from whom I have extracted this description of M. de la Faye's method observes, that it is much simpler, cheaper, and less troublesome than M. Loriot's; and that it produces a mortar less friable and more proper for works of masoury; but that it does not set so quickly, especially under water. But he adds that neither of them are superior to mortar, made of lime slaked in the common manner and mixed immediately, with equal care.

M. Rondelet's mode of making mortar.

(L) In his elaborate treatise on the art of building,* this author proposes to grind the lime and sand together, by passing a common garden roller over them upon a stone floor, which I consider much inferior to the mortar mill often used in this country, especially by Mr. Rennie, who I believe seldom dispensed with it. In 1787, M. Rondelet tried experiments upon the strength of different mortars made by him 18 months before, in the form of small bricks 15 centimetres long, 10 centimetres wide, and 4 centimetres thick, that is nearly 6 inches long, 4 inches wide, and rather more than 1½ inch thick of English measure. He mentions that in comparing these specimens, he acted upon a portion of each, forming a parallelopiped, having a surface of 4 superficial inches, equal to about 4½ of English measure, until he crushed it, by means of weights, stated in a table containing the results of those experiments, in which he did not confine himself entirely to his own new mortars, but acted also upon pieces of ancient mortar or stucco. His proceedings are very obscurely stated, but one may infer that his mortar-bricks were supported from below, as the crushing weights vary from 914 French Pounds the least, to 4738 the greatest, that is from 987 to 5116 lbs. avoirdupois, the latter weight being required to break down some mortar obtained from an ancient Reservoir at Lyons. He states, that in all cases the same experimental mortar made by him, when well beaten, was stronger than when merely mixed without beating, also that Loriot's mortar was weaker than De la Faye's the crushing weight of the latter being 1664 French lbs. or 1797 lbs. avoirdupois, but that his own mortar made of lime and sand only, even without being beaten, was stronger than either. At the end of 16 years he tried the same new mortars a second time, and found that they had all increased more or less in strength, the greatest additional resistance being in the proportion of 2-5ths of the original crushing weight.†

* Published at Paris in 1812, and entitled Traité théorique et pratique de l'art de batir, par J. Rondelet Architecte, Chevalier de la Légion d' Honneur, &c.

† The same author states in the form of a table, the names, specific gravities and resistances of 180 sorts of stone, in opposition to a crushing weight. The latter vary from 1623 French pounds, the weakest, being the resistance

(LI) He also cemented together a number of 2 inch cubes of several kinds of stone, as well as bricks and tiles, in pairs, by the same sort of mortar carefully made of lime and fine sand, the joints thus united each measuring 4 French superficial inches. At the end of 6 months, he tore them asunder by weights which varied from 64 to 123 lbs. of the old weight of France, for six sorts of stone, the first of which was tried first with a very smooth and afterwards with a rougher surface. It required 138 lbs. to separate the bricks, and 141 lbs. for the tiles. The above weights are respectively equal to 69, 133, 149, and 152 lbs. avoirdupois nearly, acting on a surface of about 4½ superficial inches of English measure. He found also on trying the same experiment with plaster of Paris, that the tenacity of plaster joints at the end of 6 months was greater by one third, than those of mortar, but he observes that the strength of mortar increases with age, whereas that of plaster diminishes if exposed to the weather or to moisture. Upon the whole M. Rondelet estimates from those experiments, that the mean force required to tear asunder one of his mortar joints is equal to 105 lbs. for 4 superficial French inches or 26 lbs. for the superficial inch, which on being reduced to English weight and measure is at the rate of 25 lbs. to the superficial inch nearly.

He states that the weight necessary for tearing asunder a prism of mortar 16 years old, and 1 inch square, by pulling from both ends was 53 lbs. French (57¼ English nearly) whilst the weight for crushing the same was 676 lbs. French (or 730 lbs. English), proving that its resistance to a crushing weight was about 12 times greater than its resistance to a tearing weight.

In plaster he estimates the resistance to a crushing weight about 9½ times greater than to a tearing weight, in mortar made with tile dust 7½ times, in puzzolana mortar 8 or 9 times, and in ancient mortars 8 times greater, all of these proportions being deduced from experiments tried by him.

When mortar has attained its full strength and hardness, he considers that its adhesiveness to stones is greater than its own resistance against a tearing weight acting upon a square prism of the same mortar from both ends, but that in plaster the reverse holds good. Yet he remarks, that in new buildings, the adhesiveness of plaster to stones or bricks is equal to one half, but that of mortar to one third only of their respective resistances to a tearing weight, acting upon square prisms of the same substances.

It is to be regretted, that M. Rondelet has not explained in detail the modes of experimenting from which he drew

of the white Tuf of Saumur, to 3000, the resistance of soft free stone proper for building, and to 56129, the strongest sort of hard stone, which are respectively equal to 752, 3239 and 60603 lbs. avoirdupois. In another table he states the same particulars for 18 sorts of marble, granite, porphyry and basalt, the resistances of which vary from 15552 lbs, French, the resistance of a veiny white marble which was the weakest, to 124416 lbs, that of the basalt of Auvergne which was the strongest, of those stones, the former being equal to 16792 and the latter to 134333 lbs, avoirdupois.

the above inferences. In speaking of breaking down cubes of stone or of mortar to ascertain their resistance to a crushing force, he uses the word machine, but leaves us in the dark as to what sort of machine he employed for this purpose, so that we do not know whether his weights were actual weights, or obtained by calculation, in reference to the leverage or power of the said machine. Hence even these experiments are by no means so satisfactory, as they would otherwise have been. But in what mode he tore asunder his square prisms of mortar by the two ends, or how he tried the adhesiveness of his mortars to stones or bricks is not explained at all, and must therefore remain a matter of mere conjecture; and consequently M. Rondelet's opinions as to the comparative tenacity and adhesiveness of his experimental mortars cannot command that sort of conviction, which belongs to principles deduced from the careful examination of the results of a series of experiments judiciously tried, and so fully reported upon in all their details, as to afford the reader the means of judging for himself. For my part, unless an author condescends to explain all his proceedings, in researches of the nature now alluded to, in such a manner as to leave nothing to conjecture, I set his alleged experiments entirely aside, and suspend my assent to his opinions, at the same time without entertaining the smallest doubt either of his talents or sincerity, for no one writer on so extensive a subject as Practical Architecture can possibly try sufficiently numerous experiments upon every branch of it, to arrive at sound conclusions. He may therefore, either from a limited number of experiments judiciously tried, or from having pursued an unsatisfactory system, be induced to adopt a hypothesis which more multiplied experiments, or a more judicious mode of proceeding, would have set aside. Hence he may deceive himself, and such of his readers as have no personal knowledge of the subject under discussion, without being at all conscious that he is in error.

(LII) *Of the Artificial Hydraulic Lime formed by M. Guyton de Morveau, about the year* 1800.

The powerful water lime of Lena in Sweden, having been analyzed by Bergman, was declared by him to consist of 90 parts of carbonate of lime, 6 of clay, and 4 of the oxide of manganese, in consequence of which M. Guyton de Morveau afterwards combined common lime stone reduced to a fine powder with clay and manganese in the above proportions, and having mixed them intimately together, he calcined the mixture, which proved to be an excellent water lime. The result of this experiment was stated in a memoir published by him in the 9th year of the French Republic (1800 or 1801), in which he also declared, that all the lean limes examined by him contained manganese in their composition, to which he ascribed the virtue of those limes, the term *lean*, which originally implied limes that absorbed the smallest quantity of water in slaking, having at that time been applied in France to all water-setting limes, most of which had been observed to possess this characteristic.

This opinion was combated by M. Vitalis in 1807, who analyzed the good hydraulic limes of Senonches and St. Catharine in the neighbourhood of Rouen, which he found to contain 12 per cent of water, 68 of carbonate of lime, 12 of alumina, 6 of silica, and 2 of oxide of iron, but no manganese; and in consequence of this analysis he stated his opinion that manganese was not an indispensable ingredient in such limes, but that clay was the chief cause of their water-setting properties. Other writers on the Continent have treated the same subject, giving various opinions, which I shall not enter into, confining myself to those who have investigated it experimentally.

Of the Artificial Stone proposed in France by M. Fleuret, Professor of Architecture at the Royal Military School.

(LIII) It appears that this gentleman published a Treatise in 1807 on the art of making artificial stones, which he declared to be of great antiquity, having been known to the Babylonians, Egyptians, Greeks and Romans, and even now in Barbary, and amongst the natives of Malabar. And he states that those stones, sometimes of enormous size, were all made in cases, by pouring in and ramming the ingredients.

M. Fleuret proposed to slake his lime by immersion, and to preserve it with great care in casks well covered up, except when a portion was taken out for use. A mixture of 2 measures of sand and 1 of tile dust, or of equal parts of sand and tile dust was next to be made. This compound was then to be mixed with lime, in the proportion of 2 measures of the former to 1 of the latter, after just wetting them. They were first to be ground together dry, and water added by degrees, until the whole was made into mortar, which was afterwards to be put into a trough and beaten by wooden rammers shod with iron, which were raised and dropped upon it by some mechanical contrivance. He says that this mortar may be improved by moistening it in the trough with one sixth part of its own mass of thin lime putty, but he does not approve of adding water alone, which he says is very injurious to its strength. He adds that hard stone ground to powder, combined with iron slag or coal cinders, are still better than sand and tile dust. When his mortar is ready, he puts it into moulds and beats and compresses it as much as possible. He has also made great use of artificial stones for water pipes, troughs, &c.

M. Fleuret actually established a manufactory of artificial stone at Pont-a-Mousson, which no doubt was good of its kind, but General Treussart from whom I have extracted this brief notice of his proceedings states, that some water pipes at Phalsburg, and the stucco used for covering some casemates at Landau and Strasburg, though made in exact conformity with his directions, failed, neither of them being water-tight, in consequence of the common lime of the neighbourhood having been used at those places; for the same author remarks that M. Fleuret had laid down no rule for distinguishing those limes, which were fit from those which were unfit for resisting the action of water.

(LIV) General Treussart also states, but without saying whether before or after the period of M. Fleuret's publication or proceedings, that the Italians made very good artificial stones at Alexandria, which were usually 14 decimetres long and 8 decimetres square, or 4 feet 7 inches long and 2 feet 7½ inches square of English measure. He states that to make 1 cubic metre of this stone requires 0·24 cubic metre of Casal lime, being a hydraulic lime, and 0·9 cubic metre of sand, which are first made into mortar, and afterwards have 0·2 cubic metre of pebbles or gravel mixed up with them, which is done in a mould. He states that the artificial stones thus formed, are covered with earth for two or three years before they are used.

(LV) *Of the Artificial Hydraulic Lime, proposed by Dr. John of Berlin, in* 1818.

The Dutch Academy of Sciences having proposed as a question for discussion 'What was the chemical cause, by virtue of which 'stone lime generally makes more solid and durable masonry than 'shell lime, and what are the means of correcting this defect in lime 'made of shells?' Dr. John of Berlin gave in a paper in reply to this query, in 1818, which was approved by them as being the best communication that had been made on the subject, and was published the year following.

Dr. John remarked that shells require to be more calcined, than is necessary for stone, which he ascribes to their being a purer carbonate of lime. Having analyzed various lime stones, he found that all those which had hydraulic properties contained clay, oxide of iron, &c. These extraneous matters which give to lime its water-setting property he called *Cement*, and said that it was possible to improve those limes which contain no cement, by adding it in the the dry way. From these considerations, he mixed pounded oyster shells, first with siliceous sand, secondly with diverse proportions of clay from one tenth to one third, and thirdly with one tenth of the oxide of manganese, after which he burned them in a kiln for 96 hours. Of these mixtures, the second produced a good hydraulic lime, but the first which failed altogether as well as the third were unsatisfactory. Hence he concluded, that it was the clay which caused limes to set under water, and that nothing can be more easy than to obtain a good hydraulic lime by adding clay either to shell lime, or to the common pure stone limes, in the manner described by him, leaving it to practical men to determine the best proportions of those ingredients according to circumstances. His memoir is said to contain also the analysis of several ancient mortars, with many important observations.*

Of the Artificial Hydraulic Lime proposed by M. Vicat, Engineer of the Royal Corps of Bridges and Roads of France, in 1818.

* I have extracted this account of Dr. John's proceedings from General Treussart's 'Memoir on Hydraulic Mortars,' which will afterwards be noticed.

(LVI) The term hydraulic lime, which is certainly more appropriate than the former epithet used in France for limes of that description, is said to have been first suggested and brought into general use by M. Vicat himself. This gentleman and Dr. John of Berlin came to the same conclusion, both working experimentally, but although M. Vicat's proceedings must have been prior in point of time, it appeared best to finish with Dr. John, in order that the account of M. Vicat's labours might be followed by the description of the proceedings of his countryman General Treussart, because these gentlemen, both of whom devoted many years of their lives to the investigation of the properties of hydraulic limes, have drawn different conclusions from their respective experiments, which appear to me to have been tried with equal zeal and industry.

Wherever good hydraulic limes are to be found on the spot, or can be procured at a moderate expense, M. Vicat recommends using them; but when only common limes or inferior hydraulic limes are to be found, he converts them into water-setting limes by the following process.

He first burns his lime in the usual manner, and allows it to fall down in powder by spontaneous slaking, exposing it for this purpose to the free action of air, in a dry place under cover, not that he considers this mode of slaking indispensable, but as being the cheapest and most convenient. He then mixes his slaked lime with a certain quantity of grey or brown clay, or simply with brick earth, into a paste with water, and forms it into balls, which are first dried and then burned in a kiln.

He states that very fat common limes, meaning those capable of absorbing a great deal of water in slaking, and pure or nearly so, like the common chalk lime of this country, may bear 20 per cent of clay, that for middling limes 15 per cent is enough, but that for limes having some hydraulic properties, 10 or even 6 per cent may suffice. When the proportion of clay is increased to 33 or 40 per cent, the lime obtained does not slake, but it is easily pulverized, after which when moistened, it forms a paste that sets very quickly under water. When the clay is mixed with stones or gravel, he stirs it up with water in a basin, and allows the fluid mixture from the top to run over into a second basin, in order to get rid of the stones, after which he mixes this liquid clay with the slaked lime powder, and he says that it is more convenient for use in this state, than if it were stiffer.

The above rules for making artificial hydraulic lime have been extracted from a Treatise published by him, at Paris in 1818, entitled 'Experimental Researches on Limes for Building,' &c, &c.*

(LVII) In a second Treatise on the same subject published by him in 1828, he describes the manner of making artificial hydraulic lime actually adopted at a manufactory set up at Meudon near Paris, by Messrs. Bryan and St. Leger, the latter being no doubt the same

* Recherches experimentales sur les Chaux de construction, let Bétons et les Mortiers Ordinaires; par L. I. Vicat, ancien èleve de l' Ecole Polytechnique, Ingenieur du Corps Royal des Ponts et Chaussées de France.

person, who took out the patent for making that article in England in 1818 (XIII). The artificial lime made at Meudon was composed of the chalk of that neighbourhood, and of the clay of Vaugirard, which were previously divided into fragments of the size of a man's fist. A mill precisely resembling the washmill of our English brick-makers, and worked by two horses, breaks down and mixes these ingredients with water, in a circular groove of two metres in extreme diameter. Four measures of chalk and one of clay are successively thrown in, which after an hour and a half form a mass of 1½ cubic metre of thin paste, which is allowed to run into a reservoir and when by continually adding more chalk and clay this becomes full, the contents run over into a second reservoir, on a lower level, and in like manner they fall from this into a third, and so on, there being four or five in all, and from the last the clear water is allowed to run off into a drain. A second set of similar reservoirs is employed in receiving more of the same fluid mixture, whilst that contained in the former is acquiring the proper consistency, which is expedited by using shallow reservoirs, in preference to deep ones.

When stiff enough, the paste is moulded into solids of a regular form, 5000 of which measuring 6 cubic metres may be made by one man tasked in the course of a day, and which after being dried are burned by a mixture of coak and coals.*

(LVIII) At the period of his second publication, M. Vicat had discovered that the finest and softest clays were the best; and although he still seems to consider the double burning as yielding rather a better lime, he had so far relaxed in his opposition to grinding soft lime stones, and burning them only once, that he admits it to be preferable, inasmuch as he states, that the cubic metre of artificial hydraulic lime, thus made of chalk and clay, would cost only 30 francs in the provinces, but that if made by the process of double burning, it would cost 40 francs.

(LIX) M. Vicat's object is evidently not to form what we call a *Water Cement* in England, which at the period of his first publication was scarcely known and seldom or never used in France, but to form a hydraulic lime. When he wrote his first Treatise, he attached so little value to water cements, that he scarcely noticed them: and even in 1828, when he devoted one entire chapter to this subject, after observing that there was a very great consumption of the natural cements improperly termed Roman cement in London, he proceeds to state 'that the use of them must infallibly be diminished 'as soon as the mortars made of very powerful hydraulic limes shall 'be better known, and consequently better appreciated in England.' In this opinion, I apprehend that he is mistaken. The blue lias, one of the strongest hydraulic limes in nature, is well known in London, and the Dorking and Halling limes also having hydraulic properties are in common use, but will never supersede cement. I conceive that the converse of M. Vicat's proposition is more likely to

* Hence each of these small prisms as he terms them, must measure about 73 English cubic inches.

be verified, namely that when the use of water cements shall be better understood and appreciated in France, they will supersede the hydraulic limes in some measure in certain districts, but never entirely, because both are good in their way, and the consumption of each must depend on the comparative expense, which may vary in different parts of so extensive a country.

(LX) After the above observation, on which I have commented, M. Vicat proceeds to state, that natural cements similar to those of England had also been found in France and Russia, which he says may be imitated artificially by burning suitable proportions of ochery or brown clay, and chalk, namely, a mixture of 66 parts of the former with 100 parts of the latter. But he adds, 'it is only 'fair to confess that none of the artificial cements, hitherto obtained 'by this process, have equalled the natural cements of England in 'hardness.' Hence considering the sort of prejudice against them in France, evinced not only by M. Vicat but by General Treussart, as will afterwards be noticed, I do not think it probable that any further attempts to imitate them have since been made in that country, and upon the whole I feel persuaded that no artificial cement equal in efficiency to the best water cements of nature was ever produced, until we undertook the task at Chatham; certainly not in England, for if Mr. Frost's experiments at that time had been successful, I would have abstained from doing any thing further than verifying them, because the field of Practical Architecture, into which my duty led me, is wide enough to induce any person who values his own time, not to waste it by endeavouring to improve upon that, which is already sufficiently perfect.

Of M. Vicat's further Proceedings as described in the Treatises published by him in 1818 *and* 1828.

(LXI) His second Treatise professes to be a compendious statement of what was positively known at that time respecting the qualities, the selection, and the most suitable combination of the ingredients proper for making common and hydraulic mortars.* On perusing both, I find this supplementary treatise a very great improvement upon the first, which is rather obscure, and therefore in giving a synopsis of his opinions, I shall generally follow his last publication in preference, which is not only clearer and more methodical, but which had the advantage of being written after the author had profited by the experience of ten years, devoted to the same important pursuit, as well as by the information derived from other French Engineers or Men of Science, whose attention was drawn to the same subject by his first publication, which was at variance with many of the doctrines then generally received in France. In the chemical analysis of the various natural hydraulic limes, puz-

* 'Résumé des Connaissances positives actuelles sur les qualités, le choix 'et la convenance réciproque des matériaux propres a la fabrication des 'Mortiers et ciments calcaires, suivi de Notes et Tableaux d'expériences 'justificatives, par L. I. Vicat, Ingenieur, en chef des Ponts-et-Chausseés, 'ancien èleve de l'Ecole Polytechnique, Membre de la Légion d'honneur.' Paris, 1828.

zolanas, and other earths experimented upon, he was indebted to the valuable labours of M. Berthier, Director-General of the School of Mines, and also to Dr. John of Berlin, whose previous proceedings in the same career of useful experiment were before noticed.

M. Vicat states in the preface to his second treatise, that in studying the nature and use of limes and mortars, he found such very discordant opinions advanced, both by writers and by practical men, as induced him to investigate the subject experimentally in 1812, in order if possible to lay down correct principles, in place of this sort of chaos, or want of positive knowledge on the subject, which had led to the premature degradation of many important works, exposed to the action of water, such as sluices, docks, bridges, wharfs, &c, in consequence of the mortar being washed out of the joints, which rendered their preservation impracticable without frequent and expensive repairs. Apparently patronized by his superiors and by the Government of France, in consequence of his exertions, for at the period of his second publication, he is styled Chief Engineer of Bridges and Roads, and a Member of the Legion of Honour, it is admitted even by General Treussart, who differed materially from him in opinion, that his labours had been extremely useful, by drawing public attention to the great importance of hydraulic mortars, and by exposing the inefficiency of common lime mortars, which does not appear to have been properly understood in France, when M. Vicat's first publication appeared. This much being premised, I shall now proceed to give an abstract of his opinions.

(LXII) *Of Lime Stones and the various sorts of Lime which they produce.* (*Section* 1. *Chap.* 1.)

After having defined the pure and mixed carbonates of lime, and explained the usual mode of testing them by diluted acids, in which the former dissolve wholly, the latter partially, with effervescence, he observes, that the most satisfactory practical mode of judging of their quality is the following.

Break the stone into pieces of the size of walnuts, and put them into an earthen vessel with holes to admit of the circulation of air, which must be placed in the middle of a kiln for burning earthenware, or of a brick or lime kiln, if the fewel used be wood: and at the end of 15 or 20 hours, take out the quicklime thus obtained, and put it whilst still hot into jars, or large mouthed bottles perfectly dry, which must be sealed up hermetically in order to preserve the quicklime in its perfect caustic state, until the time fixed for experimenting upon it. It must then be taken out of the bottle, and a quantity equal by measure, including the interstices between the pieces, to about 1 litre or 61 cubic inches English, must be put into a net or basket, and plunged under pure water for five or six seconds only, and then drawn out again, when after allowing the water to drain off for a moment, it must be emptied into a mortar of stone or cast iron, when the following effects will take place, according to the quality of the lime.

1st. It may hiss, crack and swell, throwing out a great quantity of hot vapours, and fall down in powder almost instantaneously.

2dly. It may remain inactive, but for a period not exceeding 5 or 6 minutes, after which the same effects will take place.

3dly. The lime may remain inactive for a longer period, such as a quarter of an hour, after which it will begin to emit vapours, but not so hot as in the former case, and to split and fall to pieces, but not bursting violently with noise.

4thly. Similar effects may not take place for one or even for several hours after immersion, and in this case it will crack or split, but without noise, and very little vapour or heat will be disengaged.

5thly. The slaking may commence at longer and very variable periods, the heat produced being only perceptible to the touch, and the lime either falling down into an imperfect powder, or not forming a powder at all.

He states, that in endeavouring to ascertain the hydraulic properties of any lime, the experimenter must not in any case wait for the end of the above process, but as soon as it begins to crack, water must be poured into the vessel containing it, not over the lime itself but along one side so as to run freely down to the bottom, where it will be absorbed by the parts first slaked. The lime must then be worked up with a spatula, adding water if necessary in moderate quantity so as not to drown it, and finally the pestle must be used to beat it, and the whole must be brought into the consistency of stiffish clay.

In this state the lime must be left to itself, till all the lazy particles shall be slaked.* The end of this process which may require two or three hours, or more, will be announced by the complete cooling of the whole mass. It must then be beaten again by the pestle; adding water if necessary, till it is worked up into a stiff paste of the consistency of potter's earth ready for use, which he calls an argillaceous or plastic consistency, in which state it must be put into a small vessel rather high than broad, such as a mustard pot or deep glass, so as to fill two thirds or three fourths of the vessel, bringing the paste to a level surface at top. It must then be labelled, and put under pure water, noting the hour and day of immersion.

After studying for fourteen years the most remarkable limes of France, M. Vicat has been induced to divide them into 5 classes.

1st. *Fat Limes,* whose volume may be doubled or more by the common mode of slaking, and whose consistency, after several years of immersion, is the same nearly as it was on the first day, and which will dissolve entirely in pure water frequently changed.

2nd. *Lean Limes* are those whose volume increases very little or not at all, by slaking, and which when put under water are acted upon like the fat limes, excepting that they dissolve only partially, leaving a soft residue.

3rd. *Moderately Hydraulic Limes* are those which set under water after 15 or 20 days' immersion, and continue to harden; but

* A term applied by the French to slow setting quicklimes, or portions of quicklime.

their progress becomes slower and slower, especially after the sixth or the eighth month. At the end of a year their consistency is like that of dry soap. Their increase of bulk in slaking is moderate, often equalling that of the lean, but never that of the fat limes.

4th. *Hydraulic Limes* are those which set after 6 or 8 days' immersion, and continue hardening. The progress of their solidification may extend to the twelfth month, but the greater part of it usually takes place in 6 months. Their increase of bulk in slaking is always very moderate, like that of lean limes.

5th. *Eminently Hydraulic Limes* are those which set in from 2 to 4 days' immersion. After a month they are already very hard, and altogether insoluble. In the sixth month they resemble absorbent calcareous stones, fit for ashlar masonry. They break into splinters on being struck, and present a conchoidal fracture. Their increase of bulk in slaking is very moderate.

He considers that the lime has set, when a knitting needle 1·2 millimetre, or about 1-20th part of an English inch, in diameter, with the point filed off, does not sink into it when loaded with a weight of 3 tenths of a killogramme, or about 2 thirds of a pound avoirdupois. In this state, it resists the action of the finger pushed against it with moderate force, and will not yield without breaking.

In respect to their component parts as ascertained by chemical analysis, he observes that fat limes either consist of pure carbonate of lime, or of a mixture of carbonate of lime, with a proportion not exceeding 6 per cent in all, of silica, alumina, magnesia, iron and manganese. Bituminous or fetid limes also belong to this class. He states that lean limes consist of carbonate of lime mixed with a proportion of the same substances not exceeding from 15 to 30 per cent of the whole compound: that moderately hydraulic limes contain a proportion not exceeding 10 or 12 per cent; that hydraulic limes contain a proportion not exceeding from 15 to 18 per cent; and eminently hydraulic limes, a proportion varying from 20 to 25 per cent of the same substances before enumerated; observing however that silica and alumina are the most important extra ingredients necessary for forming good hydraulic limes, especially the former, which ought to be in excess, but that the magnesia and manganese are often wanting.

(LXIII) *Of burning Lime.* (*Chap.* 2.)

He then treats of various sorts of limekilns, which he illustrates by drawings, and recommends one 7 metres high, 2 metres in diameter at top, 1·2 metres at bottom, and 3 metres in diameter at the height of about 3 metres from the bottom, having 2 or more eyes, or fire holes at bottom, as being one of the best. He recommends also another of the same height and general form, but only about three fourths of the former in extreme diameter, and having only 1 eye, as being equally good.*

* The largest of these is nearly 23 English feet high, 4 feet in diameter at bottom, 6½ feet in diameter at top, and nearly 10 feet in diameter at the intermediate height of about 10 feet from the bottom.

In France they reckon that 3 cubic metres of hard lime stone will require about 1 cubic metre of coals, but that soft lime stones or chalks require less.

(LXIV) After describing (in Chap. 3) the process of making artificial lime, which I inserted before, and will not repeat here, he explains the several modes of slaking lime.

First, the common mode of Slaking Lime. (*Chap.* 4.)

He observes that the proper rule is to use such a quantity of water as to reduce it into the state of a stiffish paste, which method the author observes is abused by the workmen of France, who run their lime with great excess of water into a reservoir in a fluid state. When thus drowned it loses the greater part of its adhesive properties.

Fat limes, slaked into a very stiff paste, attain from 2 to 3 times their former bulk. Lean limes, most hydraulic limes, and all very powerful hydraulic limes only increase, under the same circumstances, from one to one and a quarter or to one and a half of their former bulk.

In slaking fat lime by the common method, if more water be thrown upon any unslaked parts which have burst but still remain dry, a hissing like that of hot iron takes place, and what is remarkable the lime thus sprinkled slakes badly and remains lumpy, especially if the water be cold. In preparing lime for whitewashing, excess of water must be used to guard against this effect.

All limes, but especially the hydraulic ones, become lazy or slow of slaking, when rendered stale by exposure to air.

Secondly. Slaking by Immersion.

Lime is said to be slaked by this process, when it is plunged under water for a few seconds and pulled out again, after which it hisses, bursts with noise, and falls down into a fine powder. In this state it will keep a long time, if preserved from damp. It does not heat again, when mixed with water.

A hundred parts of fat lime thus slaked only absorb 18 parts of water, whilst the hydraulic limes take from 20 to 35, being in the inverse proportion of what takes place by the common mode of slaking.

Very fat limes, if broken large before immersion and laid out on a level surface, do not fall down into powder, but more than half the quantity remains in small lumps of the size of pease. This difficulty is surmounted by breaking the quicklime previously into pieces not larger than walnuts, and putting them after immersion into casks or cases, which by concentrating the heat causes the lime to fall down into powder in a satisfactory manner.

One measure of fat lime pulverized when fresh from the kiln, is only equivalent to 1·5 or to 1·7 measure of slaked lime powder; but the hydraulic limes under the same circumstances yield from 1·8 to 2·18 measures of slaked lime powder.

Thirdly. Spontaneous Slaking.

Limes may be slaked spontaneously by continued exposure to

air, which causes them to fall down in a very fine powder. In this process the fat limes increase in the proportion of 2 fifths of their original weight, and in bulk from 1 volume of pulverized quicklime to 3·52 volumes of slaked lime powder. The hydraulic limes usually absorb only 1 eighth of their weight of water, and increase in bulk from 1 of pulverized quicklime to between 1·75 and 2·55 volumes of slaked lime powder, measured without heaping or compressing.

Hereupon he observes, that on making up the slaked lime powder obtained by the second and third methods into a paste with water, of the same consistency as the slaked lime paste obtained by the first method, equal volumes of the pastes thus formed with the same sort of lime, will neither contain the same quantity of lime nor of water.

Every sort of quicklime exposed to air under cover, gradually recovers the carbonic acid which is necessary to saturate it, in a time varying according to the nature and mass of the lime.

When fat limes are laid out in layers of 2 centimetres rather more than 3 quarters of an English inch in thickness, 10 months are sufficient, after which 100 parts of quicklime will have absorbed 74 parts of carbonic acid and 17 of water. When hydraulic limes are exposed under the same circumstances, they complete the same task in the 7th or 8th month, after which 100 parts of quicklime combined with 1 fifth of clay will have absorbed 54 parts of carbonic acid and 15 of water.

Every lime slaked by immersion and then exposed to air under cover, gradually absorbs carbonic acid and water, but to a certain point only. For example, 100 parts of fat lime thus slaked, after 7½ months exposure, only acquire 36·15 parts of carbonic acid, and 23·85 of water. 100 parts of the hydraulic lime before described under the same circumstances, only acquire 16 parts of carbonic acid, and 25 of water. The weight of the slaked lime powder after this period never increases perceptibly, but only varies according to the hygrometrical state of the atmosphere.

Hence immediate immersion deprives fat limes and hydraulic limes for ever of the power of recovering the same quantity of carbonic acid that was driven out of them by calcination.

In working on a great scale, fat limes slaked by the common method are preserved in the same state in impervious reservoirs covered with 30 or 40 centimetres, from about 12 to 16 English inches, of sand or fresh earth. After being slaked by immersion or spontaneously, they may be preserved a long time in the same state in casks under sheds, or in large cases covered with cloth or straw.

The hydraulic limes soon set in reservoirs, and can only be preserved in a serviceable state, if required to be transported to some distance, in powder slaked by immersion, and packed in casks or sacks. But they may also be preserved in the state of quicklime from the kiln, by placing a mass of these lumps well packed together, on a bed of slaked lime powder obtained by immersion, on the floor of a shed, and covering the heap with a layer of quicklime applied immediately after immersion, which falling down into powder will fill the interstices of the lime, and envelope it sufficiently to protect it from air and damp.

Not having myself tried the strength of what M. Vicat calls fat, hydraulic, and eminently hydraulic limes, after slaking them in various modes, I cannot say more upon his nice distinctions in respect to the slaking of limes, than to express my doubts as to the accuracy of his opinions on this head, which do not carry conviction to my mind, except so far as his objections to excess of water extend, for both in cement and in lime I have found by experiment, that what M. Vicat aptly terms drowning is prejudicial to them.

(LXV) *Of the Hydrates of Lime.* (*Chap.* V.)

M. Vicat next treats of these, which he defines to be the solid bodies resulting from the simple combination of lime and water, the study of which he observes would be uninteresting, if it were not in its consequences connected with the most important facts in the history of mortars.

Several experiments proved, that the quantity of water employed in the slaking of lime has a great effect upon the hardness of the hydrate, that results from it; since too little water does not bind, whilst an excess of water renders the mass thus formed porous and friable, if it should not shrink properly in drying. The same effect is observed in plaster.

The consistency which produces the greatest hardness is at once ductile and firm, like that of potter's earth, ready for use, as before mentioned. He always endeavoured to give this consistency to the hydrates of lime, or lime pastes, which he employed in his experiments.

Having prepared several limes in this manner, and exposed them in the form of rectangular prisms to the air for a year, the following observations were made.

1st. The carbonic acid from the atmosphere gradually attaches itself to these hydrates, and converts them into carbonates, extending by degrees from the surface inwards. The thickness of the portion thus carbonated is at the end of a year scarcely 6 millimetres or less than a quarter of an English inch, for hydraulic limes, but only from one half to one third of that thickness for fat limes. This may be discovered after making sections of these prisms with a small saw made of watch spring, in coloured limes, by a difference of colour, in white ones by applying litmus paper slightly wetted to the surface of these sections.

The annual progress of the carbonic acid diminishes rapidly, in proportion to the depth from the surface, varying also from different contingencies, and in proportion as the mass is of an open or close grained texture.

2dly. The hardness of the hydrates of lime depends upon the

mode of slaking. In fat limes the order of superiority, is first common slaking, secondly spontaneous, and thirdly by immersion: but in hydraulic and eminently hydraulic limes, the order of superiority is first common slaking, secondly by immersion, and thirdly spontaneous.

He concludes this part of his subject by stating, 1st, that his experiments have proved that certain very fat and colourless limes may form with water alone bodies quite as hard as a number of natural stones, 2dly, that carbonic acid gradually augments the hardness of those portions, in which it fixes itself, 3dly, that hydraulic limes of every sort, mixed with water alone, only produce light bodies of moderate hardness, and 4thly, that carbonic acid also increases the hardness of these hydrates, but never renders it equal to that, which it communicates to white fat limes.

(LXVI) *Of the Action of Water upon the Hydrates of Lime.*

M. Vicat observes, that water dissolves all the non-carbonated parts of the hydrates of fat lime, however great their cohesion may be.

If from the same fat lime, but slaked by the three different modes before described, three hydrates of a firm and perfectly equal consistency be obtained; and if they be immediately immersed under pure water, the hydrate obtained by common slaking will absorb 4 per cent of water after one month, the hydrate obtained after slaking by immersion will absorb 10·8 per cent of water in 2½ months, but that obtained after spontaneous slaking will absorb 24·6 per cent in 2½ months, after which they will be saturated.

He observes that after slaking fat lime by the second and third modes, and making it into a stiff paste with water, the hydrates thus produced will absorb more water until saturated, which however only produces increased density and consequently hardness, without any perceptible change of bulk.*

He states that hydraulic and eminently hydraulic limes, when immersed in the state of a very soft paste, contain superfluous water which they afterwards reject in setting; but in the state of a very stiff paste, they set more quickly, absorbing an additional quantity of water in this process, and in the end obtain a greater degree of hardness than the softer pastes.

Every hydraulic lime that has set in the air may be immersed without being perceptibly acted upon by water.

He observes that the hydrates of fat limes can be of very little use in Practical Architecture. The principal difficulty arises from the shrinking, which takes place in hardening in the air, especially after the common mode of slaking, so that in large masses, of complicated forms, made in moulds, the paste adheres to the sides, cracks take place, and only fragments are obtained, and he says

* I cannot comprehend these statements in which there may be some typographical error, or something left unexplained by the author, as I have always found the hydrates of chalk lime go to pieces when immediately immersed.

he has found that beating the lime paste well does not prevent it from going to pieces, especially in using very fat limes, which shrink the most. But the hydrates of very fat white limes, when made into small squares and allowed to dry on a surface to which they do not adhere, become extremely hard, and capable of receiving a fine polish resembling beautiful white marble, and may be prepared at a moderate expense for various useful purposes, such as mosaic pavements, &c.

M. Vicat observes, that hydraulic limes in the state of hydrates or mixed with water alone, can only be employed successfully under ground or in water. But as they would not under those circumstances answer better than the mortar made by mixing them with sand, it would evidently be improper ever to employ them with water alone.

Upon the foregoing parts of M. Vicat's Treatise, I shall remark, that some of his statements are in direct contradiction to the results of my own experiments, upon which I naturally place greater confidence. For example he states, that fat limes slaked by the common mode produce a paste of 2 or 3 times their former bulk, whereas the chalk lime experimented upon by me rather lost than gained in bulk, when made into a paste (see Table XIV, Article 297); also that the hydrates of fat limes are much harder than those of hydraulic limes, which is contradicted by my experiments, recorded in Table XII (277), in which the resistance of the chalk lime of Chatham one of the fattest, was scarcely one fourth of that of the blue lias, one of the most eminently hydraulic limes in nature. Indeed the small prisms made by me of the former, so far from rivalling the hardness and beauty of white marble, never attained sufficient consistency to resist the foot of a man treading upon them, which invariably crushed them to pieces.

(LXVII) *Of the Ingredients, which combine with Lime in the formation of Common or Hydraulic Mortars.* (*Section* 2, *chap.* 6.)

Of Sands. Here M. Vicat commences by defining sand, which he distinguishes from *dust*, by observing that the finest sands fall down in clean water without perceptibly altering its transparency. He observes that calcareous sand is seldom found in nature, because the hard lime stones do not break into small grains by disintegration, like the granitic rocks. If soft they crumble into dust; if hard they splinter.

Arenes and *Psammites.* He describes the first of these substances as being a sand generally quartzose or siliceous, but mixed with clay of various colours, in the proportion of from one fourth to three fourths of the whole. This argillaceous sand usually occupies the summits of round hills of moderate height, but is also sometimes found in veins in the crevices of calcareous rocks. He next defines psammites, consisting of grains of quartz, schist and feld spar, and flakes of mica, united together mechanically by a variable medium. Most of these enter into the class of rocks, but the schistous psammites of a yellow red or brown colour, fine grained, unctuous and forming an argillaceous paste with water, are a separate species, which he thinks deserving of attention. These are found in beds or veins intermixed with schists, of which they are only a decomposition.

(LXVIII) *Of the Natural and Artificial Puzzolanas.*

After remarking that the natural puzzolanas in the neighbourhood of Mount Vesuvius were known to the Romans, &c, he observes that M. Faujas de St. Fond found some in France in the extinct volcanoes of the Vivarais. He observes that few volcanic regions are without puzzolana, but it presents itself in various forms, sometimes in powder, sometimes in large grains, in scoriæ, in the form of pumice, tufs, &c. Its colour generally brown may pass into yellow, grey or black.

Puzzolanas are essentially composed of silica and alumina, united with a little lime, potash, soda and magnesia, and iron is found mechanically mixed with them in its magnetic state.

He states that clays, arenes, schistous psammites, or grauwackés and schists, if properly calcined, as well as the residue obtained by burning peats and coals, and finally the fragments of tiles and pottery may be used as artificial puzzolanas. And he observes that these substances, though generally composed of silica and alumina act very differently, some uniting well with fat limes, others with middling or powerful hydraulic limes, some resisting equally well the action of air, weather and water; others only holding together under continual immersion, and finally others losing all cohesion when immersed.

(LXIX) *Of the qualities of the Substances which unite with Lime, in making Common or Hydraulic Mortar.* (*Chap.* 7.)

1st. He defines *very energetic substances* to be those, which mixed into a stiffish paste with very fat lime slaked in the common mode, produce a mortar capable of setting from the first to the third day of immersion, and of attaining in one year the consistency of good bricks, and of yielding a dry dust when sawed in two by a watch spring saw.

2dly. That *energetic substances* are those which under the same circumstances set in from 4 to 8 days after immersion, and in one year attain the hardness of soft stone, and which under the spring saw yield a moist dust.

3dly. That *substances of little energy* are those which under the same circumstances set in from 10 to 20 days after immersion, and in one year attain the consistency of dry soap, and which clog the spring saw.

4thly. *Inert substances* are those, which when mixed with fat lime, and put under water, produce no change or improvement in that lime.

He observes, that of the two tests, namely the time of setting and the hardness finally attained, the second point is the most important and characteristic, for some substances produce a mortar that sets in a few days but always remains soft, whilst others that set more slowly at first become very hard in the end.

From his experiments he declares sands generally to be *inert.*

That arenes, psammites and clays are commonly possessed of *little energy*, and rarely *energetic.*

That the natural and artificial puzzolanas may be of all degrees of energy, from the most moderate to the most powerful.

That nothing in the physical characters of arenes, psammites and clays, can enable one to judge beforehand what their action upon fat lime will be.

He states that the natural and artificial puzzolanas offer some negative symptoms, in this respect, those which are hard grained and compact or of a vitrified appearance, and which do not act upon the tongue, and in short all those which possess great cohesion, being inferior.

(LXX) ***M. Vicat's Rules for judging of the quality of Natural or Artificial Puzzolanas.***

He recommends testing them by acids and in lime water, and observes, 1st, that those which are not acted upon by acids, and have no action on lime water are inert. 2dly, that those which are moderately acted upon by acids, and which act very moderately on lime water have little energy. 3dly. That those which are powerfully acted upon by acids, and which act powerfully themselves on lime water, are energetic.

He explains that the action of acids in this case does not only imply effervescence, which is the symptom of the presence of lime, but the solution or separation of the oxide of iron, or of the alumina, contained in those substances, which may be judged of by the proper chemical tests. Also in reference to this action on lime water, that sand and the other substances, styled inert, produce no effect when thrown into it, but that very energetic substances combine with the lime contained in it, which falls down to the bottom of the vessel, leaving the water above it pure.

So says M. Vicat, but on trying this experiment myself, I found that excellent puzzolana both natural and artificial produced no effect of this sort on lime water.

Of the best mode of making Artificial Puzzolanas. (*Chap.* 8.)

M. Vicat observes that the substances chosen for this purpose, whether clays, schistous psammites of a yellow or brown colour, arenes rich in clay, or certain sorts of schists, must all be subjected to the action of fire, to such a degree as to obtain, first, that the substance shall no longer form a paste with water, secondly that it shall attain the minimum of specific gravity, and maximum of absorbing power possible; thirdly that it shall be more easily acted upon by acids than before. These objects are attained by a moderate degree of burning, so regulated that the air shall reach every part of the matter during this process.

For this purpose he recommends previously pulverizing the substance, and spreading it out in a layer one centimetre, or about a tenth of an English inch thick, on an iron plate brought to a red heat, and subjecting it to the same heat for 20 or 25 minutes, stirring the powder continually in the mean time, that every part may be equally acted upon. This method being only suitable for experiments on a small scale, he proposes making the substance porous by mixing it up after pulverizing it, into a stiff paste, with combustible substances in a state of minute division, such as saw-dust, chopped straw, &c, and burning it when dry enough, in the upper part of a lime kiln, or where the heat is moderate.

If these methods cannot be used, he recommends burning the substance in its natural state, but with the precaution of first breaking it into small pieces less than a man's fist, exposed to air and with moderate heat.

He observes that any very fine and soft clay composed principally of silica and alumina, whether it contain little or much oxide of iron, or little or much carbonate of lime, will make a very energetic artificial puzzolana, if burned by the two first methods, but only an energetic one if burned by the third method, and if burned to the hardness of strong bricks, it will form one of little energy.

He observes that the substances usually, but as he thinks improperly, termed *cement* by the builders of France, namely old tiles, fragments of bricks and pottery, ought never to be used, because the refuse materials of brickworks, tileworks and potteries, consist of overburned as well as of underburned fragments produced by the burning of earths of different sorts often of an improper quality, so that the chance of success with such incongruous substances must be too precarious.*

* Both M. Vicat and General Treussart, who attached still more importance to artificial puzzolana than the former, have been anticipated in this part of their researches, not only by the Dutch manufacturers of this article, who are said to have succeeded so well as to pass it off for genuine trass, but also by a number of men of science, who experimented with the same view, of whom M. Baggé of Gottenburgh was the first, as mentioned in Article 258. This gentleman made use of a very hard black schistous rock, which he burned until it lost its hardness, and afterwards pulverized it, in which state, when mixed with lime, it formed a good hydraulic mortar.

Afterwards Count Chaptal calcined the ochery clays of Languedoc in 1786, and in a memoir published the year following he declared, that if properly managed, they were not inferior to the natural puzzolanas of Italy.

About the same time, M. de Cossart tried similar experiments on the basalts of the upper Loire at Cherbourg, which are said to have yielded results no less satisfactory. M. Gratien senior also succeeded with the schists of Haineville, by burning which he formed an artificial puzzolana, inferior to the natural puzzolanas or trass, but by no means bad.

Afterwards M. le Masson, in concert with M. Vitalis a chemist of Rouen experimented upon the ochery earths in that neighbourhood in 1806 & 1807, converting them by Count Chaptal's process into artificial puzzolanas, which when tried as betons, by sinking casks filled with them in the Seine, were found to have become extremely hard, when taken out at the end of six months.

The above notice of the various attempts to make artificial puzzolanas of a better quality, than was likely to be produced from fragments of tiles, bricks and pottery, has chiefly been extracted from General Treussart's Treatise.

Finally he observes that coal and peat ashes are sometimes energetic, but sometimes quite inert; and that the slag or cinders of forges, &c. usually yield very weak artificial puzzolanas.

The particular cement known by the name of *Ciment d' eau-forte*, which is a combination of clay and potash, resulting from the moderate calcination of nitre and moistened clay, is stated to be a very energetic but expensive artificial puzzolana.

(LXXI) *Of the best combinations of various Limes, with the other Ingredients for the composition of Mortars.* (*Chap.* 9.)

In order to obtain hydraulic mortars capable of acquiring great hardness under water, or under ground, or in situations always moist, M. Vicat is of opinion, that weak hydraulic limes must be combined with energetic puzzolanas, that hydraulic limes may be combined with puzzolanas of little energy: but that eminently hydraulic limes may be combined with inert substances such as sand, whether silicious or calcareous, and also with the slag of forges, &c.

To obtain mortars or cements capable of acquiring great hardness in the open air, and to resist rain, heat and severe frosts, he is of opinion, that fat limes cannot effect this object at all, nor can weak hydraulic limes accomplish it in a satisfactory manner; but that hydraulic and eminently hydraulic limes will succeed if mixed with any clean sand, quartzose dust, or with the powder of hard lime stones, or other inert substances.

He states that by other combinations, passable or good mortar may be obtained, but that those which he has pointed out above are the best.

(LXXII) *Of Mortars to be used under Water.* (*Sect.* 3, *Chap.* 10.)

He states that the proportions of the sand and other ingredients, used in combination with different sorts of lime, must vary according to circumstances, but to give some general notion he observes:

That arenes, psammites and clays, will answer with a smaller proportion of lime, than other substances generally, for in measuring them in the state of dry powder they require, of fat lime paste slaked by the common method, from 15 to 20 per cent; of moderately hydraulic lime from 20 to 25 per cent; and of hydraulic lime from 25 to 30 per cent. That the energetic or very energetic puzzolanas require under the same circumstances, of fat lime from 30 to 50 per cent, and of moderately hydraulic lime from 40 to 60 per cent; and that silicious and calcareous sands require of hydraulic or eminently hydraulic lime from 50 to 66 per cent.

As a general rule, he states that it is better to err by using too much than too little lime in such mixtures: for he adds that excess of lime makes them adhere better to stone; but if to be used by themselves, that the just proportions enable them to attain the greatest hardness.

Mode of Slaking. He states, that when artificial puzzolanas are mixed with fat or weak hydraulic limes, the best order for obtaining good mortar is 1st, spontaneous slaking, 2dly, by immersion, 3dly, the common mode.

For all mortars to be obtained from hydraulic or eminently hydraulic limes, the best order of slaking is 1st, the common mode, 2dly, by immersion, 3dly, spontaneous. In using these limes the two first modes do not produce any great difference in the result, but the third is very prejudicial.

Mode of mixing the Mortar. He observes that after every mode of slaking, the lime must be brought to the state of a homogeneous paste for receiving the ingredients that are to be mixed with it. If these consist of grains like sand, it should be a stiff paste: if in the form of impalpable powder such as puzzolanas, arenes, clays or psammites, it should be softer; but that the mortar or cement obtained should always have the plastic consistency before described.

As he objects in all cases to drowning lime as he terms it by excess of water, he gives a rule for slaking the hydraulic limes by a moderate quantity of water in a water-tight reservoir or basin, into which it is thrown by the shovel in successive layers of from 20 to 25 centimetres, or from 8 to 10 English inches in thickness, each of which he sprinkles with a moderate quantity of water gradually applied. He considers 2 basins necessary, one to be filled each day, in order that the lime paste may always have 24 hours to slake thoroughly, and to admit of all the lazy particles being perfectly divided. Next day as the hydraulic lime will become very hard, it must be cut out by the pickaxe or spade, and beaten with rammers to restore its moisture and toughness.

He says that the greatest care must be taken to avoid making the lime paste too fluid, which the workmen will always do if left to themselves, because it saves them labour, but it injures hydraulic mortars destined to be used immediately under water, by reducing their strength to one half, one third, or even one fifth only, of what it would be if mixed up into a proper consistency, and therefore this practice ought not to be tolerated.

He states that the hydraulic mortar for beton, must be beaten and mixed properly, before the small stones or gravel that are to form it into beton are introduced, and that these also must be mixed by the rammer.

(LXXIII) *Of letting down Beton into its place.*

As beton is to form foundations and walls under water, without previously pumping the spot dry, he says that it must be let down into its place by some contrivance, such as a box, which on reaching the bottom must be turned over and emptied, by pulling a string or otherwise as proposed by Belidor; but must on no account be thrown down by the shovel or even poured through a hopper. He observes that on the surface of each course of beton, which should not exceed 40 centimetres or about 16 inches in thickness, a milky scum is usually formed, which must be got rid of, being prejudicial to the solidity of the work, as it has a tendency to collect in strata of from 3 to 5 centimetres, or of an inch or two in thickness, between successive courses of the beton, which it does not unite properly and tends to produce settlements. M. Vicat states that in running water, very small openings on both sides to admit the current to pass through and change the water within the inclosure, and that in still water one or two powerful pumps, for draining off this sort of milky scum from the surface of the beton, in each portion of the inclosure, will remedy the evil.

He disapproves of ramming the beton under water, saying that it may be spread or arranged by compression, but not by percussion. And he also reprobates a practice which was formerly generally approved, that of using hot lime for beton, it being his opinion, that none but perfectly slaked lime ought to be used for this purpose.

He does not approve of Belidor's rule of always allowing Beton to become dry before it is used, which ought to be carefully avoided, but if this effect should have taken place owing to a burning sun or a hot wind, he recommends that it should be remixed and beaten again with a small addition of water.

From perusing this part of M. Vicat's work with attention, I infer that Beton is a very precarious mode of building under a mass of water, and subject to greater difficulties than he seems to suppose. From my own observation of the concrete in Chatham Dock-yard when disturbed by land springs, I would not recommend it at all; and the author himself admits, that when the water can be pumped out regular masonry is preferable; and in this case he states, I apprehend correctly, that the final resistance of the hydraulic mortars employed may be doubled, by letting them set as much as possible before the water is let in again, but without waiting till they become absolutely dry.

(LXXIV) *Of the Action of Water on the parts of Mortars in immediate Contact with it.*

He observes that the parts of several mortars in immediate contact with water, after having acquired at various periods a certain degree of hardness, sometimes lose the consistency that they had at the moment of immersion, and that a sort of soft envelope is formed at the surface, continually increasing in thickness and tending to gain the center; which if scraped off, is succeeded by another. These effects which are very remarkable, if fat limes, and weak puzzolanas be used, are imperceptible when very energetic puzzolanas, or eminently hydraulic limes are employed, in combination with the proper ingredients. This observation is of great importance in respect to wharfs or basins frequented by boats, and also exposed to the action of a current or of water in a state of agitation; but in still water, he observes that this sort of decomposition has its limit, for a black or white crust is formed by degrees, which he ascribes to carbonic acid, upon the soft envelope before mentioned, and protects the parts below, which by degrees may recover their consistency; but he thinks that more than six days may elapse, before they become as hard as the center of the same mortar.

(LXXV) *Of the Influence of Time.* After premising that a difference of the time of setting is only a good criterion of the future strength of hydraulic mortars, when the same ingredients are combined together in different proportions, or in the same proportions but with different quantities of water; he observes that the excess of fat or moderately hydraulic lime, in a hydraulic mortar, retards its setting; and that the second and third modes of slaking are favourable to this process. Also that hydraulic mortars, composed of a mixture of the above limes with energetic puzzolanas continue to increase in hardness or strength during the third year after immersion; but that those made of a mixture of powerful hydraulic limes with silicious or calcareous sand, do not improve perceptibly after the second year of immersion.

(LXXVI) *Of Mortars exposed to Air, and to all Weathers.* (*Chap* 11.)

Having remarked, that in a previous part of his work, he had proscribed the use of common mortar altogether for the walls of buildings of any importance, exposed to air and weather, he states the following remarkable circumstances observed by him in respect to mortars generally.

1st. That the hydrates of lime, which acquire the greatest hardness in the air, are those which when mixed with clean sand produce the weakest mortars.

2dly. That the intervention of clean sand does not contribute, as was formerly believed, to increase the cohesion of which every sort of lime is susceptible; but is injurious to fat limes, very useful to powerful hydraulic limes, but neither injures nor improves some intermediate qualities.

Of the Effect produced by the Size of Sand.

He describes the most essential qualities of sand to be the cleanness, and hardness of the particles. He considers all particles much smaller than 1 millimetre in diameter as only deserving the name of *dust*. He defines *fine sand* to be that which has grains varying from 1 to 1½ millimetre in diameter, or from the 25th to the 16th part of an English inch, and *coarse sand* that which has grains varying from 1½ to 3 millimetres in diameter, or from the 16th to the 8th part of an English inch.

He then observes, that with eminently hydraulic and hydraulic limes, fine sand makes the best mortar, secondly a mixture of fine and coarse, lastly coarse sand; that with moderately hydraulic limes mixed sand is the best, secondly fine sand, thirdly coarse sand; but that with fat limes, coarse sand is the best, secondly mixed sand, and lastly fine sand.

He observes that the greatest difference, which the use of different sands produces in fat lime mortars does not exceed one fifth, but that in hydraulic mortars it may amount to one-third, the proportional strength of each thus varying from 100 to 80 in the first and from 100 to 60 in the second sort of mortar.

The dust of very hard cohesive lime stones such as marble, and even silicious dust makes a good mortar with powerful hydraulic limes, but all dust consisting of mud or clay is to be reprobated.

(LXXVII) *Of the Proportions of the Ingredients for Mortars exposed to Air.*

The resistance of common mortar, produced from fat limes slaked in the common method, increases from the proportion of from 50 to 240 parts of sand, mixed with 100 of stiff lime paste, after which it decreases indefinitely. When slaked by immersion or spontaneously, 220 parts of sand to 100 of lime yield the maximum of resistance,

In hydraulic limes slaked by the common method, 180 parts of sand to 100 parts of lime paste produce the strongest mortar; but when slaked by immersion or spontaneously, 170 parts of sand to 100 of lime paste produce the strongest mortar; and in both cases more or less sand than the above proportions yields inferior mortar.

In respect to the manipulation, M. Vicat observes, that long exposure to air with repeated beatings and remixings improves the mortars of common fat lime; but that the improvement which may thus be obtained is not commensurate with the extra expense of labour. The same process is very injurious to the mortars of hydraulic limes.

He observes that mortar should always be made under cover, to avoid the too rapid drying which takes place in summer, or the still more inconvenient effect of winter rains. In wet wintry weather he observes, that the slaking of hydraulic limes by immersion may be best, to soak up wet sand, though not to be recommended in working with dry sand, also that in summer, lime paste of the proper consistency may not be capable of combining with very hot sand: but he observes that great caution must be used in adding more water, not to drown the lime.

Mode of using Mortar. It being evident, that very stiff mortar cannot combine well, with dry absorbent materials, he lays down a rule to wet the materials incessantly, and to keep them permanently soaked, the proper maxim for masons being, according to him, '*stiff mortar and wet materials,*' whereas he observes that the masons of France adopt the contrary maxim, using *wet mortar and dry materials.* To comply with his rule he remarks, that a sufficient quantity of water should be put under every stone, in order that on beating it down upon its bed, the mortar below may be made to work up and fill the vertical joints on each side of the stone thus placed.

He observes that the mortars of buildings generally become friable, when exposed to a very rapid drying, which is always injurious, but chiefly to those of eminently hydraulic limes. Therefore in hot weather he recommends watering masonry.

In respect to the age of mortar, after quoting a proverb of the

French masons, that *mortar a hundred years old is still a baby*,* he remarks that it is only in foundations or in very massy walls 400 or 500 years old, that good fat lime mortar is to be found;† and if so, he justly observes, that mortar which requires so long a time to harden is of no more use, than if it never became strong at all. In respect to the mortars of powerful hydraulic limes, he observes that when exposed to air in small masses, they attain in the short time of about 18 or 20 months, if not their maximum degree of hardness, at least a state differing so little from it, that one may foresee with certainty what they will eventually become.

(LXXVIII) *Of Hydraulic and Common Mortars constantly exposed to Damp Earth.* (*Chap.* 12.)

In respect to hydraulic mortars, he considers that the proportions, mode of slaking, and of mixing them ought to be the same as for constant immersion, and in regard to common mortars, the same as if they were to be exposed to air and weather.

Reprobating the use of dust, especially that of soft lime stones, he prescribes certain proportions of sand both for fat and hydraulic limes, as being proper for mortars intended to be kept in constant contact with damp earth, different from those which he before recommended for making mortar with the same limes to be exposed to air and weather; but his rules for these proportions are so very obscure, that I shall not attempt to state them, as I think it probable that there may be an error of the press in this part of his Treatise.

He observes that excess of lime makes the mortars thus used adhere better to stones and bricks, but at the expense of their own cohesion.

(LXXIX) *Of the Vicissitudes, to which Hydraulic and Common Mortars may be exposed.* (*Chap.* 13.)

Some of these, which set tolerably well under water, may lose a part or the whole of their cohesiveness, when exposed to a dry hot air. This remark applies particularly to those which are not well suited to each other, such as a mixture of fat limes, with artificial puzzolanas of little energy, or of weak hydraulic limes with sand. The surface of such mortars if partially decomposed in water, as before described, deteriorates still more rapidly, when afterwards exposed to air. Mortars that have hardened under damp earth usually hold good, if afterwards exposed to water, but not always when exposed to air. But good hydraulic mortars will bear any of these changes without injury.

Common mortars of fat lime and sand which have rather dried than hardened in air, become completely decomposed under water.

* Le mortier à cent ans, est encore un enfant.

A similar distich or proverb of the Scotch masons ascribes more virtue to it at this age.

When a hundred years are past and gane,
Then gude mortar grows into stane.

† I do not think that good fat lime mortar is to be found any where. I have never seen any in the oldest buildings that I have examined.

M. Vicat considers, that frost is particularly injurious to mortars made with a small or moderate quantity of sand. For example, in respect to fat lime mortars made in April, he observes that they will not stand the next winter's frost if severe, if less than 220, 160 or 100 parts of sand respectively be used with 100 parts of lime paste, according as it may have slaked by the first, by the second, or by the third mode. After two years the danger ceases, and the severest frosts will have no effect.

In respect to the mortars made of hydraulic limes, he states, that their preservation against frost also depends upon the proportion of sand, but after 6 or 7 months he considers them out of danger.

(LXXX) *Of the Effect of beating Mortar.* (*Chap.* 14.)

Considered as a plastic substance, mortar may be fashioned in moulds to any form you please, and when made with colourless sand or the dust of hard white lime stones, it may be made to take the appearance of stone.

In such moulds it may be worked up, beaten or rammed, and thus acquire great density, but as this process does not always increase its strength, it becomes a question under what circumstances it does good or harm.

M. Vicat observes, that mortar cannot be rammed, whilst it remains so soft as to rise in one part when beaten down in another. It must therefore first be allowed to dry to a certain degree. He observes that the ramming of mortar operating always in the same direction must cause it to become laminated, although this effect may not be perceptible to the eye, and he is of opinion that the greatest resistance of mortar thus foliated or stratified is in a direction perpendicular to the strata, when under ground, but in a direction parallel to the strata in the air. He observes that beating is prejudicial to the hydrates of hydraulic limes, when used any where excepting in air. But he remarks, that he has himself often made artifical stones of hydraulic mortar in ornamental forms, which resisted frost, and hardened progressively, and that in time they were covered with a sort of varnish of such a nature, that the most experienced observer would take them for natural stone. He adds, that his efforts in making ornamental vases with hydraulic mortar had been so successful, that the Society for encouraging Arts in France had presented him with their gold medal on this account, in 1823.

Of the Natural Cements. (*Chap.* 15.) He treats of these in a manner which I have before alluded to (LIX), and which sufficiently proves his total want of practical knowledge of the use of those cements.

(LXXXI) *Of the Comparative Strength of Ancient Mortars, those of the Middle Ages, and Modern ones.* (*Chap.* 16.)

He observes that mortar, first used by the Egyptians, was brought to considerable perfection by the Greeks, who not only used excellent mortar for walls, but also made good inside plaster, as well as stucco for flat roofs and floors. In respect to the Romans, who had the advantage of all the knowledge acquired by the Greeks, he remarks that there was a generally received prejudice in France, that they possessed some secret for making superior mortar unknown to the Moderns. He combats this opinion, which he considers perfectly erroneous, not only by reasoning, but from his own experiments on the comparative strength of mortars. But before I notice the results of the numerous experiments tried by him, from which he deduced the opinions and laid down the rules, of which I have given an abstract, it will be proper to explain the process pursued by him in his researches.

(LXXXII) *M. Vicat's mode of trying the comparative Strength of Limes and Mortars.*

The manner in which he experimented upon the various hydrates of lime, mixed with water only, in order to ascertain whether they had hydraulic properties or not, has already been explained. He also combined each sort of lime tried by him with sand, or with the natural or artificial puzzolanas, or trass, or with a mixture of sand and of those puzzolanas or trass, using always the same proportion of lime paste, namely 100 parts by measure, but varying the proportions of the sand, puzzolana and trass; and whether he experimented with the lime paste alone, or added to it the other ingredients, which have been mentioned, he always brought them to the same degree of plastic consistency before defined, taking the greatest care not to add water rashly, so as to make a fluid paste, and thus to drown his mixtures, which he considered equivalent to destroying the lime. And in preparing them, he states that he always beat or rammed the ingredients.

It was before mentioned that he ascertained the time, that each experimental specimen of hydraulic lime putty or mortar took in setting, which he noted in days from the period of immersion, that is from the day in which they were made or mixed. At the end of a year or sometimes more, he endeavoured to ascertain their hardness and consistency, which he did by two different processes.

(LXXXIII) First, he used a steel stem slightly conical or tapering downwards, and measuring at bottom 1·66 millimetre, about one 16th of an English inch, in diameter, which being loaded with a weight of 0·9961 kilogrammes, or 2 lbs. 3 oz. avoirdupois, was allowed to fall upon the specimen, like a little pile driver from a constant height of 5 centimetres, rather less than 2 English inches. A scale of millimetres and parts attached to this little apparatus indicated the penetration of the steel point into the specimen after it was allowed to drop. In his first publication in 1818, he estimates the *comparative hardness* of each specimen experimented upon, as being in the inverse ratio of the square of the penetration thus indicated: and he also sometimes estimated their *comparative resistance*, as he termed it, by the same rule, but using a different and much higher scale of proportional numbers. For example, he estimated the resistance of a good well-burned brick, as being equal to 5960, and its hardness as being equal to 0·096, which numbers when both found by the steel point, ought to have been alike, if different scales had not been used. But he also sometimes used another mode of estimating the resistance of his experimental specimens of hydraulic mortars, &c, which appears to me to be much more conclusive, being a matter of direct experiment, and not of hypothetical calculation, like the former. Having formed rectangular prisms of indeterminate length, but measuring 25 by 40 millimetres in section or 1 by 1½ inch nearly of English measure, he fixed one end firmly down, leaving the other projecting horizontally, and to the latter he applied a small iron stirrup large enough to pass over the end of the prisms with ease. By this stirrup which was placed at 30 millimetres or nearly 1·2 inch, from the point of support, he suspended a wooden box to hold fine sand, which was poured gradually into it from a sort of hopper, until the specimen broke down under the weight, and this breaking weight was assumed as the value of the resistance, which in his Tables is stated in decagrammes, the number 5960 for instance, by which he estimates the resistance of a good well-burned brick, implying that a rectangular prism of this sort of brick measuring 1 by 1½ inch would break down under 5960 decagrammes, equal to about 131 lbs. avoirdupois.

It is to be observed, in reference to his first publication, that in 12 of his tables of experiments on various limes and mortars, the resistances are expressed by numbers deduced from direct experiment, being the actual weight in decagrammes, under which his specimens broke down; but in 10 other tables of similar experiments, they are expressed by other numbers calculated as before described, the inverse proportion of the squares of the penetrations of the steel point.

In consequence of those incongruities in the mode of experimenting, and in reference particularly to the steel point apparatus, which appears to me almost nugatory, for I entirely coincide in the reasons urged against it by his countryman General Treussart, and considering also the obscurity of his style, I must confess, that I was disappointed in M. Vicat's first publication. His second Treatise, on the contrary, which I did not see till lately, though published nearly ten years ago, is less exceptionable, his opinions being stated clearly, and illustrated by notes, and those relating to the comparative strength of various limes and mortars being deduced from a new series of experiments, the results of which are contained in Tables at the end of his book, to which I am sorry to say, that I cannot attach that confidence, which the author places in them. In his new steel point experiments, pursuant as he says to the advice of his friends, he has recorded the actual penetration made in each specimen in millimetres, which so far is more satisfactory than his original system; but with this mode of estimating the hardness of his specimens, he has mixed another and a different mode by boring into them,* which if better should have been used exclusively, if not better, not at all. In respect to the breaking apparatus, he has adopted a much less satisfactory method of operating, as well as of recording the results of this process, than he did before: for in his second series of experiments he worked with square prisms of unequal dimensions, from 25 to 40 millimetres in height and from 40 to 50 millimetres in width, that is from about 1 to 1·6 inch in height and from about 1·6 to 2 inches in width; and which may consequently be of the same strength, though broken down by different weights, and of different strengths though broken down by the same weight,† and in adopting this injudicious system he does not fairly state the actual dimensions of his several specimens, and the actual weights under which they broke down, instead of which he gives their ABSOLUTE RESISTANCES in kilogrammes, reduced to the superficial centimetre by some sort of calculation, in reference to certain dimensions and to their relative or comparative resistances, determined by his experiments, which dimensions he has not stated, and which mode of calculation he has not explained.‡ Hence his new Tables of Experiments are still in my opinion ex-extremely unsatisfactory, being incongruous with each other, and excepting so far as the water-setting properties of various hydraulic limes or mortars are investigated, having no analogy to the mode in which the mortars of walls either act upon the stones or bricks with which they are combined, or are themselves acted upon by the pressure or forces, to which as a part of those walls they are exposed.

In his second Treatise, in reference to the course of experiments above alluded to, he estimates the absolute resistance of mortars of lime and sand to vary from 0·75 to 18·53 kilogrammes, and that of building stones to vary from 20 to 77 kilogrammes, per superficial centimetre, being equivalent to from 11 to 264 and from 285 to 1095 lbs. avoirdupois to the superficial English inch, observing at the same time that the soft stone, commonly used at Paris, has scarcely half the resistance of the weakest of the above.

* He bored into each of his specimens by an equal number of turns of a steel borer under a constant pressure, and stated the comparative hardness of each, in numbers reciprocally proportional to the depths of the respective holes thus formed; but he has not stated the actual depths, which would have been much more satisfactory, than the results deduced from them, by some sort of calculations, of which he has only explained the principle.

† In trying these unequal rectangular prisms, he has not stated any fixed distance from the point of support, at which he applied his breaking apparatus, as in 1818. Hence we may infer that the distances may have varied in different experiments of this new series; so that the breaking weights may have acted with unequal leverages, another source of uncertainty.

‡ The only explanation which he attempts to offer is this 'the absolute 'resistances of the prisms buried or exposed to air have been deduced from 'the relative resistances of these same prisms, by determining by a suitable 'series of experiments, and for specified dimensions, the particular value 'assumed by the coefficient, which is represented by $\frac{1}{3}$ in the formula of 'Galileo.'

(LXXXIV) He concludes his Treatise by stating, that as a matter of calculation one may reckon upon the following absolute resistances for mortar one year old, made of various sorts of lime; namely for eminently hydraulic lime 12, for hydraulic lime 10, for inferior hydraulic lime 7, and for fat limes 3 kilogrammes per superficial centimetre; which reduced to English weight and measure are equivalent to 171 to 142 to 100 and to 43 lbs, avoirdupois, per superficial inch; provided that the proper mode of slaking the several limes, and of mixing the ingredients in the most suitable proportions, as recommended by him, have been adhered to. But he states that the resistance of the mortar made with fat lime by the workmen of France, who drown their lime in mixing it, is only three quarters of a kilogramme per centimetre, or about 11 lbs, to the superficial inch, which mortar he observes scarcely deserves the name. And he further states that the best cements and mortars of the same age, kept immersed in water, or buried under earth constantly damp, will not exceed 10 kilometres to the superficial centimetre, or 142 lbs, avoirdupois to the English inch.

I must confess that I really do not understand, in what manner the absolute resistance of stones or mortars can be reduced to any just proportions, founded on superficial measure. The resistance of a mortar or cement joint against a force, tending to tear asunder the two stones or bricks united by it, which peculiar resistance I have designated by the term *Adhesiveness*, depends in some measure upon the superficies of the joint, although as I have shewn in the foregoing Treatise (220), it cannot be estimated accurately in reference to such small dimensions as the superficial inch or centimetre, to which this quality of mortars, &c, bears no fixed proportion. But I cannot see, in what way experiments made as M. Vicat did upon rectangular prisms fixed at one end and broken down by weights, to ascertain the comparative or absolute resistance of stones or mortars, can have their results estimated by superficial measure; for these small prisms having been acted upon like beams fixed in the same manner, their resistance ought to be estimated like that of beams, which no one has ever thought of reducing to the superficial centimetre or inch.

I have thus attempted to give an abstract of the text of M. Vicat's second Treatise, but not of the numerous explanatory notes annexed to it, in support of his statements and opinions, which I could scarcely have included also, without making almost a complete translation of his work, which was not my intention.*

* M. Vicat's second Treatise has recently been translated by my friend Captain Smith of the Madras Engineers, who has added valuable notes of his own upon various parts of the subject. Having made the foregoing abstract of M. Vicat's work before I saw Captain Smith's translation, with which I afterwards compared it, I can certify to the extreme fidelity of his translation, excepting in what I admit to be a mere trifle, his having invariably changed the French phrase *fat lime*, used by the author into *rich lime*, which I do not think by any means so appropriate as the former; because fat limes are the worst of all that can possibly be used, and the term *fat*, not inaptly

Of the Proceedings of General Treussart of the Royal Corps of Engineers, in France.

(LXXXV) In the first chapter of a work published in Paris in 1829, on hydraulic mortars and common mortars,* this Officer states that he was employed at Strasburg from 1816 to 1825. On his arrival at that place, he found that nothing but common lime was used there, and that almost all the masonry of that Fortress, which was exposed to or connected with the manœuvres of water, required to be reconstructed, having been executed badly in the time of Vauban. Twenty years experience at former stations having convinced him of the great superiority of hydraulic limes, he tried all those which were to be procured in the neighbourhood, and finding some of them very good, he used them in the reconstruction or repair of 1500 metres or 1640 yards, of revetment, as well as in the reconstruction of the sluices, floodgates, &c, for admitting the water into the ditches. He afterwards states (in Chap. 4), that being required in 1818, by a circular order of the Minister of War, to repeat M. Vicat's experiments, he employed successively two Officers of Engineers, both of whom failed in trying to produce an artificial hydraulic lime from common lime twice burned, and mixed with brick earth previously to the second burning, according to M. Vicat's directions. He then took the matter into his own hands, and succeeded at last, by using a very fat clay, but being still dissatisfied with the strength of the best of his own mixtures, he sent for some artificial hydraulic lime made at the manufactory near Paris, established on M. Vicat's principle, and this also proving unsatisfactory; he was afterwards induced to try experiments for improving common lime himself, by adding artificial puzzolana to it, according to the old system before mentioned, as having been generally practised in the south of Europe, and especially in France, from the time of the Romans (257). These experiments were commenced by General Treussart in 1821, and in a few years afterwards he published a little pamphlet on the subject, in which he combated several of M. Vicat's opinions.

(LXXXVI) Both agree in one point, that the natural hydraulic limes should always be used, when they can be procured at a moderate expense, but where this is impracticable, M. Vicat recommends

adopted, in reference to their imbibing much more water than other limes in slaking, does not imply any sort of superiority, like the term *rich*. M. Vicat's reputation stands so high in France, and his work upon calcareous mortars and cements is so very much superior to all those which preceded it, with the exception of Smeaton's only, that I do not wonder at Captain Smith prizing it so highly as he does, and it is a matter of regret to me, that I have found myself compelled to notice the very unsatisfactory nature of M. Vicat's experiments, upon which Captain Smith has made no comments, and if I had not experimented so largely myself upon the same subject, the strong objections to those experiments might perhaps have escaped my observation also.

* Mémoire sur les Mortiers Hydrauliques, et sur les Mortiers Ordinaires, par le Général Treussart, Inspecteur du Génie.

his own artificial hydraulic lime, whilst General Treussart contends that it is much better to mix common lime having no hydraulic properties, in combination with natural or with artificial puzzolana prepared for the purpose.*

In respect to the failure of M. Vicat's method, when first tried at Strasburg by General Treussart's order, it appears to me to have proceeded from the error of M. Vicat himself, in recommending any sort of clay or brick earth as being equally good. If instead of attempting to form an artificial hydraulic lime, he had endeavoured to form an artificial water-cement, which does not seem to have been contemplated by him, M. Vicat would have found, like me, that brick earth was perfectly unfit for that purpose. I am therefore not surprised, that General Treussart should have found his first attempts to make artificial lime unsatisfactory, owing to this error in M. Vicat's specification : but at the same time, I am perfectly convinced, from my own experiments, of the great perfection to which M. Vicat's artificial hydraulic limes might be brought by using proper clay, selected for the purpose, instead of using any argillaceous earth indiscriminately, as he in his evident inexperience at first directed.

(LXXXVII) *Of the slaking of Lime.* (*Chap.* 2.)

First, or common Mode. General Treussart describes first, the common mode of slaking lime in France, by throwing so much water on it, as to reduce it to a thin paste or milky consistency, in which state it is run into pits or reservoirs, sometimes reveted with masonry on the sides, and after some time, when it becomes thicker, by covering it with sand or earth to preserve it from the air. He remarks that the common opinion is, that the longer lime has been thus run the better. This he considers to be erroneous, as he found by experiment, that common lime thus slaked made bad mortar when mixed with sand alone.

Second mode, by Immersion. Having described M. de la Faye's method of slaking lime in this manner, which I shall not repeat here as it was fully explained before, General Treussart states, that having found this process inconvenient, he adopted a different mode in his works at Strasburg, tending however to obtain slaked lime powder of the same quality. For this purpose he measured his quicklime in a bottomless cubical measure containing 3 tenths of a

* M. Vicat in his publication of 1818, prior to General Treussart's proceedings, had foreseen that the question would be agitated, whether his new method or the old French system of improving common lime by adding tile dust or what they called cement to it, was the best, in reference to which he observed by anticipation, 'it must not be supposed that clay 'burned separately and afterwards added to common lime can produce the 'same results, as when these two substances are mixed previously and burned 'afterwards, because,' he asserts, 'that the fire modifies the ingredients 'composing the mixture, and produces a new compound, having new properties, of which he considers the difference of colour, between artificial 'hydraulic lime, and a mixture of common lime with the same clays when 'burned separately, to be a proof.'

metre or about $10\frac{1}{2}$ cubic feet of English measure, and laid it out in a small heap. He then measured the proportion of sand intended to be used, in the same measure, and laid it out round the lime, after which a quantity of water equal to one fourth of the same measure was sprinkled out of tin watering pots over the lime, so as to moisten every part of it equally, and it was allowed to slake until it ceased to emit vapour. It was then examined by turning it over or poking it with an iron shod staff, to see if any pieces still remained unslaked, upon which a little more water was sprinkled, and then the mass of lime was covered with the sand; which being done over night, it was left till next morning, and no more of these little heaps were prepared than were necessary for the next day's use. For mixing it afterwards with the sand and water, he recommends when the works are considerable, a mill, worked by two horses, which from his description is the common mortar mill having two vertical rollers, that has generally been used in this country, as well as in France.

Third method of slaking Lime spontaneously, or by Exposure to Air. After quoting M. Vicat, who considers the strong objections of all practical men to the use of lime slaked in this manner, as a vulgar error repeated by one author after another, but destitute of any foundation in truth, he states that his own experiments do not confirm this opinion of M. Vicat.

In his first Table, General Treussart gives the results of experiments tried with various sorts of lime, which after having properly burned and pounded in a mortar, and sifted them, and measured 1 litre of each in the state of impalpable quicklime powder; he then added water sufficient to change them into a slaked lime powder, and finally by adding more water, he reduced them into a paste, noting the successive quantities of water, and the quantity by measure of each state of the lime. Proceeding in this manner it appears from the Table, that 1 measure of the quicklime powder of white marble, sprinkled with half a measure of water produced $2\frac{1}{2}$ measures of dry slaked lime powder, and when more water was added so as to make $1\frac{6}{10}$ measure of water in all, it produced $1\frac{1}{2}$ measure of paste.

In like manner 1 measure of quicklime powder of Altkirk lime, which he considered a good hydraulic lime, sprinkled with half a measure of water, produced $1\frac{5}{6}$ measure of dry slaked lime powder, and with $\frac{4}{5}$ths of a measure of water in all, produced 1 measure of paste.

In the same Table he gives the results obtained by experimenting upon seven other limes, besides the above, and on the *Galets de Boulogne*, cement stones found there (similar to the Sheppy cement) of which he says that the properties were first discovered by an Englishman. This cement calcined and pulverized would not bear more than one third, and the limes would not bear more than one half, of their respective volumes of water, without being reduced to a wet powder. He therefore kept within those limits, that he might obtain a dry slaked lime powder, but he observes that this effect might have been produced on all the limes by one fifth part of their own volume of water only. He observes further, that the strongest hydraulic limes require the least water, and occupy the least space, first as slaked lime powder and afterwards as a paste.*

General Treussart's Mode of experimenting upon the Hydraulic Properties and Strength of Limes and Mortars. (*Chap.* 3).

(LXXXVIII) He first reduced his quicklime to a fine powder by using about one fifth of its bulk of water. After 12 hours he made his slaked lime powder into a paste with water, in which state he measured it and mixed it with trass, or puzzolana only, whether natural or artificial, or with sand and trass or puzzolana, also combined with it by measure, in various proportions, sometimes also experimenting upon the lime paste or hydrate of lime alone. He worked up these ingredients with water to the consistency of honey, passing them 7 or 8 times under the trowel. He then put them into little fir boxes 15 centimetres or about 6 English inches long by 7 centimetres or 2¾ inches in width and depth, and exposed them to the air for 12 hours, after which he compressed them a little by the hand and the trowel, they having at this time acquired what he calls a half consistency, and in this state he put them into a large tub of water in a cellar. He observed how many days they were in setting, which he judged of, by their not yielding to the pressure of the thumb.

(LXXXIX) At the end of a year he took his experimental pieces of lime or mortar out of the tub and box, and by a stone cutter's chisel and afterwards by grinding them on a stone, he reduced each specimen to a prism of the same length as before, but only 7 centimetres or 2¾ inches square. He then broke them down by weights suspended from a collar or inverted stirrup, passing over the center of each, whilst their ends were supported by a couple of iron stirrups fixed to a beam of wood, at the clear distance of 1 decimetre, or very nearly 4 English inches apart. This will be understood by a reference to Article 199, in which I have described a similar apparatus used by me, but copied from General Treussart. He states that his object, in scraping off part of the surface all round,

* General Treussart makes his white marble lime expand from 1 measure of quicklime powder to 1½ measure of slaked lime paste. In my experiments on chalk lime, the same quality of lime, but from a less compact stone, produce nearly 1¼ measure of slaked lime paste from 1 measure of quicklime powder. See the Abstract of Table XIV (297). M. Vicat says that 3·1 measures of paste of the same description of lime may be obtained from 1 measure of quicklime, which he states as a fact, without describing the experiments from which he deduced this opinion. Hence as this assertion is in direct contradiction to General Treussart's experiments and mine, which very nearly agree, provided that allowance be made for the General's mode of measurement, which is the only thing he has not explained, and which may have differed from mine; I naturally place more confidence in his experiments, than in the opinions of M. Vicat.

was to subject to the breaking apparatus, that part of each experimental prism, which had not been exposed to the direct action of water, for he remarks that he often found his mortars harder outside than inside, and sometimes the contrary. This method appears to me much more satisfactory than M. Vicat's breaking down process, as before described (LXXXIII), and infinitely better than his steel point apparatus, to which no value can reasonably be attached, for as General Treussart remarks, in trying the hardness or resistance of a prism of beton by this apparatus, if the steel point happened to impinge upon one of the hard broken stones or flint gravel, which form a component part of this substance, its effect would be null, but if it happened on the contrary to impinge upon the lime or fine mortar, by which those stones are united, its penetration might be considerable.

(XC) *Of the Natural Hydraulic Limes.*

General Treussart estimated the strength of these, as well as that of the mortars produced from them, by the actual breaking weight in kilogrammes, as was before explained. To give an idea of their comparative resistance in proportion to that of bricks, he reduced well burned bricks of two sorts to the same dimensions as his experimental square prisms of mortar, and on trying them in the same manner, he found that one sort broke down under 210, and the other under 269 kilogrammes, or under weights respectively equal to 463 and 593 lbs. avoirdupois. He found that his natural hydraulic limes were generally stronger as hydrates, than when made into mortar with sand alone, but that their strongest form was when 1 measure of lime paste, 1 measure of sand and 1 of trass or of puzzolana, were mixed together.

For example, Altkirk lime in the state of putty broke down under 122 kilogrammes, or 269 lbs, avoirdupois; when 1 measure of this lime was made into mortar with 2 measures of sand, it broke down under 79 kilogrammes or 174 lbs; but when made into mortar with 1 measure of trass and 1 of sand, it required 245 kilogrammes or 540 lbs, to break it down. The times which these mixtures had previously required to set under water, were respectively 10, 12 and 4 days, the trass thus appearing not only to add to the strength of the mortar, but to accelerate its setting. Similar results were produced, though not in the same proportion, by the like combinations of the other hydraulic limes with water, sand and trass. In these experiments, the lime having been slaked fresh from the kiln, and used immediately, he afterwards tried what difference would be produced by allowing the slaked lime powder to remain until it got stale. With few exceptions, when these limes were kept for various periods not exceeding three months, and then experimented upon, the mortars produced became gradually weaker and weaker when sand was used; but not always when sand and trass were both used, for the addition of trass appeared to correct the injury thus caused by the staleness of the lime. He also tried the effects of spontaneous slaking; and found that the mortars produced were much worse, than those obtained from the lime slaked into powder when fresh from the kiln, and used immediately, excepting in the Metz lime alone, which when slaked in air, and mixed 4 months afterwards with sand and trass, required a breaking weight of 200 kilogrammes or 441 lbs, avoirdupois. The Boulogne cement (equivalent to the Sheppy cement) as I said before, broke down when used pure under an average weight of 60 kilogrammes or 132 lbs, but when mixed with sand it did not on an average bear more than one fifth of that weight. When calcined strongly, the same cement used pure was so much increased in strength that it bore 130 kilogrammes, or 287 lbs, even in this state however being inferior to his hydraulic limes, two of which mixed with water alone required 220 kilogrammes or 485 lbs to break them down. This apparent inferiority of a cement, which I have no doubt is equal to our Sheppy cement or nearly so, can easily be accounted for by the circumstance of General Treussart having experimented upon it in the same manner that he did in trying his limes, which must necessarily have spoiled it.*

Of the Burning of Hydraulic Limes.

In respect to this point, having observed that the lime of Obernai, one of the best in the neighbourhood of Strasburg, during the process of calcination assumed first a yellow, then a grey and afterwards a blue tint, the latter of which was a proof of its being rather overburned, General Treussart found on trial that the yellow verging on grey, or buff colour, was the best state for yielding a strong mortar, and his experiments on the Boulogne cement also led him to attach importance to the degree of burning of hydraulic limes. But from my own experiments I am of opinion, that these varieties in the apparent strength indicated by our experiments, may have arisen from some accident; for I consider that when a calcined stone first ceases to effervesce in acids, it is burned enough, and that any further burning may injure but cannot improve it.

Of Artificial Hydraulic Lime. (*Chap.* 4.)

(XCI) General Treussart next relates the results of his first attempts to make this sort of lime according to M. Vicat's system, which when composed of common lime reburned and one fifth of common brick earth, required 20 days to set under water, and when mixed with 2 measures of sand broke down under 20 kilogrammes or 44 lbs, but when made of pipe clay and mixed with sand alone it required 85 kilogrammes or 187 lbs, and when made with pipe clay and mixed with sand and trass also, it required 140 kilogrammes or 309 lbs to break it down. The artificial hydraulic lime, which he sent for from Paris, when mixed with 2 measures of sand, set on

* General Treussart states, that when chemically analyzed their component parts were the same and in the same proportions nearly, excepting that the Sheppy stone had some thousandths parts of carbonate of magnesia and of manganese not found in the Boulogne stone, a difference too insignificant to be worthy of notice.

an average of nine trials in 16 or 17 days, and broke down under an average of 101 kilogrammes or 223 lbs, but when allowed to become stale, by keeping it one or two months, before it was made into mortar, it either lost a great deal of its strength or broke down by the apparatus alone, that is when sand only was used; but on adding trass also, this defect of staleness was considerably corrected, for the same lime though kept 2 months, when mixed with sand and trass in equal parts set in 6 days, and required 135 kilogrammes or 298 lbs, to break it down. He afterwards tried artificial hydraulic lime made with reburned lime and pipe clay, which had proved the best of his former mixtures, but now mixing these ingredients with soda water and potassa water, and on making it into mortar with 2 measures of sand, he found that the soda did some good, but that the potassa, increased the strength of the mixture still more considerably, in one experiment requiring a breaking weight of 185 kilogrammes, or 408 lbs.

Of Hydraulic Mortars made with Common Lime, and Trass or Puzzolana. (*Chap.* 5.)

(XCII) His trass was of the best quality obtained from the village of Brohl near Andernach on the Rhine, and his puzzolana was from Naples. Hè found that the puzzolana used by him, in combination with common lime, whether with or without sand, produced rather stronger hydraulic mortar, than the like mixtures of trass with the same substances. For example the average breaking weight of lime and puzzolana was 206 kilogrammes, or 454 lbs; whilst that of lime and trass was only 183 kilogrammes, or 404 lbs; and in like manner the average breaking weight of lime, sand and puzzolana, was 203 kilogrammes, or 448 lbs; whilst that of lime sand and trass was 191 kilogrammes, or 421 lbs. He also found, on trial, that mixtures of common lime, sand and trass, were not materially injured by allowing the lime to remain slaked for several months, or even for a year, until it got completely stale. In short when trass or puzzolana was one of the ingredients, he found that hydraulic mortar might be formed with common lime with or without sand slaked in any way whatsoever, and no matter how stale, and he adverts to experiments afterwards tried, which proved that common fat lime run in the state of fluid lime putty, and kept in a reservoir for 5 or 6 years, formed a good hydraulic mortar even at the end of that time, if mixed with the ingredients before specified. But he thinks that the best hydraulic mortar is obtained from common lime slaked into a powder, by a moderate quantity of water, and then allowed to become stale by exposure to air under cover.* This he considers to be a very great advantage of this sort of hydraulic mortars, over that formed of sand alone mixed with any of the hydraulic limes natural or artificial, since these limes are deteriorated and may be spoiled by becoming stale.

* He observes that some of the hydraulic limes make good mortar with sand and trass, although stale, but not all of them, others being much deteriorated if not used fresh. He always designates what had been termed *fat Lime* in France, by the more appropriate epithet of *common Lime.*

In all my own experiments with chalk lime, puzzolana and sand, I used the lime thoroughly slaked, but never stale. Hence I have no right to contradict General Treussart in this point, from my own personal knowledge; but as slaked lime by degrees approaches in its chemical qualities to the original carbonate of lime from whence it was obtained, and we know that the paste of pulverized carbonate of lime has no cementing properties at all, I cannot help suspecting that there must be some fallacy in supposing, that any sort of lime will not be materially injured by continued exposure to air before it is made into mortar for use. I conceive that all must suffer more or less, although fat limes may be less deteriorated than the hydraulic limes, by being allowed to become stale.

Of Artificial Puzzolanas. (*Chap* 6.)

(XCIII) The next part of General Treussart's subject relates to artificial puzzolanas, which are of great antiquity in Europe as before mentioned, that is, if tile dust or brick dust be considered such, and in this part of his work he notices the labours of M. Baggé in Sweden, and of other gentlemen who preceded him, in this investigation, which I abstain from repeating.

In using bricks as a substitute for puzzolana, he observes, that some sorts require to be more burned than others. In order to judge of this point, he recommends a trial of three bricks to be chosen in any brickfield, one rather underburned such as we would call in England *Place Bricks*, another well burned, and a third rather overburned but not vitrified. Let each of these be reduced to the state of impalpable powder. Then let them be intimately mixed with common lime which has been slaked for some time, in the proportion of 2 measures of brickdust to 1 measure of lime paste, adding as much water as may be necessary to effect the perfect incorporation of the ingredients. Put the mixtures thus formed at the bottom of a glass, and if stiff, you may pour water over them immediately, if not, let them dry for some hours till they obtain what he calls a half consistency, and then add the water, taking care that at this period they shall all be in the same state. Label the glasses, and after two or three days, examine your mixtures daily, until they set under water so as to resist the pressure of the finger, and that which soonest attains this state will be the best. If there be several brickworks in the neighbourhood, let three bricks of each be selected and tried in the same manner, and the result of your experiments will prove, which of the several earths made into bricks is the best for an artificial puzzolana, and to what degree it ought to be burned. General Treussart further remarks, that bricks, which in burning have had a strong current of air passing through them, make a better artificial puzzolana, than bricks of the same earth equally well burned but not subject to a current of air during this process. He also observes, in reference to the degree of burning, that marly clays, or those which contain a small portion of carbonate of lime, and which therefore effervesce in acids, should be less burned than the purer clays not acted upon by acids, which are not so liable to become vitrified as the former. He states that the plastic clays fit for forming earthenware are the best, which although I have not experimented in the same way appears to me to be a sound opinion, since I found that none but pure plastic clays would make an artificial cement, and I conceive that the same clay, which is the best for this purpose, will be the best for an artificial puzzolana also. Hence I would exclude bricks and brick earth altogether, as M. Peyronnet did in his specification for the artificial puzzolana for the Bridge of Neuilly, for which he only allowed pulverized tiles, which are always formed of a much finer clay, than bricks (LXVII).

(XCIV) General Treussart states that he has found, that a mixture of common lime, with sand and brickdust in equal parts, will form a hydraulic mortar requiring a breaking weight of from 100 to 150 kilogrammes, or from 221 to 331 lbs, avoirdupois, which he considers sufficient for large masses of masonry: but for important works, such as the masonry of the locks of canals, floodgates, &c, the foundations of batardeaus, the plastering of the ridges of vaults or casemates, &c, and for making artificial stone to be used for the facing of beton walls, he thinks that a resistance of from 150 to 200 kilogrammes or from 331 to 441 lbs, is desirable, for which purpose the clays fit for pottery should be chosen, which should be made up into balls the size of an egg, and burned first experimentally in crucibles having holes at bottom, to ascertain the degree of burning proper, and afterwards on a larger scale in a kiln exposing them to the action of air, but not of cold air. In using wood he recommends kilns having fire places below, and in all respects like a common brick kiln in this country, but with the addition of being vaulted over, whilst in using coals or peat, he recommends the inverted-cone-shaped lime kiln to be applied to the burning of his clay balls, intermixing them with the fewel in alternate layers.

(XCV) The results of his experiments seem to prove that generally his artificial puzzolanas, when made of fine clay, were equal to the natural puzzolanas. Having tried salt in a few experiments he thinks it probable, that it might be beneficial, especially as he found that potassa had produced a considerable improvement, and as it was well known that much of the trass sold in Holland was not natural as the dealers pretended, but artificial trass obtained from the calcination of a clay found under the sea on the Dutch coasts. He states that he finds that magnesia mixed with clay previously to calcination produced no effect, and therefore he considers it passive or inert, in which opinion I do not agree. He also tried a mixture of sand with his clays, and as he says that these sands were reduced to a very fine powder before they were used, they seemed rather to do good than otherwise. Some of the clays on the contrary when deprived of a part of their original sand were improved.*

* I do not consider this as a contradiction. Clay is chiefly composed of silica in its finest state, which in its coarse state is sand. My own first experiments proved that coarse sand is prejudicial to an artificial cement, and by analogy I conceive that it must be prejudicial to an artificial puzzolana.

(XCVI) *Of the Metallic Oxides, Coal Ashes, &c.*

General Treussart combats the opinions of those, who think that the oxides of iron and of manganese act a principal part in communicating hydraulic properties to limes and mortars; in respect to which I admit that they are secondary to clay, but my experiments will not allow me to consider them absolutely useless, as the General maintains.

In respect to the utility of coal ashes, as an ingredient for a hydraulic mortar, concerning which the opinions of Engineers in France were divided, he considers that some may be very good, such as the ashes of Tournay, of which he experienced the benefit at Lille in 1815 and 1816. Others may be bad, such as the coal ashes of Strasburg, which he afterwards tried and found useless. He considers this difference to arise from the first sort of coals containing a small proportion of clay, which being incombustible forms the chief residue of the coals burned in the lime kilns of Tournay, and being calcined corresponds in its properties with an artificial puzzolana. He observes that some sorts of basalt when burned properly answer the same purpose.

(XCVII) General Treussart remarks that artificial puzzolana, formed of bricks exposed to a draft of air whilst being burned, will set in 3 or 4 days, whereas that formed of the same sort of bricks equally well burned, but not so exposed, may not set for 10, 20, or sometimes even 30 days, and yet may form good hydraulic mortars in time. He says that he has observed, that the strength of artificial puzzolana may generally be judged of by its quickness in setting; but that the quickest setting hydraulic limes are not the strongest. I do not agree with him in this last supposition, which is at variance with the results of my own experiments on the hydraulic limes of England.

(XCVIII) *Of Mortars kept under Water.* (*Chap.* 7.)

Here General Treussart returns to the subject of slaking, and proposes for hydraulic limes to throw about $\frac{1}{5}$ or $\frac{1}{4}$ of their bulk of water over them; and to allow them 12 hours for thoroughly slaking, before they are made into mortar and used, but not longer. He states however that some Obernai lime experimented upon by him, when slaked with its own bulk of water, which reduced it to a dampish powder, and mixed with 2 measures of sand immediately, formed a mortar having a resistance of 110 kilogrammes or 243 lbs, whilst the same sort of mortar, made from the same lime slaked with only $\frac{1}{5}$ lb of its bulk of water, had a resistance of 80 kilogrammes or 176 lbs, only. On allowing several portions of the lime, slaked as above with its own bulk of water, to remain various periods, from 12, 24, 36, &c, to 84 hours, before it was made into mortar with 2 measures of sand, its resistance gradually diminished to 95 and from thence to 30 kilogrammes, or to 210 and from thence to 66 lbs, thus losing nearly three fourths of its original resistance, by allowing it to remain only 7 days from the period of slaking to that of using it.

(XCIX) *Of working up Mortar again by remixing and beating it.*

He next relates experiments upon a mixture of common lime, sand and trass, in equal measures, tried in order to ascertain whether mortar made but not used for some days, on account of bad weather, would be injured by the delay. He found that this mortar when put under water immediately had a resistance of 132 kilogrammes, or 291 lbs. When allowed 12, 24, 36, 48 and 60 hours to set in air before it was immersed, its resistances were 149, 140, 145, 145 and 140 kilogrammes, or 329, 309, 320, 320, and 309 lbs, respectively, all being superior to that obtained by immediate immersion, but on remixing and rebeating the same mortar, at the periods of 12, 12 a second time, 24, 36 and 48 hours after being first made, and putting it into water 12 hours afterwards, its resistances were 150, 148, 165, 165 and 150 kilogrammes, or 331, 326, 364, 364 and 331 lbs, respectively, in one of which rebeatings, namely the first, no additional water was used, whilst in all the others a little was added. General Treussart further states that Obernai lime made into mortar with 2 measures of sand, and immersed immediately, had a resistance of 70 kilogrammes, or 154 lbs, but that when allowed 12 hours before it was immersed, its resistance was increased to 75 kilogrammes or 165 lbs; and that on being rebeaten in 12 hours and immersed after 12 hours more, its resistance was increased to 90 kilogrammes or 199 lbs. Also that a mortar composed of equal measures of the same lime, sand and trass, had a resistance of 175 kilogrammes or 386 lbs, when immersed immediately; but that when rebeaten in 12 hours and immersed after 12 hours more, its resistance was increased to 240 kilogrammes, or 529 lbs.

Upon this part of General Treussart's experiments, although the circumstance, of mortar subject to immediate immersion proving weaker, than the same when allowed some time to set in air, agrees with all my own experiments, yet I cannot understand in what manner remixing and rebeating any mortar can possibly improve its strength, unless it should have been imperfectly slaked when first mixed.

(c) He tried other experiments upon trass, and found that mortar made of 1 measure of common lime, mixed with 2 measures of sifted trass, was nearly twice as strong as the mortar of the same lime mixed with coarse trass in the same proportion, their resistances being 210 and 105 kilogrammes, or 463 and 232 lbs, respectively; and he also ascertained that wetting trass did it little or no harm. In short his opinion, and I believe a correct one, is, that neither trass nor any sort of artificial puzzolana, to which he might have added natural puzzolana also, can possibly be injured by exposure to weather.

M. Minard, an Engineer of Bridges and Roads, having announced as an important fact, that by burning lime imperfectly, so as to retain a proportion of carbonic acid gas, very good hydraulic lime might be obtained, General Treussart tried the experiment with common lime, which he made into mortar with sand and put it under water; but at the end of a year it still remained soft. For my part, I do not think that I should have taken the trouble to repeat the experiment, as General Treussart did, since from our present knowledge of the nature of limes, one might at once pronounce M. Minard's scheme to be a fallacy like that of Dr. Higgins.

(CI) *Of the Cause of the Induration of Mortars.*

General Treussart observes, that M. Vicat in his publication in 1818, having ascribed the solidification of mortars to a chemical action of the lime upon the sand, this opinion was objected to by M. Berthier and Dr. John, with whom the General himself concurs, inasmuch as he asserts that hydraulic limes set under water, whether mixed with sand or not. But he conceives that a chemical action must take place between lime and trass, or lime and puzzolana, whether natural or artificial, since common lime obtains the property of setting under water, when mixed with those substances, but not when mixed with sand.

(CII) General Treussart observes, that in making mortar, there is always a diminution of volume, after the mixture of the ingredients. For example, 4 heaps of 0·3 cubic metre of quicklime of Obernai, and of 0·6 cubic metre of sand, equal in all to 3·6 cubic metres, when mixed into mortar measured only 2·878 cubic metres: and in another experiment, 60·9 cubic metres of quicklime and sand, when mixed produced 50·338 cubic metres of mortar only.

(CIII) He mentions a proposal of M. Lacordaire Engineer of Bridges and Roads, who obtained superior hydraulic mortar, by half burning hydraulic lime, slaking it by immersion, grinding the portion that will not slake, and mixing the whole with sand. This is called cement of Pouilly from a place where a manufactory of it has been set up, and it is said to be used to great advantage in combination with fat limes. I cannot help doubting the merit of this system, which General Treussart considers to be very promising, so far as regards hydraulic limes; though as he previously stated, the same mode of improving common lime by half burning it, as proposed by M. Minard had completely failed.

(CIV) General Treussart combats the opinion advanced by many Engineers and Chemists, that the setting of mortars is caused by the lime gradually recovering the whole of the carbonic acid gas driven out of it in the kiln; for he asserts on the contrary, that the oldest mortars never recover enough of this gas, thus to be reconverted into carbonate of lime. In support of his own opinion, he gives a Table of the component parts of various old mortars analyzed by Dr. John. On examining this Table, I cannot see that General Treussart's objection holds good, for from the component parts of several of the mortars, stated in the Table, if we deduct the quantity of quartz, sand and moisture, which could not have formed any part of the original lime, the remainder contains nearly as great a proportion of carbonic acid gas, as any pure carbonate of lime possesses.*

* For example, the first mortar in his table 100 years old was from St. Peter's Church at Berlin. Besides 8000 parts of quartz and 330 of water, it consisted of carbonic acid 600 parts, lime in the state of carbonate 800, lime combined with other substances 170, combined silica 100, total 1000. Hence if the quartz and water be deducted, the remainder of this old

(cv) *Of Sands and Arenes.* (*Chap.* 8.)

After remarking on the various situations in which sand is found, and on its nature, as being silicious, granitic, calcareous, &c. he justly remarks, that one of the chief requisites in sand for making mortar is that it should be clean and free from all earthy particles.

Afterwards he treats of arenes, a sort of sand recently brought into notice in France by M. Girard de Caudemberg, Engineer of Bridges and Roads, who published a paper on them in 1827, in which he describes them as fossil sands of a reddish brown, reddish yellow, and even of a yellow ochery colour, found on the banks of the River Isle in the Department of the Gironde, which when mixed with lime made a very good mortar for the walls of buildings, and possessed also hydraulic properties, which had caused them to be used by the proprietors of the mills situated on that river. Having no hydraulic lime in his possession, M. Girard executed several sluices with mortar made of common lime and these arenes, which proved very good, and which required the pickaxe to cut into beton formed of it, when a year old.

Captain Le Blanc of the French Royal Engineers, employed at Peronne, addressed an interesting memoir on the 30th of November of the same year, in which he remarked, that in demolishing some of the old walls of that fortress, which had been built from 150 to 600 years, they were generally found to be extremely hard, and were evidently built of the common sand of that neighbourhood, which in more modern specifications for the works of the same fortress had been rejected, as of an earthy quality. This induced Captain Le Blanc to make experimental cubes of mortar, three of the clean sand recommended in the specifications for the construction of the new works, and three with the earthy or argillaceous sand, with which the old masonry had evidently been built. The latter set under water in a month to such a degree, that it was impossible to make an impression on it with the thumb, but the former was still soft after several months. In the mean time M. Girard's pamphlet appeared, by which Captain Le Blanc found that this earthy sand of Peronne was of the same description, that had been treated of by that author under the name of arenes; and since that time arenes have been noticed in several other parts of

mortar consists almost entirely of carbonic acid and carbonate of lime. Again, in the 5th mortar described in the same table, said to have been obtained at Cologne from a wall built by Agricola in the first century, the proportions are carbonic acid 900, lime in the state of carbonate 1194, lime combined with other substances 322, combined silica 25, quartz and sand 6884, alumina and oxide of iron 275, and water 400, making a total as before of 10000. General Treussart pointedly remarks that this last mentioned mortar had only recovered 13 per cent of carbonic acid, whereas he observes that lime stone usually contains 33 per cent of that gas. Consequently, I apprehend, that in stating the proportion of 13 per cent, he must through mistake have included the quartz and sand, which could not have formed any part of the original lime. If we strike out these, the carbonic acid absorbed by this ancient mortar cannot be less, than what the lime stone originally contained before it was burned.

France, of which it appears however that the properties have generally been well known to the masons of the country, and used by them in building mills, &c.

From the above it must be evident, that the clay contained in these arenes must rather be in the solid state of clay but broken down, than in the plastic state, in which last state it would ruin mortar, instead of improving it. In short it must be a sort of natural puzzolana probably combined with sand; and it is possible that something similar to it may be found in the British Islands also, in those districts where softish clay stone abounds, the disintegration of which would probably form a sort of arenes, having the same properties as those of France.

Of Betons. (*Chap.* 9.)

(CVI) After remarking upon the mode of making foundations by throwing down stones and hydraulic mortar in the time of Belidor, when the term beton was applied to the mortar only, General Treussart observes that the term beton is now applied to the mixture of this sort of mortar, with small stones broken to the size of an egg. Therefore the French beton as he observes is in reality masonry composed of small materials, the goodness of which depends upon the quality of the hydraulic mortar used in making it, of which puzzolana frequently forms a part.

According to this definition, beton is not precisely the same as our English concrete, which generally consists of a mixture of sand and gravel grouted with some sort of hydraulic lime, though they may eventually attain nearly the same resistance; but it may fairly be compared to the concrete in the sea-wall at Brighton, which was a mixture of mortar and shingle from the beach, excepting that the present French beton has its small stones rendered angular by breaking, like those used for road-making in this country, instead of being rounded.

In using beton for the foundations of walls commenced under water, General Treussart states, that two rows of sheeting piles should be driven, and after clearing out the earth from the intermediate space the beton may be thrown in; but instead of doing this as soon as the ingredients are mixed, he recommends, contrary to the opinion of M. Vicat, that it shall be allowed to remain from 12 to 36 hours in the air to consolidate, according to the season of the year, and that it shall be let down in a sort of spoon until it reaches the bottom of the space in which it is to be used, at first supposed to be full of water, when by pulling a string the spoon is turned over, and the contents discharged into the proper spot. In this manner the whole foundation should be covered with a course of about 30 or 40 centimetres or from 12 to 16 inches in thickness. He observes, that beton thrown down after acquiring a half consistency in the air, becomes soft again under water, but that some time afterwards it begins to set once more. He then recommends ramming it at the end of 12 hours, first gently and afterwards more strongly by a flat rammer, which being done a second layer of beton may be applied. He observes, that in spite of every precaution, part of the lime detaches itself and forms a liquid paste on the upper surface, which he says ought to be removed before another course of beton is applied.* He states that hydraulic mortar made with common lime and artificial puzzolana will answer as well for beton as any other ; but in this case he considers lime that has been slaked two months to be the best, whereas hydraulic lime should be slaked fresh from the kiln, and mixed as soon after slaking as possible, and within a period not exceeding 10 or 15 days at the utmost.

He observes that after some experience, he finally adopted 1 measure of Obernai lime mixed with 2½ measures of sand, for the mortar both of some of his beton foundations, and of his masonry at Strasburg, but that for more important works he used mortar composed of 1 measure of Obernai lime, 1 of trass and 1½ of sand. To convert each of these mortars into beton he added from 2 to 2½ measures of gravel and broken stones in all, the proportion of the former being one half of the latter. Observing that the chief use of the gravel and broken stones is to diminish the expense, he remarks that the proportion of them used in beton should be such, that all the pieces of stone shall be united by a sufficient quantity of mortar.

He states that one day's labour of 4 men is sufficient to mix 1·8 cubic metre, or about 23 English cubic feet of the above ingredients, which when properly mixed will compress into about 4 fifths of their original volume ; and to sink this quantity of beton into its proper place. Also, that if 40 men be so employed, 2 intelligent workmen and 1 carpenter in addition will be required, the former for slaking the lime and proportioning the ingredients, the latter for repairs of tools, scaffolding, &c, besides 2 labourers more for unforeseen contingencies.

(CVII) He does not propose beton for the whole height of the walls of wharfs, &c, but for the foundations only, and when these are finished, he recommends a third row of sheeting piles to be driven into the beton foundation, and the space between this and one of the former rows to be filled with clay to make a regular coffer dam, and then to pump out the water from the remaining space, in which the wall shall be built upon the beton foundation with stones and hydraulic lime.

This arrangement he considers indispensable in a current, but in still water and when the materials for the mixture are remarkably good, he thinks it admissible to carry the beton up as high as the low water mark, and to commence the regular masonry above that level. Yet in concluding this chapter, he expresses his doubts as to the propriety of the employment of beton except in foundations, in which doubts I fully concur, but in a stronger degree, being con-

* He considers the case of springs being found at the bottom, which would wash away the beton as fast as it was applied, and suggests a remedy which I think would prove quite unequal to the object, from my own observation of the effect of land springs on foundations, under much less difficult circumstances.

vinced of the impropriety of using beton, or concrete, or artificial stone, for any thing but the foundations and backing of walls exposed to water.

(CVIII) *General Remarks on Hydraulic Mortars.* (*Chap.* 10.)

Having greatly abridged the first section of General Treussart's treatise, which relates to hydraulic mortars, I shall not follow him through his recapitulation of this part of his work, except in a few observations not before noticed.

He observes, that all hydraulic limes set twice as quick in summer as in winter; and considers any hydraulic lime to be good, which either as a hydrate or mixed with sand, sets in summer in 8 or 10 days after immersion so as to resist the finger; but if it should require from 15 to 20 days to attain this consistency, he considers it only middling.

He observes that hydraulic mortars made of lime and artificial puzzolana ought to set in summer in from 3 to 5 days, if the latter should be of good quality and have been burned in a current of air: but if the same substance should not have been burned in a current of air, he says that the like mixture of it with lime will require three times as long to set.

Both General Treussart and M. Vicat appear to me to attach more importance to this circumstance than it deserves, for it may be remarked, that there is no sort of kiln now in use, in which clay for an artificial puzzolana can possibly be burned, in which there would not be a free circulation of air; and unless some part of the kiln be blocked up afterwards, which no one would think of doing, except in making bricks or tiles which require to be *annealed*, or cooled gradually to prevent them from becoming too brittle, this free circulation or thorough draft of air will continue, until all the fewel shall be consumed and the clay perfectly burned. Even in experiments on a small scale, M. Vicat and General Treussart consider a current of air so essential, that in using crucibles they recommend boring numerous air holes through the sides and bottom; whereas in all my crucible experiments I used close crucibles with covers, not of course absolutely air tight, though fitting well over the top, and never found the experimental cements thus burned to be inferior to those burned without crucibles, in a common fire place, and having a free circulation of air on all sides.

In his second section General Treussart treats of mortars that are to be exposed to the air.

Of Mortars for Air made with Lime and Sand or Puzzolana. (*Section* 2, *Chap.* 11.)

(CIX) General Treussart is of opinion, that the Romans had no mystery or nostrum for making superior mortar, and he states expressly that M. de la Faye's peculiar mode of slaking and M. Loriot's method of making mortar before described, which they both declared to be rediscovered secrets of the Romans, were not of that importance which they held out, not being able as they asserted to produce superior mortar from common lime.*

(CX) He observes, that age cannot be considered as the chief cause of the goodness of mortars, for some very old or ancient mortars have been found perfectly moist below the surface. He therefore tried some experiments on this subject, and adverting to the opinion of most builders, that common or fat lime should be run into reservoirs in a fluid state, and allowed to remain there a long time before it is used, he mixed old slaked lime putty of this description with 5 several proportions of sand, varying from 3 to 4 times the volume of the lime, and having made square prisms of each sort of mortar thus composed, and put them by in a cellar, he took them out at the end of a year, and experimented upon them in the usual manner, when he found that they all broke down under the iron collar and scaleboard alone, without any additional weight. Afterwards he tried the same old slaked lime putty mixed with trass alone, in proportions varying from 2 to 2¾ measures of the latter substance, when he found that after keeping these mortars also one year in a cellar their resistance was from 110 to 155 kilogrammes or from 243 to 342 lbs, increasing in proportion to the quantity of trass; but after keeping them in a dry apartment without a fire for the same length of time, their resistance was from 200 to 240 kilogrammes, or from 441 to 529 lbs; but in both cases the resistance diminished, on increasing the proportion of trass from 2¾ to 3 measures.

On mixing 1 measure of the same slaked lime putty with from 2 to 3 parts in all of sand and trass in equal quantities, he found that the resistance in the cellar was from 130 to 150 kilogrammes, or from 287 to 331 lbs; but in the dry room only from 65 to 95 kilogrammes, or from 143 to 210 lbs, being the contrary of his former results. Not being able to account for this difference otherwise, and not having time to repeat the same experiments, he suspects, that the trass used in part of those experiments may have been of a different quality from that used in the others, for an interval of one year had intervened between them.

(CXI) He found that common lime slaked and measured in paste, and made into mortar immediately, acquired a resistance of 31 kilogrammes or 68 lbs, when mixed with 2 measures of sand, but that the strength of common lime mortar was diminished by an increase of sand, and became null, when mixed with from 2¾ to 3 measures of sand. These results are not confirmed by my experiments on chalk lime mortars (See Table XII, Article 277).

* In his second publication, M. Vicat has not noticed Loriot's system at all, though he attached some value to it in his first publication of 1818. General Treussart observes that the quicklime powder applied to stale mortar, according to this system, causes the mortar to dry very rapidly, which he says gives it the appearance without the reality of setting, for it does not communicate greater hardness or resistance.

Of Mortars made of Hydraulic Lime, mixed with Sand or Puzzolana. (Chap. 12.)

(CXII) He states that under exposure to air, 1 measure of hydraulic lime combined with 2 measures of sand only has much less resistance, than 1 measure of hydraulic lime, 1 of sand and 1 trass, the former in three instances having broken down under 45, 52 and 100 kilogrammes, or 99, 115 and 221 lbs; whilst the latter required 105, 122, and 120 kilogrammes, or 232, 269, and 265 lbs, to produce that effect. Here also his results differ from those of my experiments upon the blue lias lime, with the like proportions of sand alone, and of sand and puzzolana, which are equivalent to sand and trass. See the same Table before quoted (277).

He states further, that on reducing hydraulic limes to a slaked lime powder, by adding one fifth of their volume of water, and mixing them afterwards with sand alone, the resistance diminished considerably, on allowing the powder to become stale before it was mixed with the sand; but when combined with both sand and trass, the deterioration from this cause either did not take place, or in a very moderate degree, within the period of four months after the slaking of the lime. He observes also that the strength of the same mortars made with hydraulic limes was usually a little greater under water than in air: and moreover that common lime, sand and the puzzolanas, make a stronger mortar, not only for water but for air also, than the hydraulic limes mixed with sand alone. This is confirmed by my own experiments, so far as the resistance of these limes is concerned, being the only quality to which General Treussart's experiments apply.

He tried several hydraulic limes exposed to air in the state of hydrates, namely those of Obernai, Verdt and Metz, and the Boulogne cement, the breaking weights of which were 147, 45, 54 and 55 kilogrammes, or 324, 99, 119 and 121 lbs, avoirdupois, respectively; and it appears that the first mentioned lime in the state of hydrate exceeded in resistance the mortars made by mixing it with sand. I have already remarked, that he must have spoiled the Boulogne cement, in his former experiments, by an injudicious mode of using it (XC). The same observation applies here also.

On trying 1 measure of old stale common lime paste mixed, 1st with 2 measures of puzzolana, 2dly, with 2 measures of trass, 3dly, with 2 measures of artificial puzzolana of the calcined clay of Suffleheim, and 4thly, with 2 measures of forge cinders, he found that they broke down respectively under weights of 195, 205, 195 and 25 kilogrammes, or of 430, 452, 430 and 55 lbs. A mixture of 1 measure of the same lime paste, 1 of sand and 1 of the same artificial puzzolana, broke down under 165 kilogrammes, or 364 lbs. In respect to sand, he found that common lime mixed with 2 measures of common sand produced a mortar that broke down under 35 kilogrammes, or 77 lbs, but on mixing it with the same sort of sand ground fine, it required 55 kilogrammes or 121 lbs, to break it down.

On trying the same experiment with Obernai lime, the weights which broke down the coarse sand mortar and the fine sand mortar were respectively 85 and 125 kilogrammes, or 187 and 276 lbs. Finally on mixing 2 sorts of lime successively with twice their volume of earthy sand, he found their resistance null; but on afterwards mixing them with the same proportion of sand that had been properly washed, they each required a weight of 80 kilogrammes, or 176 lbs, to break them down.

(CXIII) *Observations on Mortars exposed to Air.* (*Chap.* 14.)

In recapitulating his observations on mortars exposed to air, he remarks that the system of preserving common lime in the state of putty or paste is so ancient, and has been so generally used, that though his experiments were unfavourable to that system as destructive of the strength of mortars made of that lime and sand alone, he is unwilling to condemn it without more numerous and careful trials, as he admits his own to have been few. I am rather inclined to consider General Treussart's experiments on this point, likely to prove decisive; but he himself recommends that it should be tried with various common limes, in several countries. He is of opinion, that Vitruvius recommends the running of lime, as it is termed, into reservoirs in the state of putty, for interior plastering only, not for the general purposes of building. He is so strongly impressed, from some of his experiments, that no common lime with sand alone can make a durable mortar, that he declares his opinion, that none of the ancient Roman buildings in a state of preservation in the present day could possibly have been built of that lime without a mixture of puzzolana or of tile dust. He observes that the occasional want of the brown or reddish colour in the mortar, which indicates those two last named ingredients, does not prove that the lime used was common lime mixed with sand only, he being persuaded, that in this case, it must have been one of those hydraulic limes which are colourless from the absence of the oxide of iron. Under the same impression, he recommends that common lime and sand mortars should never be used at all, even for walls exposed to air alone; but that the mortar of such walls should either be of hydraulic lime and sand, or of common lime and sand mixed with the natural or artificial puzzolanas, and he reprobates very justly the mode which had hitherto generally prevailed in France, of building the arches of military casemates with common lime and sand mortar, and covering them with a mixture of the same lime and of tile or brick dust, which owing to the precarious nature of that sort of artificial puzzolana and the badness of the mortar, almost always allowed the rain water to filter through. Instead of this defective mode of construction, he recommends the whole of such casemates to be executed with good hydraulic mortar.

Whilst I fully agree with General Treussart in believing that no common lime, or fat lime as M. Vicat would term it, such as the common chalk lime of the South of England will ever make a strong mortar with sand alone; yet his opinion as to the perishable nature of such mortar even in air, unless a proportion of puzzolana be added to it, appears to me to be erroneous. If no ancient building, constructed with common lime mortar alone could be expected to stand, as General Treussart has asserted, Rochester Castle ought now to be a mass of ruins, for it has undoubtedly been built with common chalk lime mortar, unimproved by any sort of puzzolana natural or artificial; and in consequence of the inferiority of this mortar, all the external joints have been corroded in the course of centuries to a considerable depth, but there is not the least appearance of the lofty walls of this ancient edifice ever falling to pieces, according to the General's theory, from mere exposure to the vicissitudes of the weather.

Of Artificial Stones, and of Betons exposed to Air. (*Chap.*15).

(CXIV) After quoting M. Fleuret and M. Rondelet, respecting the great antiquity of artificial stone for buildings in air, General Treussart observes that artificial stones may be made to great advantage of any sort of good hydraulic mortar; but in making them with common lime and artificial puzzolana, he recommends using pipe clay to obtain the latter, in order that it may have a handsome stone colour, because the common clays assume a red colour.

He now declares an opinion in favour of artificial stone or beton, so much stronger than he seemed to entertain in the first section of his treatise, that I suspect he must have printed the first part before the second was written: for he not only observes that it may be cast in moulds, for coins, copings, door and window jambs, sills and lintels, cornices, gutters, water pipes, &c, and even for objects of the largest dimensions, but that artificial stone might be made for the piers of the bridges over the ditches of fortresses, for the walls of sluices, for columns, obelisks, &c, &c, and as it would not be easy to bury such masses, as is done at Alexandria in Italy, and to place them afterwards in their proper situations, he recommends constructing them on the spot, and states that he thinks a regular casing to form the beton in, might be dispensed with: but that it ought to be secured by always protecting the top with wet straw, on leaving off at night, and by surrounding the mass with something to keep it moist in summer; and he recommends making such masses somewhat thicker than is ultimately intended, in order to pare off some centimetres from the surface next year, which from having dried too soon may be softer than the interior. He states that these artificial stones may be coloured, by mixing diverse metallic oxides with the surface mortar.

(CXV) In treating of this subject he mentions, that in 1827, about two years before he published his book, Colonel Finot had constructed a vault of beton of 4 metres, about 13 feet $1\frac{1}{2}$ inch English, in span, with piers of masonry, which answered very well, and that a similar vault had also been constructed at Schélestadt.

He observes that some vaults at Strasburg, the floors of which were lower than the high water level, of the river and of the wet ditches after floods, were made available for use by having the floors and walls secured by beton.

(CXVI) General Treussart is of opinion also, that for aqueduct bridges to pass canals over rivers, beton may be used to the greatest advantage, not only to cover arches built of stone or brick, but even to supersede them. He likewise thinks that it may be used to widen a wharf wall by projecting over the usual water way, when that must not be encroached upon, because he considers that the over-hanging portion of beton would form as it were one immense stone, with the remainder of the wall, of which it made a part. He thinks it also well adapted for forming air and water-tight granaries for preserving corn, which may be of great importance in fortified places.

(CXVII) After observing that in some departments of the North of France, they are obliged to build the mass of the revetments of their Fortresses with chalk, and to protect the outside by bricks, which often give way and require continual and expensive repairs; he proposes under such circumstances, to construct the whole revetments in future with beton, and to stucco the outside with a strong coat of the same hydraulic mortar, used for the body of the beton. He thinks that revetments of this description, being as it were in one piece, might be made thinner than those of common masonry, under the same circumstances. He is also of opinion that vegetation attacks chiefly the joints of common lime mortar, and that beton revetments stuccod as he proposes, would not allow vegetation to attach itself to the surface at all.

(CXVIII) Adverting to the circumstance of houses in the South of France being built of *Pisé*, or rammed earth, which has not stability enough for Northern climates, he thinks that beton may be used with great advantage, building it in frames exactly on the same system as the pisé, and making all the coins, as well as the contours of the doors and windows, with artificial stones formed of the same beton cast in moulds. And as he observes, that his experiments have proved that good hydraulic mortar approximates to the strength of common bricks, he conceives that the walls of such buildings, which he would stucco with a thick coat of stone coloured hydraulic mortar, would be very solid.

(CXIX) He is also of opinion that beton may be used for military bomb-proofs, such as powder magazines, casemated barracks, &c, &c. If it should not be deemed expedient to construct the whole of such buildings of beton, he recommends that the stone or brick arches, which should on no account be built with common but with hydraulic mortar, should be covered at top with beton, which he considers the only means of preserving them from moisture; and although the stuccoing of the upper surface, consisting of the ridges and valleys of the vaulted roofs, if formed of beton, might perhaps be dispensed with, yet he recommends it to be done, taking care to prevent a hot sun from acting upon this stucco, by covering it with wet straw for the remainder of the summer, after previously bringing it to a very smooth or polished surface; that is when earth is not to be laid over it, after it is finished.

(CXX) General Treussart is further of opinion, that beton is better for securing the extreme ends of the chains at the abutments of a suspension bridge, than common masonry; so much so, that he thinks that an accident which occurred in a suspension bridge, opposite to the Hotel of Invalids at Paris, which gave way owing to the chains fracturing the masonry of the abutments, would not have taken place, if beton had been used and allowed one year to consolidate, instead of stones and mortar. He states that autumn is the best season for building with beton in the air, and concludes by remarking, that there is more real economy in going to the expense necessary for good workmanship, than to execute bad at low prices.

(CXXI) Here I shall remark, that as far as his own experience and experiments on limes and mortars go, I consider General Treussart's opinions to be of great importance; but in the whole of his observations on beton and on artificial stone, which are almost entirely speculative, he seems to have been led away by the same sort of enthusiasm, which prevailed in this country a few years ago, and which caused concrete and artificial stone to be valued so far above their real merit, that many persons seemed to think that they would supersede stone altogether: but the failures of this substance in the Docks and Wharfs in Her Majesty's Dock-yards at Woolwich and Chatham, which at first were to have been built with con-concrete exclusively, have brought it to its proper level, by reducing it from the rival to the humble companion and assistant, of stone and of brickwork laid in cement; and in respect to the former in particular my experiments have proved, rather at first to my own surprise, for I confess that I partook to a certain degree of the favourable impression alluded to, that the resistance of the best concrete after being allowed a year to set, is not one tenth part of that of Portland stone.

(CXXII) In speculating beforehand on the probable result of untried measures or expedients, having repeatedly observed, that the most promising failed, whilst others, that seemed to have no probability in their favour, have occasionally succeeded beyond expectation, I have for many years laid down a rule not to give an opinion under such circumstances, upon any practical question, capable of being experimented upon, without stating at the same time that it was an untried and therefore a doubtful point; and the propriety of this rule has been most strikingly confirmed, by the failure of the only expedient that I ventured to recommend in the foregoing Treatise, before I had tried it, and of the efficiency of which I felt as certain beforehand, as a person can be of any thing that he has not seen tried. I mean the artificial stones for the steps of geometrical staircases proposed to be formed with bricks and cement, which even with the powerful aid of bond iron in the joints, proved to possess much less resistance than any prudent architect would consider sufficient for the purpose in view. If General Treussart had followed the same rule, from which I soon found cause to regret my deviation in the solitary instance, that has just been mentioned, he would have left out far the greater part of his Observations on Beton and Artificial Stone, which he has recommended for purposes, to which our subsequent experience in this country has proved them either to be altogether incompetent, or of very inferior value.

(CXXIII) It was before stated, that General Treussart was ordered officially to report on M. Vicat's artificial hydraulic lime in 1818, and that he differed from that gentleman's opinions in several respects. It appears that he published a pamphlet in 1824, stating his objections to M. Vicat's doctrines; and from various passages in the General's more elaborate Memoir of 1829, of which I have now endeavoured to give an abstract, I infer that M. Vicat must have replied to his objections in the mean time, either by a pamphlet or in some of the periodical scientific publications of France. I am not in possession of either of these pamphlets or papers. In M. Vicat's Second Treatise published in 1828, in which he entirely remodels the whole of his subject, and tacitly abandons several of his former opinions, he has abstained from every thing in the shape of controversy, and does not even allude to the opinions of General Treussart where they differ from his own. The General, on the contrary, after objecting to M. Vicat's mode of estimating the resistance of mortars by the penetration of the steel point as being liable to error, observes that he himself has compared the strength of his experimental mortars by direct experiment, entering in his Tables the number of kilogrammes required to break down each specimen, and modifying nothing by calculation. Hence he declares, that whenever M. Vicat has advanced opinions contrary to the results, which he obtained from his own experiments, he would invariably notice them, which he has accordingly done from time to time throughout the whole of his work; and as M. Vicat published both his treatises, which made a very strong impression in France, before General Treussart's memoir, the system adopted by the latter, of stating every point in which he differed from the former, was no doubt much more useful to the public, than if he had stated his own opinions and experiments, without noticing M. Vicat at all. For in a subject of such immense practical importance to mankind, in those particulars, in which two men of talent investigating it experimentally both agree, the chances are that they must be right; but when they draw directly contrary results from their different modes of experimenting, *one must be and both may be wrong*; and therefore it is proper to draw the attention of the public pointedly to these controverted matters, in order that more conclusive experiments may decide the question.

(CXXIV) Upon the whole, General Treussart appears to have taken up the subject of hydraulic mortars as a matter of experiment, some years later than M. Vicat, but before the appearance of the first treatise published by the latter. Their opinions do not after all materially differ, for M. Vicat admits the efficacy of artificial puzzolanas upon which he also experimented largely, whilst General Treussart gives the preference to these over M. Vicat's artificial hydraulic lime, without denying the efficacy of the latter. They both agree in recommending that the fat or common limes, equivalent to the chalk lime of the South of England, should never be used in combination with sand for any building of importance, in which opinion I entirely concur: and the practice of all our most eminent British Architects of the present day, though probably not influenced in any way by the French publications alluded to, is in conformity to the same opinion, for they have for many years past abandoned the use of chalk lime altogether, experience having proved its inferiority not merely for hydraulic purposes, but for walls exposed to air only.

(CXXV) Though willing to place confidence generally in those opinions, in which these two able and laborious investigators of the properties of calcareous cements agree, whether in accordance with received opinions or not, there is one point, in which I have already stated, that I entirely dissent from both. I allude to those passages in M. Vicat's last work, in which he depreciates the excellent water cements of England, and to the like conclusion drawn by General Treussart from his experiments, in which he asserts that a water cement, equivalent to the Sheppy cement, was inferior to certain hydraulic limes, although the latter would not set under water in less than 8 or 10 days from the period of immersion. These conclusions are in such direct contradiction to the daily experience of men in this country, that every English Architect or Builder would reject them at once, and they must be ascribed solely to the entire want of experience and of practical knowledge, in the proper mode of using such cements. By this observation, I mean no disrespect either to M. Vicat or General Treussart, for I confess that, until very recently, I was myself equally ignorant of the proper mode of experimenting with hydraulic mortars, and with the natural or artificial puzzolanas, and in fact after my own first experiments with excellent natural puzzolana, tried from 1826 to 1830, had entirely failed from this very cause, namely that I did not know how to use it, I felt, though I did not express it, the same sort of unfavourable opinion of this important substance, which those two French authors have recorded against the valuable water cements of England, the wonderful strength of which, for that epithet applies to them, far exceeds that of all other mortars. Setting aside therefore their opinions on water cements, properly so styled, as erroneous, and General Treussart's opinion of the value of beton and artificial stone, as being much too favourable, I conceive that the labours of these two eminent Frenchmen in investigating the properties of common and hydraulic limes, mortars and puzzolanas, are of the highest importance, and that they will be useful not merely to the Service of France, but of all other nations. In respect to those points in which the results of their experiments are at variance with each other, I have no hesitation in saying, that I consider General Treussart's direct mode of conducting his experiments and his straightforward system of recording them, as so much more satisfactory than M. Vicat's mode of proceeding, that I cannot help placing much more confidence in the former than in the latter. In respect to the comparative hydraulic powers of limes and mortars, they are both equally satisfactory, and General Treussart's are perfectly conclusive as to the comparative resistance of square prisms of all the limes and mortars experimented upon by him; but even his experiments, though also perfectly satisfactory as to the comparative resistance of beton or of rubble masonry having large irregular joints, afford but a very imperfect criterion of the resistance of the same mortars in the joints of ashler masonry or of regular brickwork, which are necessarily thin. So far as regards the comparative adhesiveness of the various mortars, which they took into consideration, neither M. Vicat's experiments, nor even those of General Treussart, afford any means of forming a proper judgment, for my recent experiments have sufficiently proved, that this quality, which is perhaps of greater importance than any other, is neither commensurate with, nor proportional to, the resistance of mortars, as has generally been supposed.

(CXXVI) In respect to the chief difference of opinion between M. Vicat and General Treussart, whether it is better to form an artificial hydraulic lime as proposed by the former, or an artificial puzzolana, as proposed by the latter, when neither of these natural substances can be procured, but common lime may be had in abundance (LXXXVI), I am inclined to give the preference to General Treussart's system, as being equally efficient with M. Vicat's, but much less expensive; for it is evidently simpler and cheaper to burn clay alone, which forms an artificial puzzolana at once, than to burn a hard lime stone, slake it, and then reburn it mixed with clay; or to pulverize a soft lime stone, mix it with clay and then burn the compound; without one of which troublesome processes an artificial hydraulic lime cannot be obtained. Lest it should be supposed on a hasty view of the subject, that what I have just stated militates also against the manufacture of artificial cement, I have only to suggest, that artificial puzzolana mixed with common lime, is quite as good for building under water, as the best hydraulic limes mixed with sand; but neither the one nor the other are by any means equal to cement, under difficult and important circumstances, whether for building in air or in water. In fact as I said before, cement renders many works easy or practicable, which without it would be difficult or impossible. (See Articles 58, 59 and 235.)

*Of the Asphaltic Cement of Seyssel, recently introduced in France.**

(CXXVII) I had heard this cement described with great approbation by a French General not connected with any building speculations; and had accordingly ordered some specimens of the raw materials, and a small sample of the artificial stone made from them, to be sent to me from Paris. In the mean time I am enabled to give an accurate description of it, from a pamphlet recently published by Mr. F. W. Simms, the author or compiler of several very useful works on Surveying and Civil Engineering.

The materials from which this sort of cement is made are found in great abundance at Pyrimont, situated at the foot of the Eastern side of Mount Jura on the right bank of the Rhone, one league North of Seyssel. These consist partly of calcareous asphaltum, composed of carbonate of lime nearly pure, intimately combined with from 9 to 10 per cent of bitumen, which gives it a deep brown colour beneath a whitish surface: and partly of what they term *Molasse*, which is a micaceous sand combined with from 15 to 18 per cent of the same bitumen, which contains from 60 to 70 parts of resinous petroliferous matter, and from 30 to 35 of carbon; and to this last ingredient M. Millet who analyzed it, believes that its black colour, and its important quality of hardening in the air, are to be ascribed.

The asphaltic cement is said to have been known in France before the middle of the last century, and even occasionally to have been used in works of some importance, having been employed by the celebrated Buffon in lining a large basin in the garden of plants, for which he vouched as being perfectly water-tight 36 years afterwards, and having also been used for the principal basin in the King of France's garden in 1743, &c.† But it is only recently that it has been generally used, for the foot pavements of the Eastern side of the Pont Royal, and on both sides of the Pont de Carrousel, and on the North side of the Palace and Gardens of the Thuileries, and recently in a more ornamental form for paving the Place de la Concorde. It has also been used for the upper as well as lower floors of some Barracks, and for the floors of the Cavalry Stables at the Quai d'Orsay, and for the floors of the *Abbattoirs* or public Slaughter-houses at Paris.

It is stated that in that city, they import the calcareous asphaltum

* Which has also been called Asphaltic Mastic, as I believe that in France the word mastic signifies any sort of strong cement not coming under the definition of a water cement, or hydraulic mortar. But as this word has been naturalized in England, and exclusively applied to the oleaginous cement, described in Article XXXIII of the Appendix, it appears desirable, for the sake of precision, not to use the term mastic, in speaking of the asphaltic cement.

† Asphaltum or bitumen was also used by the ancient Babylonians as a cement for building. I do not know that it has recently been proposed for this purpose at Paris, but Mr. Simms relates experiments tried there, by which the asphaltic cement appeared to have a power of uniting bricks superior to that of the best mortars composed of lime and sand.

in rough blocks of about a cubic foot each, and the bitumen, previously extracted from the molasse in casks, in the state of mineral pitch. They roast the calcareous asphaltum upon an iron trough about 10 inches deep, below which a brisk fire is made, in a sort of furnace 10 feet long and 3 feet broad, which causes a good deal of moisture to pass off, and the asphaltum falls down in powder, in about half an hour, and after sifting it, the lumps are pounded, and thus the whole is pulverized.

In preparing the asphaltic cement for use, 7 kilogrammes of bitumen are first put into a cauldron over a fire, to which 93 kilogrammes of the powdered asphaltum are gradually added, and finally a bucketful of very small clean heated gravel is put in, and the contents of the cauldron are continually stirred from the commencement of this process, until the whole compound shall begin to simmer and become rather more fluid than treacle, emitting a light white smoke, when it will be ready for use, and must be taken out of the cauldron in buckets or ladles. When the asphaltic cement is intended for roofs, no gravel must be put into the cauldron.

When used for superior pavements, a beton foundation of about 3 inches thick is prepared for receiving it, which is covered about one inch thick by the asphaltic cement, that has just been described, after which some fine gravel is sifted over it whilst still fluid, and beaten down with wooden bats as it cools, so as to produce a very smooth surface, which by using gravel of proper colours may be made to resemble a pavement either of granite or of ornamental mosaic. For common pavements, I have been informed by an Engineer Officer who inspected this operation at Paris, that it is usually laid in successive breadths of about 18 inches at a time, on a common road surface of gravel, the beton being dispensed with.

(CXXVIII) The Director of the Royal French Artillery at Douai, and several Officers of the French Engineers, have certified in favour of this asphaltic cement as having been used in their respective departments, and the Chevalier de Pambour, author of a work of merit upon locomotive steam engines and railways, gives the asphaltic cement a high character, not only for paving, as having proved better than a stone pavement laid down in its vicinity, and for the flooring of stables, but also for floors and roofs. For this latter purpose it was used for three extensive roofs of buildings in the Artillery Arsenal at Douai, the rafters of which had a slope of about 1 foot in 3, over which were nailed transverse laths for receiving plain tiles, each about 19 centimetres or nearly 7½ inches square, having their joints filled and united together by the asphaltic cement. Canvas, similar to that of paper-hangers, was stretched over the tiles and nailed down between the joints, after which the asphaltic cement was spread to the thickness of from 10 to 12 millimetres, rather less than half an inch over the whole surface, which was immediately covered with hot sifted gravel, applied whilst the cement was in a semifluid state, and the surface was then beaten by wooden bats, until it became perfectly smooth. In the upper floors of Barracks also, it was applied over a course of plain tiles previously laid, and properly supported by woodwork attached to the girders or joints.

(CXXIX) In addition to the certificates of the French gentlemen before mentioned, Lords Elgin and Lincoln, and Sir John Hay, made a favourable report of the asphaltic cement of Seyssel from their personal observation of its various applications at Paris in November 1837; and we are told that preparations are now making for applying it on a part of the Greenwich railway, and as a foot pavement in many of the metropolitan parishes, as well as at Liverpool.

The Architects and Engineers of the Metropolis and its vicinity will therefore soon have an opportunity of judging of this bituminous cement, which is said to possess greater resistance to wear and tear than some sorts of stone, combined with sufficient elasticity to prevent it from cracking, in consequence of any small settlement or crack in the work over which it is applied; at the same time, that it is alleged to be proof against a heat of 100 degrees of Fahrenheit, in which respect it is superior to Lord Stanhope's composition, which contained the same or at least similar component parts, but combined with a greater excess of bituminous matter. I ought to have mentioned before that the casemates of the Citadel of Quebec were coated by order of the Commanding Royal Engineer, under whose direction they were built, about 12 years ago, by a composition of the same ingredients as Lord Stanhope's cement, mixed in the proportion of 3 barrels of tar, 2 cwt. of whitening and 8 bushels of sand: but I have never been able to ascertain the precise proportions of his Lordship's composition. Sir Robert Smirke informs me, that on giving it a trial, he considered that composition better adapted for covering vaults protected from the sun by a sufficient thickness of rubbish, than for the flat roofs of dwelling houses, which do not admit of being thus loaded.

(CXXX) In the pamphlet published by Mr. Simms, it is stated that a ficticious (query, facticious or artificial?) asphaltic cement, also proposed at Paris by other parties, had failed. But from the chemical analysis of the calcareous asphaltum, and of the bitumen obtained from the molasses of Seyssel, and the mode of preparing and using them described in the same pamphlet, I cannot comprehend in what way it is possible to fail in rivalling the asphaltic cement of that locality by artificial means, excepting that the preparation and combination of the natural ingredients found there, may perhaps be cheaper than a modification of Lord Stanhope's mixture of pulverized chalk and tar, which will no doubt be the case, if the use of a harder carbonate of lime than chalk, such as the Plymouth marble, should be found necessary to produce a superior article.

I have not the smallest doubt of the great utility of this sort of cement, but when any one material on being first proposed, is declared as this has been, to be equally efficient for several different purposes, and as being likely to supersede several other sorts of materials long used and approved as the most applicable for those respective purposes, it is the safest mode to wait the test of experiments judiciously tried, and not to give way to that sort of enthusiasm in favour of novelty, which prevailed in this country a few years ago in respect to concrete, which has entirely failed in superseding the more expensive materials of stone, and even of bricks, which some persons were sanguine enough to suppose that it would do ; and yet notwithstanding this entire failure, concrete when confined within its proper limits, may be considered one of the most important of all the modern improvements in Practical Architecture.

I shall conclude with some miscellaneous information, respecting the subject of this Treatise, introducing first some memoranda with which I was favoured by Colonel Fanshawe on sending him a copy of my lithographed Treatise on Practical Architecture. (See the Note to Article XXV of the Appendix.)

(CXXXI) *Colonel Fanshawe's Memoranda and Remarks on Hydraulic Mortars.*

Description of the different sorts used in the new works at Water Port, Gibraltar, executed under the direction of the Commanding Royal Engineer at that station, in the years 1790 and 1791.

1st. *Coal Ash Mortar.* This consisted of lime $2\frac{1}{2}$ measures, sand $2\frac{1}{2}$, coal ashes $2\frac{1}{2}$, puzzolana $1\frac{1}{2}$, and smiths' danders $1\frac{1}{2}$, the proportion of lime to the other ingredients thus being as 1 to $3\frac{1}{5}$.

2d. *Dutch Tarras Mortar.* This consisted of equal parts of lime and trass by measure.

3d. *Puzzolana Mortar.* This also consisted of equal parts of lime and puzzolana by measure.

4th. *Puzzolana Mortar for lining Cisterns and coating the Ridges of Casemates.* This consisted of slaked lime 16 measures, puzzolana 8, sand $5\frac{1}{4}$, brickdust $5\frac{1}{4}$, beaten glass 4, and smiths' danders 4, the proportion of lime to the other ingredients, thus being as 1 to $1\frac{5}{8}$ nearly.

It is almost unnecessary to observe, that the whole of these hydraulic mortars were adopted, before the properties of the excellent water cements of England had been discovered.

5th. *Hydraulic Mortar, &c. used by the Dutch Engineers for coating the Casemates of the new Citadel at Ghent.*

This also was observed by Colonel Fanshawe, who describes it as follows. First, trass mortar was laid on over the ridges of the casemates, which was paid over with coal tar. Upon this a series of continuous gutters were formed, by bricks laid on the sloping sides of each ridge, and leading towards the valleys. These gutters were covered with bricks set in mortar, so that any water that oozed through would run freely off, whereas in the usual mode of laying pebbles or dry bricks on the sloping sides of the ridges as filterers, small particles of earth are apt to get through, and in time choke the valley gutters.

Upon the above construction, I may be permitted to remark, that unless these bricks were laid in cement mortar, which probably was not the case, stalactites of lime will in time be formed in all the joints.

(CXXXII) *Experiments on the Strength of various sorts of Harwich and Sheppy Cement Stones, by Mr. J. Mitchell of Sheerness Dock-yard.*

Mr. Mitchell formed all the specimens of cement experimented upon by him in a mould 14 inches long 6 inches wide and 1 inch deep. He supported them at the ends, by wooden props at the clear distance of 12 inches apart, and broke them down by weights applied vertically over them, and resting on a sort of blunt knife edge, pressing on the center of the top of each slab of cement.

The following specimens were all made one day after the cement was manufactured, and those kept dry were broken down 11 months afterwards, but those immersed in water, 10 months afterwards.

Of the specimens kept dry, and formed of nett cement, the Sheppy cement from Minster Manor broke down under 155 lbs, the Harwich cement under 116, and the mixture of 1 fourth of Sheppy with 3 fourths of Harwich cement, being the same usually burned in the cement kilns in the Dock-yard, of which Mr. Mitchell has charge, broke down under 123 lbs. Of the specimens kept under water, the Sheppy cement broke down under 215 lbs, and the Harwich cement under 172 lbs. Their resistances, when mixed with equal parts of sand and kept dry were 81 lbs, for the Sheppy, and 67 lbs, for the Harwich cement.

The following specimens were made of nett cement exposed to air 7 days, and experimented upon 2 months afterwards. Their resistances when kept dry, and estimated on the average of 2 trials each, were as follows.

Harwich cottage stone 113½ lbs, Hills' pebbles from Whitstable Bay 111½ lbs, Harwich dredged stone 106½ lbs, Sheppy pebbles from Minster Manor 88 lbs, Sheppy pebbles from the South-east part of the Island of a yellowish tinge 85 lbs, and Harwich Manor stone 67 lbs.

When immersed their resistances estimated by one trial only, except the Harwich manor stone which had two, were as follows. The Whitstable pebbles 123 lbs, Harwich cottage stone 109, Harwich manor stone 106, Harwich dredged stones, and Sheppy pebbles from Minster, each 102, and Sheppy pebbles from the South-east of the island 91 lbs.

When mixed with equal parts of sand and kept dry, their resistances, estimated generally on an average of 2 trials were as follows. Harwich cottage stone 76 lbs, Sheppy pebbles (S.E.) 61½, Sheppy pebbles from Minster manor 60, Whitstable pebbles 57½, Harwich manor stone 56½, and Harwich dredged stone 32 lbs.

When kept under water their resistances were as follows, on an average of one trial only. Harwich cottage stone 76 lbs, Whitstable pebbles 71, Sheppy pebbles from Minster 60, from the S.E.

of the island 59, Harwich dredged stone 55, and Harwich manor stone 46 lbs.

(CXXXIII) *Mr. Griffith's Report upon the Gases contained in the Blue Clay of the Medway, in reference to Article* 85.

Having ascertained in concert with Mr. Howe, in June 1832, that the blue alluvial clay of the Medway yielded by a sort of distillation, first what appeared to be carbonic acid gas, and afterwards hydrogen gas, but we not being sufficiently expert in chemical analysis, to come to a satisfactory conclusion, Mr. Griffith formerly of the Royal Institution had the goodness to repeat and record the results of the same experiment, which he tried in concert with us on the 22d of February 1833, by an apparatus consisting of an iron oxygen bottle serving as a retort, from which the gases were led off through a bent copper tube, into a pneumatic water trough, and collected in the usual manner. Mr. Griffith's minute of the experiment is as follows.

'Weighed 4000 grains of blue clay fresh from the River Medway, which we made up into small balls, and put them into an iron oxygen bottle, the neck of which and the copper tube were then connected by a luting of pipe clay, worked up stiff with water.* As the bottle was placed in the center of the fire, blue clay was put round the iron to protect it.'

'The steam, which was generated at first, was allowed to escape into the atmosphere.'

'A transfer jar was then placed over the beak of the tube, which rested on the shelf of the pneumatic trough, and the gas was collected therein. It came over at first cloudy, of a whitish colour, and the surface of the water in the jar appeared greasy.'

'The first air we got appeared to contain carbonic acid gas, as it extinguished a taper. It was obtained in a very small quantity, before the steam was generated, as before mentioned. We considered it at first as common air, and did not pay any attention to it. The quantity was about 15 cubic inches.'

'As the experiment proceeded, the cloudy gas turned to a milk white colour. We collected it in several successive jars-full. That collected in the first jar-full appeared to contain carbonic acid gas, in consequence of producing absorption when caustic was applied to it. On putting a lighted taper into it, the taper was extinguished and the gas burned with a lambent blue flame, analygous to the flame produced by the burning of carbonic oxide.'

'On trying the second jar-full, the cock at the top of the jar was opened, and a lighted taper applied, the jar being immersed in the trough, when the gas burned briskly with a blue flame.'

'Towards the close of the experiment when the bottle became red hot, the gas came over very nearly clear; and latterly it was perfectly clear.'

* The 4000 grains of blue clay measured about 11$\frac{1}{3}$ cubic inches. The clay was made into balls of the size of school boys' marbles, in order that after being burned hard in the retort, they might be got out of the neck of it, in that form, which in larger masses would have been very difficult.

'On trying a third jar of the gas, which was also inflammable burning with a blue flame, an explosion took place, which was supposed to be caused by a little atmospheric air mingled with the gas at the time the taper was put into it.'

'Afterwards 1 volume of gas and 2 of atmospheric air were mixed together in a cylindrical jar, which exploded upon the application of the taper. The effect was due to the presence of hydrogen, which was proved by burning a jet of the gas from the top of the transfer jar, under a dry glass, which was placed over the flame in an inverted position, and the surface of which was moistened during this process.'

'The whole quantity of gas collected during the operation amounted to 355½ cubic inches. The clay balls, at first perfectly moist, were burned generally into hard red globules of the quality and consistency of well burned bricks, some of them being a little more burned and vitrified at the surface, and consequently of a darker colour.'

(CXXXIV) *Analysis of the Brown Pit Clay of Upnor, and of the Blue Alluvial Clay of the Medway, by Michael Faraday, Esq. L.L.D. F.R.S. &c. &c.*

The circumstance, of both these clays forming an artificial cement, when calcined in combination with chalk, and an artificial puzzolana, when calcined singly, made me desirous of having a correct analysis of each, partly to ascertain what difference caused the blue river clay to be superior to the brown clay for both these purposes, and partly also to determine and record the actual proportions of the component parts of the former, which appeared to me to have varied between the years 1829 and 1836 (153), and which might possibly vary again.

Having sent specimens of each in earthen jars well secured against the action of air, to Mr. Faraday at the Royal Institution, he was kind enough to analyze them at my request, as detailed in the following extract from his letter to me of the 9th of March 1837.

'The pit clay of Upnor has a specific gravity of 2·07. This you required to know, though I do not see of what use it can be to you.* It contains a trace of carbonate of lime, but it also contains little calcareous concretions, like small pebbles, which would render a specimen carelessly taken very uncertain in its composition in its moist state, as sent to me.'

'A hundred parts of the brown pit clay of Upnor contain the following proportions very nearly.

Water		19·0
Sand		30·5
Finer particles	Silica	29·8
	Alumina	16·5
Peroxide of Iron		3·7
Carbonate of Lime		0·5
		100·0

* My reason was, that in mixing this clay with chalk by measure, to form an artificial cement, I previously reduced its specific gravity, by working it up with additional water, to that of the blue clay, which being in a fluid state, would otherwise have contained much less solid matter in the same measure.

'The Medway clay in its dark coloured and moist state had a specific gravity of 1·46, but this of course would vary as the quantity of water. One hundred parts gave

Water		50·9
Sand		14·0
Finer particles	Silica	14·8
	Alumina	10·8
Peroxide of Iron		3·4
Carbonate of Lime		1·5
Fragments of Wood		1·5
Organic Matter		3·1
		100·0

'I have put down all the iron as peroxide, because it is the best state in which to estimate it, but in the clay whilst dark coloured, a portion of it is in the condition of sulphuret (the greater portion being even then peroxide), and the presence of this sulphuret causes the dark colour, and also the evolution of sulphuretted hydrogen, upon affusion of acids. A little protoxide may also be present.'

'The gases you speak of, as existing in the clay do not exist in it really, but are produced from the wood and organic matter, the carbonate of lime, and the sulphuret of iron, &c, &c, by the action of heat.'

Remarks on the above. If we strike out the water from both, which evaporates, as well as the organic matter and the accidental fragments of wood, from the blue clay, which are burned out, in the act of calcination, the solid matter of these two clays chiefly differs, in the pit clay containing more sand than the blue clay, so that the former is in a much coarser state, which I have always found prejudicial to the quality of an artificial cement (91), and no doubt the same circumstance must injure an artificial puzzolana also. In respect to the gases, which are found combined with the blue clay, when fresh from the river, but which evaporate after long exposure to air, and which therefore, as Mr. Faraday observes, are no inherent part of the solid matter of that clay, their presence is certainly of some importance as our experiments have fully proved (84 and 155). But to return to our comparison of the solid matter of each, that of the brown pit clay contains little more than 4½ per cent of iron, and one half per cent of lime, whilst that of the blue river clay contains rather more than 7½ per cent of iron, and 3 per cent of lime; but I do not attach much importance to the difference in the proportions of these two substances, the iron and the lime, being persuaded that the superiority of the blue over the brown clay, as an ingredient for an artificial cement or puzzolana, depends chiefly on the superior fineness of the former.

(cxxxv) *Analysis of Trass and Puzzolana.*

The chemical component parts of trass and puzzolana are so nearly alike, that the one cannot possess any property not belonging to the other, though it may in a stronger or weaker degree; but it ought to be understood, that puzzolanas and trasses even of the same neighbourhood, may differ a little from each other in the proportions of their component parts, and those of different localities still more; and the same remark applies to hydraulic limes and water cements.

The trass used by Mr. Stevenson on commencing his works at the Bell Rock Lighthouse, for want of puzzolana, is stated by him to have contained in one hundred parts, silica 57,* alumina 28, lime 6·5 and oxide of iron 8·5.

The puzzolana used by him for the same Lighthouse contained silica 55 parts, alumina 20, lime 5 and oxide of iron 20.

The following Table contains the component parts of the trass and puzzolana, used by General Treussart at Strasburg, as analyzed by M. Berthier, together with a former analysis of each of those substances, which as he observes may be somewhat different in their quality.

ANALYSIS OF TRASS AND PUZZOLANA.

Component Parts.	According to some former Analysis.		According to M. Berthier.	
	Trass.	Puzzolana.	Trass.	Puzzolana.
Silica	570	350	570	445
Alumina	230	400	120	150
Lime	65	50	26	88
Magnesia	....	....	10	47
Oxide of Iron	85	200	50	120
Potassa	....	....	70	14
Soda	....	....	10	40
Water	....	....	96	92
Total..	950	1000	992	996

(CXXXVI) *Analysis of some Water Cements.*

General Treussart states that the calcined Sheppy cement stone, after having had the water and carbonic acid gas driven off by the fire, when analyzed by M. Drapier, was found to contain in one thousand parts, quick lime 554, clay 360, oxide of iron 86, and some traces of carbonate of magnesia and of manganese: and that the Boulogne stone after having been treated in the same manner, was found to contain quick lime 540 parts, clay 310, and oxide of iron 150, when analyzed by M. Berthier.

Mr. Porrett chief clerk of the Storekeeper's Department of the Ordnance favoured me with the analysis of two specimens of the Sheppy cement stone, and one of the Southend stone, analyzed by himself in 1811, after previously drying them at a heat of 400 degrees. He also gave me an analysis of the Yorkshire or Atkinson's cement, and of the Harwich cement stone, which he had copied without knowing the authorities, but which he considered likely to be correct.

* It is stated as 37 parts of silica, which I have corrected to 57, the former being evidently an error of the press.

ANALYSIS OF VARIOUS CEMENT STONES.

Component Parts.	Sheppy, No. 1.	Sheppy, No. 2.	Southend.	Yorkshire.	Harwich.
Water....................	....	3·00	....	....	3·875
Carbonic Acid	31·00	29·00	29·77	31·00	22·750
Lime	30·20	35·00	34·08	30·50	29·250
Silica	18·00	17·75	12·00	24 00	9·375
Alumina	3·10	6·75	13·00	6·75	9·500
Magnesia	0·20	0·50	1·52	1·04	....
Black Oxide of Iron....	5·25	6·00	8·80	1·31	17·750
Brown Oxide of Manganese...........	6·75	1·00	....	....	....
Sulphate of Soda, and Muriates of Lime, of Magnesia and of Soda	....	....	....	....	7·500
Loss	5·50	1·00	0·83	5·40	....
Total..	100·00	100·00	100·00	100·00	100·000

(CXXXVII) *Analysis of some Hydraulic Limes.*

Captain, now Major, Savage, R.E. informed me, that the blue lias lime stone of Lyme Regis, as analyzed by Dr. Carpenter the Mayor of that town in 1826, was found to contain in 100 parts, carbonate of lime 65, sulphate of lime 18, silica 10, alumina 5 and oxide of iron 2.

The lime of Obernai used at Strasburg by General Treussart, as analyzed by M. Berthier, was found to contain in one thousand parts, carbonic acid and water 380, lime 422, silica 15, alumina 43, magnesia and iron 50.

The lime of Metz, also analyzed by M. Berthier, was found to contain in one thousand parts, carbonic acid and water 412, lime 445, silica 53, alumina 13, manganese and iron 67.

(CXXXVIII) *Of the superior Mortars or Stuccos made in India.*

The natives of India seem to have had no knowledge of natural water cements, similar to those of England, and I believe that Lieut. or Cadet W. W. Saunders of the Madras Engineers then on a visit to Bengal, and since retired from the Service, first introduced my method of making an artificial water cement in Calcutta in 1831, which he explained to the Asiatic Society, as stated in their Journal for October 1831, that intelligent young Officer, whilst previously under my command at Chatham, having attended at and assisted me in most of the experiments on cements, tried by me during that period. He also found that some of the kunkurs, or lime stones of that country, which I believe are only obtained in nodules near Calcutta, required less clay than common lime, to make a water cement, they having hydraulic properties, especially the kunkur of the salt water lake.

In making flat roofs in Bengal, strong beams are first laid 3 or 4 feet apart, and over these light horizontal rafters, to support tiles about a foot square and 1 inch thick, over which they lay fragments of bricks or tiles, lime and water, which are well beaten with small mallets, until the mixture becomes very stiff. Over this a finer coating is applied, consisting of pounded bricks or tiles mixed with lime, which is also well beaten, and made quite smooth, and sometimes rubbed over with oil. In the upper Provinces, long stones are used instead of the small wooden rafters and tiles. This information I received from Major Hutchinson of the Bengal Engineers.

Captain Smith, in the notes to his translation of M. Vicat's Treatise states, that the Natives of India mix the common lime with Jaghery water, that is with a solution of a very small proportion of coarse sugar in water, which he says produces an admirable effect, inasmuch as the common mortars made of calcined shells, when well prepared in this manner, resist the action of the air for centuries; and he thinks that the same expedient might be adopted with benefit in this country.

(CXXXIX) Persons who have been in India generally praise the handsome external appearance of the houses, produced by a stucco, which is termed Chunam. Captain Smith describes that used at Madras, which is said to be the most celebrated, as being a stucco laid on in three coats, the first of a common mixture of shell lime and Jaghery water half an inch thick, the second made with sifted shell lime and fine white sand also sifted; the third and last coat, which receives the polish, is prepared with the greatest care, of lime of the purest and whitest shells, mixed with from one fourth to one sixth part of the finest white sand; and the ingredients of these two last coats, which are mixed with pure water, as Jaghery water would spoil their colour, are ground with a roller on a granite bed to a perfectly smooth uniform paste, which should have the feel and appearance of white cream. In every bushel of this paste are mixed the whites of ten or a dozen eggs, and half a pound of ghee or clarified butter, to which some also add from a quarter to half a pound of powdered balapong, or soap stone.

The last coat is laid on exceedingly thin, and before the second is dry. It dries speedily, and is afterwards rubbed with the smooth surface of a piece of soap stone (steatite) or agate, to produce a polish, an operation which is sometimes continued for many hours, after which it is necessary to wipe it from time to time with a soft napkin to remove the water which sometimes exudes from it for a day or two after completion.

Captain Smith observes, that every Master Bricklayer has his own favourite recipe, but he considers the essential ingredients, in addition to the lime and sand, to be the albumen of the eggs and the oily matter of the clarified butter, for which oil is sometimes substituted.

(CXL) *Proceedings in India relative to the Carbonate of Magnesia, which has been proposed to be used as a Cement, in the Madras Presidency.*

Being informed that a communication had been made to the Court of Directors of the Honourable East India Company in 1837, recommending that the carbonate of magnesia, which was said to be very abundant in some parts of the Madras Presidency, should be used as a water cement in public works, under that Government, I naturally concluded, that the hint must have been taken from my pamphlet on cements, printed but not for general sale, in 1830, of which several copies had been sent out officially to all the Presidencies in India, and which soon after its circulation in that country, was republished in a periodical work entitled Gleanings of Science at Calcutta: and under this impression I was gratified by the reflection that an experiment of mine, from which I never anticipated any practical result, owing to the great expense of magnesia in this country, should have proved useful in that distant part of the globe.

On inquiry, however, I ascertained, that the water-setting properties of the carbonate of magnesia, had been discovered by Mr. J. Macleod a Surgeon in the Company's Service, some years before the same property became known to me, in the course of my experiments, who as it appears had been employed by the Government of Madras several years prior to the death of Sir Thomas Munro, to endeavour to discover a hydraulic cement, that might be used in India as a substitute for the water cements of England, which could not be procured there conveniently, or at a moderate expense.

On the 11th of February 1826, Mr. Macleod wrote to the Secretary of the Madras Government, communicating the results of his experiments, and recommending that pure magnesia, of which abundance was procurable, should be used as a hydraulic cement, instead of the calcareous water cements, hitherto considered as the only efficient cements for the purposes of Hydraulic Architecture. The Military Board was accordingly directed to institute such experiments, on a moderate scale, as might enable them to judge of the propriety of Mr. Macleod's suggestion.

These experiments having been performed on the walls of Fort St. George, the result was reported to the Military Board, by the superintending Engineer, in a letter dated the 2d of February, 1827, in which he stated, that the magnesian cement seemed, as far as a judgment could at that time be formed, to be in every respect equal to Parker's cement, and it had the advantage of being of a handsome white colour. On the 14th of February, after receiving this report, Sir Thomas Munro is said to have inspected the experimental cements in person, and to have expressed a confident hope of their ultimate success.

I have extracted the above information from a letter of Mr. Macleod to the Right Hon. Sir Frederick Adam, K.C.B, then

Governor of Madras, in which he requests that Sir Frederick would bring his labours under the notice of the Court of Directors. This letter is dated from Bellary the 1st of October 1835, and it appears that he was induced to write it, from the circumstance of having read in the Fort St. George Gazette, that the Government of Madras intended to employ the magnesian cement on their public works.

This measure does not appear to have resulted from Mr. Macleod's original proposition, or from the Report of the Engineer before-mentioned, but in consequence of the Report of a more recent Committee, consisting of Lieut.-Colonel Garrard Chief Engineer, President, Lieut. Colonel Monteith also of the Madras Engineers, Surgeon James White Medical Storekeeper, and Assistant-Surgeons G. Malcolmson and Dr. P.M. Benza, Members.

This Committee met on the 25th and 29th of September 1835, in obedience to a Garrison order of the 24th of the same month, to examine a supply of calcined magnesia received from Salem, and to report on its fitness for building purposes.

They ascertained by chemical analysis, that the native carbonate of magnesia contained in one hundred parts, carbonic acid 51·5, magnesia 48, water 0·3 and silica 0·2, with slight variations of the latter substance in different specimens.

To quote the whole of their report would be unnecessary, but they conclude by an opinion, that the calcined magnesia from Salem forms an excellent cement suited for any building purposes to which it may be applied, and capable of bearing 1½ measure of sand to 1 of magnesia. As it absorbs moisture rapidly even in a close room, and is thereby deteriorated, they recommend that it shall be calcined as near as possible to the spot where it is to be used, and that it shall be preserved in air-tight casks, and as the weight of the carbonate or raw material is double that of the calcined magnesia, they recommend that it shall be obtained in its natural state, from those situations, where it can be conveyed to Madras by water carriage.

In the Appendix to his translation of M. Vicat's Treatise, Captain Smith has introduced an additional Article of his own (IV), in which he states that 'extensive beds of the native carbonate of 'magnesia have been recently discovered in the South of India, 'near Salem and Trichinopoly in the Madras Presidency, and the 'supply is so abundant, that measures have already been taken for 'turning this valuable material to account as a cement, for which 'purpose the experiments of Colonel Pasley have shown it to be 'admirably adapted;' and he adds further particulars respecting it, according to information received from Dr. Malcolmson, who states that it is proper to allow it 12 hours or more to set, previously to immersion, which agrees with my experiments, and who observes that it is the most beautiful of all stuccos, excelling even the celebrated chunam of Madras. Captain Smith does not notice the researches of Mr. Macleod, which must therefore have been unknown to him, as they were to me, although I have no doubt of the authenticity of the documents relating to them, as before quoted.

(CXLI) *More detailed Description of Mr. Brunel's Semi-arches, in reference to Articles* 59 *and* 319.

The elevation of the North side only of Mr. Brunel's semi-arches having been given in Article 59, I herewith subjoin the elevation of the South side of the same wonderful piece of workmanship, showing seven small arches, which on this side only projected about 1 foot from the spandrels of the main semi-arches or great ribs, as they may be termed. The pier from which these semi-arches sprang measured in plan, 10 feet in extreme length and 4 feet in width, at the level of the ground, of which two dimensions the last only appears in elevation; and its extreme height above the same level was 8 feet. It was founded on 3-inch Yorkshire paving stones, let into the ground only 8 inches below the surface, which was the cause of its downfal as described in Article 319, in consequence of a much deeper excavation having been made near it.

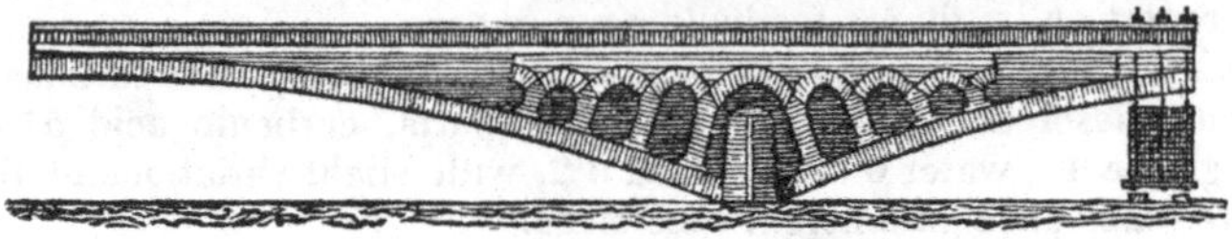

South Elevation of Mr. Brunel's Experimental Semi-arches.

The semi-arches springing from the central pier were 3 feet 6 inches wide, and 13 inches in depth or thickness; and the cornice, with which the one coincided and the other nearly so, at their outward extremities, being also the respective crowns of these semi-arches, was of the same width also; but in the spandrels or intermediate spaces on each side of the central pier, above the spring of each semi-arch, and below the cornice, the width of brickwork measured transversely was only 18 inches, that is within the openings of the small arches, and not including the additional width caused by the projection of them or their piers.

The brickwork was bonded with 10 tiers of 3 wooden pantile-laths each, commencing at the height of 4½ feet above the level of the ground, and laid horizontally in the joints of the remaining courses of the spandrels, above which it was bonded with 12 tiers of hoop iron, also laid horizontally, in the joints of the remaining courses of the spandrels and in those of the cornice, in the following order as to number, viz; in the first or lowest joint 2 pieces of hoop iron, in the second 3, in the third 4, in the fourth 5, in the fifth, sixth, seventh, eighth and ninth each 6, in the tenth 4, and in the eleventh 6 pieces, above which there was only one course of bricks more to complete the extreme height.

The above dimensions and details have been taken chiefly from the 6th number of the Civil Engineer and Architect's Journal, which I believe to be correct, excepting that nett cement, supplied by

Messrs. Francis and Co., was used, instead of mortar composed of cement and sand as therein stated, and excepting also, that the elevation of the first explanatory figure given in that number (Page 119) has been reversed by mistake, the short semi-arch represented there being placed on the wrong side of the central pier. In correcting these mistakes in the journal before quoted, I have no wish to depreciate that work, which promises to be very useful and interesting. In those dimensions or points, in which I differ from it, I have not trusted entirely to my own observation, but made a reference to Mr. Brunel, who confirmed some particulars in the description of his semi-arches, as stated in that journal, of which from my own recollection, I had doubted the accuracy.

Mr. Brunel has often informed me, that he considered wooden laths, properly secured against decay, to be quite as good as laths of hoop iron for brick and cement bond. From what I have seen of the ruins of the broken down experimental beams, &c, in which the iron always came out very bright, I do not think that hoop iron is liable to rust, when laid in cement joints, because the cement in setting absorbs the water, with which it has been mixed, without allowing it time to corrode the iron. Mr. Howe is of opinion, that previously heating the hoop iron, and allowing hot linseed oil to dry upon it, would preserve it, if liable to rust. Should this expedient be adopted, no oil in its fluid state must be allowed to remain on the surface of the iron, because it would injure the cement (100).

(CXLII) *Further Information respecting Asphaltic or Bituminous Cements.*

In a former article of the Appendix (XXXIV), I noticed Mr. Fitz Lowitz's resinous cement, and Lord Stanhope's bituminous cement, the latter of which I described more particularly as having been actually used for roofs in London, both of which were brought forward nine or ten years ago, or perhaps sooner, as I can only vouch for the periods when they came to my knowledge. In a note to the same Article, I also noticed Mr. Cassell's bituminous cement, for which I have since learned that he took out his patent in 1834, but none of these inventions appear to have made any impression on the public in this country, until after Mr. Claridge had taken out a patent and established a Company in London, for the sale of this article, which was described by Mr. Simms the Engineer of that Company, in a pamphlet, from which I made ample extracts in some former articles of the Appendix. (See from Article CXXVII to CXXX, inclusive.) In so doing, it was my object to describe a cement, which like concrete, may be superior for some purposes, and therefore very useful, if kept within proper bounds; but I had no wish of course to recommend the asphalte of Seyssel in preference to any other bituminous cement, for as I hinted before (CXXX), I cannot see any thing in its natural chemical component parts, as described in Mr. Simms's pamphlet, that can render it superior to similar compositions, that may be made by combining the same substances found any where else. It will therefore be a question of economy for those, who wish to use a bituminous cement, whether the cement of Seyssel is sold cheaper than others of equal quality or not, and the public will now have ample means of making a selection ; for although Mr. Claridge's patent was only advertised in March 1838, about the time when those Articles of the Appendix in which I have described this cement were printed, the circumstance of bituminous cements having become fashionable in Paris, and consequently having been seen by a great number of English travellers as applied to pavements, &c, in that metropolis, has led to the formation, still more recently, of numerous rival asphaltic and bituminous companies in this country, all claiming the superiority of quality and economy for their own respective compositions.

For example, having had the curiosity to look back to the Times Newspaper, which is filed at the United Service Club, Pall Mall, for the months of April and May 1838, I found that in this brief space the following Companies were advertised, or their pretensions if of prior date, which applies to two only, were recapitulated.

1st. Cassell's patent of 1834, of which I have not seen the original advertisement, but which in May 1838, is stated as having a Capital of 100,000£ in 10,000 shares at 10£ each.—2nd. Claridge's patent Asphalte of Seyssel Company, Capital 200,000£ in 10,000 shares of 20£ each.—3rd. The London Bitumen and Asphalte Company, Capital 60,000£ in 3000 shares of 20£ each, who, as it appears by their advertisement, are to obtain their Asphalte from Egypt, and have sent an Envoy to negociate with the Pacha for that purpose.—4th. The Bastenne and Gaujac Bitumen Company, Capital 250,000£ in 2500 shares of 10£ each.—5th. The British Asphaltum and patent Coal Company, Capital not stated.—6th. Robinson's patent Parisian Bitumen paving Company, for roads, footways, bridges, &c, Capital 100,000£ in 5000 shares of 20£ each.—7th. The Savoy Iron, Coal and Asphalte Company, Capital 200,000£ in 10,000 shares of 20£ each.—8th. Polonceau's patent elastic Bitumen Company, Capital 200,000£ in 10,000 shares of 20£ each.—9th. The Zante Asphaltum Company, Capital 150,000£ in 3000 shares of 50£ each, whose spring of Asphaltum is said to have been at work for more than 2000 years.—10th. The Netherlands Asphalte Company, Capital 20,000£ in 2000 shares of 10£ each. 11th. The United States Asphalte of Seyssel Company, Capital 200,000£ in 10,000 shares of 20£ each.—12th. The British Colonial Asphalte Company, Capital 100,000£ in 5000 shares of 20£ each, to be supplied with bitumen from the Island of Trinidad, where they assert the supply to be inexhaustible, and of a much purer quality than that of Switzerland, France, &c.—13th. The London, North America and West India Asphaltic Company, Capital 100,000£ in 10,000 shares of 10£ each.—14th. The Swiss Asphalte Company, to be supplied from the celebrated mine of Travers, Capital

100,000£ in 5000 shares of 20£ each.—15th. The Liverpool Seyssel Asphalte Company under Claridge's patent, for the Northern Counties of England and the Isle of Man, Capital not stated.—16th. The Scotch Asphaltum Company, Capital 125,000£, in 5000 shares of 25£ each, who propose to use the native bitumens of Scotland only.—17th. The Manchester Asphaltum Company, licensed by Mr. Cassell, Capital 100,000£ in 10,000 shares of 10£ each, the asphaltum to be found near that city.—18th. Claridge's patent Scotch Asphalte of Seyssel Company, Capital 200,000£ in 10,000 shares of 20£ each.—19th. The Birmingham, Foreign and British Asphaltum Company, Capital 50,000£ in 2500 shares of 20£ each, proposing first to use foreign, and afterwards native asphaltum found in the neighbourhood of Birmingham.—20th. The London, Paris and Hamburgh Asphalte Company, Capital not stated. —21st. Asphalte Lobsanne Association, Capital 300,000 in 10,000 shares of 30£ each.—22nd. The Dublin, Cork and Belfast Asphalte Company, Capital 100,000£ in 4000 shares of 25£ each.

The amount of Capital above stated is no less than 2,655,000£, to which the three Companies who have not yet advertised their respective Capitals, may probably add at least 300,000£ more. Those who have advertised their Capitals required deposits of from 10 shillings to 5£. per share, which I did not think it worth while to insert.

In the advertisement of one of the Companies licensed under the patent of Mr. John Henry Cassell, who is stated as being a manufacturing chemist at Mill Wall London, after quaintly observing, 'that in the present rage for Asphalte Companies, which many of the 'sober-minded are disposed to class among the bubbles of the day,'

"The earth has bubbles as the water hath,
"And these are of them."

The vengeance of the law is denounced against all the other Companies without exception, if they shall presume to vend their asphaltic or bituminous cements in this country, unless by Mr. Cassell's permission, within the period of the next ten years, during which his patent right for the exclusive sale of all cements of this description, is declared to hold good.

Having tried a specimen of the cement of Seyssel prepared for use as part of a foot pavement, I find that it melts and partially burns when held over a lighted candle, but not with a clear flame, as Mr. Fitz Lowitz's resinous cement did.* I have no doubt, that all the other bituminous cements enumerated in this article will do the same.

Specimens of a foot pavement actually made with Claridge's Asphalte of Seyssel may now be seen in the metropolis, one opposite to the Horse Guards on the other side of the way, and another at the Zoological Gardens, Regent's Park, near the vaulted passage, by which the two portions of those gardens communicate with each other, under the public carriage road.

* See the Note to Article 42.

(CXLIII) *Observations on the Proportions of Blue Alluvial Clay, and of Brown Pit Clay, proper for making the best Artificial Cement in combination with Pure Chalk, in reference to Article* 363 *and the first Paragraph of Article* 366.

Before I received the Analysis of those two clays with which I was favoured by Mr. Faraday (CXXXIV), I thought it probable, that the proportion of each, required to form the best artificial cement with chalk, would depend upon the quantity of solid mineral matter contained in it, exclusive not only of water and gases, but also of animal and vegetable matter. If this conjecture had been correct, as 100 parts by weight of the blue clay, contained only about 46 parts, whilst 100 parts of the brown clay contained 81 parts, of solid mineral matter, one might suppose, after we had ascertained by numerous experiments that 100 parts of chalk to 137½ parts of blue alluvial clay formed the best artificial cement with that sort of clay, that 100 parts of chalk to 79 of the brown pit clay of Upnor would form the best artificial cement with the latter, because 79 parts by weight of the brown contain as much mineral matter as 137½ of the blue. This rule however did not hold good, as we found by our experiments, for the difference was by no means so great as the above; but we did not try a sufficient number of experiments, with the brown pit clay of Upnor, to fix its best proportion with the same accuracy, as we had done in respect to the blue alluvial clay of the Medway.

Hence I consider, that in using different sorts of clay as ingredients for an artificial cement, one must have recourse to experiment for determining the best proportion of each, as I stated in the first paragraph of Article 366, instead of attempting to determine this point from their comparative chemical analysis.

(CXLIV) *Of the Artificial Cement, made by Major White of the Royal Staff Corps, at Hythe.*

This zealous and intelligent Officer, who, either as Adjutant or latterly as Senior Officer of the above Corps, has had the charge of the Works of the Royal Military Canal in that neighbourhood for many years, informs me that after receiving my pamphlet of 1830, printed for official circulation, he made a considerable quantity of artificial cement, according to the rules therein laid down, which he used in various parts of the canal exposed to the action of water, and which he considers equal to the best natural cements of this country. But as our own subsequent operations of 1836 recorded in Article 146, and in those immediately following it, fully proved that the proportion of blue clay laid down in that pamphlet was too small, and ought not to have led to such satisfactory results as Major White experienced, I think it probable that the clay he used, which was pit clay obtained in that neighbourhood, may by its greater density and specific gravity have corrected the error in my original proportion, and rendered that an excellent cement in the hands of Major White, which, if he had used alluvial clay like me, might have been an inferior one.

(CXLV) *Remarks on the comparative Weight, &c, of Cement Stone, in reference to Article* 372.

Mr. Mitchell, whom I have quoted as my authority for that article, informs me that there are from 26½ to 27 cubic feet of broken Harwich cement stone, and about 25 cubic feet of Sheppy cement pebbles, to the average ton weight of each. Also that the Harwich cement stone loses less weight in burning than the Sheppy; and that any given quantity of calcined cement powder, if as closely packed, as is the custom in Sheerness Dock yard, will occupy only about two thirds of the space, previously required for containing the quantity of broken cement stone, from which it was produced. Messrs. Francis consider 26 cubic feet of broken cement stone, per ton, to be a just average, both for the Sheppy and Harwich cements.

(CXLVI) *Proportions of Fewel, necessary for burning Cement on a great Scale, in reference to Article* 377.

Very small kilns require a much greater expenditure of fewel than larger ones, as was proved by our experiments in 1830, when we found it necessary to use 1 measure of coals to 5 measures of our raw artificial cement cubes (113). Mr. White informs me that Frost's artificial cement, being always perfectly dry, may be burned by 1 measure of coals to 10 of cement, but that he uses an average of 1 measure of coals to 8 measures of broken cement stone in burning the Sheppy and Harwich cements. Messrs. Francis consider 1 measure of coals to 10 of broken stone to be a fair average for the Sheppy and Harwich cements, except when the stone is very wet, when rather more coals are required. Mr. Mitchell also considers one measure of coals to 10 of cement stone to be a just average, but under favourable circumstances, and with a skilful and careful foreman to manage the kiln, he has found that 1 measure of coals has burned even more than 10 of cement. In the common lime kilns in the South of England, 1 measure of coke or coals is generally used to burn from 5 to nearly 7 measures of broken chalk, the former proportion being seldom exceeded in using coke.

(CXLVII) *Probable comparative Expense of Artificial Cement, in reference to Article* 378.

Messrs. Francis and White, after having purchased Mr. Frost's works in Swanscomb parish in 1833, were in the habit of selling their artificial cement for one shilling per bushel, at the same time that they sold their natural cement for eighteen pence, the former being thus sold at 33⅓ per cent less than the latter. But as they invariably recommended their customers to purchase their natural cement in preference, they may be supposed to have had more profit upon the sale of this than of the other; although they must of course have had some profit on their artificial cement also, because having no claim, like Mr. Frost, to whatever merit there may be in the invention or first introduction of this article in England, they would of course have discontinued their manufactory of artificial cement altogether, if it had been a losing concern. But artificial cement, made according to the rules deduced from our experiments at Chatham, and consequently equal to the Sheppy and Harwich cements in efficiency, would fetch a higher price than Frost's, which is acknowledged to be of much inferior quality, even by the manufacturers themselves, who have dealed in it (XVII).

From these circumstances, I consider myself warranted in believing, as I stated in the article before quoted, that the excellent artificial cement, proposed by me, if offered at 25 per cent lower than the natural cements of England, would find a ready sale in the markets of the metropolis, and that at this difference of price, the former would yield the same profit to the manufacturer, as the latter.

(CXLVIII) *Observations on Coal Ashes, and on the Peroxide or Rust of Iron, as Ingredients for improving Mortar.*

Every country bricklayer in England knows, that an addition of coal ashes improves common lime mortar, and therefore a mixture of this kind is often used for pointing outside joints, which mixture is distinguished by its dark colour. The degree of improvement communicated by coal ashes to mortar of this description is however so insignificant, that I did not think it worth while to notice it in the body of my essay, nor would I have done so even in the Appendix, but from a wish to omit nothing, that has come to my knowledge.

In respect to the peroxide of iron, it is well known that the rust formed in wet weather from the lower part of iron posts, or from iron water pipes or gas pipes laid in the ground, gradually consolidates the gravelly soil near them into a compact substance, of which the rust or peroxide of iron is the cement. Hence I believe, that some of our Civil Engineers, from observing this circumstance, have occasionally ordered their mortar for wharf walls, &c, to be made with water impregnated with iron. I tried for a long time to cement sand and gravel together by the peroxide of iron, but entirely failed, owing I believe to having always used clean siliceous matter only; for on reflection since, it has occurred to me, that wherever I have seen gravelly soil cemented in the manner above described, it has never been clean gravel, but always a mixture of that substance with clay. But I have not thought it worth while to make any farther experiments for ascertaining this point, by mixing clay and gravel together, and subjecting them to the action of the peroxide of iron in a moist state; because even if this should gradually form a sort of compact substance, it would be of very little use as a mortar, without lime: and every practical man knows, that clay in its natural state spoils all limes and calcareous cements.

(CXLIX) *Confirmation of Opinions previously advanced, that the Mortar of weak hydraulic Limes ought not to be used for Wharf Walls, &c, &c.*

In various parts of the foregoing Treatise, it has been laid down as a rule, that chalk lime ought not to be used at all, for wharf walls or others exposed to wet or moisture; and also that even the weaker hydraulic limes ought not to be used, unless protected at the surface against the action of the water, by pointing all the external joints with cement; and it was stated as one proof amongst others of the necessity of this rule, that the wharf wall adjoining to the new Baths at Rochester, which was built with Halling Lime Mortar, had given way for about an inch in depth in all the external joints. (See Article 280). Having again carefully examined the same wall more recently, I find that this process has been going on rapidly, and that the mortar of all those joints, which have been alternately exposed to the Medway at high water, and to the air at low water, has been wasted away to a much greater depth than the above, since the late severe winter of 1837-8; so that it appears certain, that this wall which is built with rubble Masonry, of Kentish Ragstone, and of excellent workmanship, will go entirely to pieces in time, unless protected by pointing all the external joints with Cement Mortar. This circumstance and others before noticed confirm me, in the opinion, that the weaker hydraulic limes, such as the Halling, Dorking, Warmsworth, &c. ought not to be used even for the body of wharf walls, unless improved by a proportion of natural, or of artificial puzzolana of good quality.

Having heard lately that Messrs. Gladdish, had built a rubble wharf wall on their premises, which are situate between Gravesend and Northfleet, with flints and mortar made of 2 or 3 measures of sand and 1 measure of the blue lias lime from Lyme Regis, which was said to have remained at the end of 2 or 3 years in a serviceable state, without any perceptible degradation of the external joints, though it had been exposed to the alternate action of the Thames and of air at high and low water, during the above period; I went to see it, and found that the above statement was not exactly correct, for the lower part of the wall so exposed had been newly pointed, and where this precaution had not been taken, the mortar had been washed out of all the joints for about two inches in. Hence even this very strong hydraulic lime, of the quality of which I have no doubt, for all the blue lias lime used in my last experiments, which I found to be excellent, was supplied from Messrs. Gladdish's kilns, is unfit for the facing of wharf walls; as it goes to pieces unless pointed from time to time. Moreover all the mortar of the lower part of this wall, whether old or new was soft, whilst that of the upper part above the level of the high water of spring tides seemed nearly as hard as the flints, with which it was combined. This circumstance confirms me in the opinion, which has been repeatedly advanced in the foregoing Treatise, that the mortar even of the strongest hydraulic limes requires to be protected in all the external joints by cement, without which it is incapable of resisting the action of water in mass, when exposed to immediate immersion, which is necessarily the case in walls built by tide work.

(CL) *Final Remarks on Common Chalk Lime Mortar.*

I had ascertained more than 12 years ago, that the pure limes, such as chalk lime, Carrara marble, &c, were perfectly unfit for the purposes of hydraulic architecture, as they dissolved away on the outside, and never set at all in the inside, of walls exposed to the action of water. But I was of opinion at that time, that the mortar of these limes was good for dry situations and for inside work, provided that the external joints were protected against the effects of beating rains, by pointing them either with cement or with some superior sort of lime, which opinion I expressed in Article 10, near the commencement of the foregoing treatise. Having, however, acquired much more experience, partly from experiments, and partly in consequence of continual observation of new buildings in progress, or of old buildings being pulled down, since the first sheets of this work were sent to press; I am compelled to retract the above opinion in favour of chalk lime, which I now consider bad under all circumstances, even in the driest situations; as it never attains any great degree of adhesiveness even when only exposed to the atmosphere, and its resistance is so insignificant, that it rather dries, than sets, in air (280). All that can be said in favour of chalk lime mortar is, that it is better than none; and that walls built with it will not fall to pieces in process of time, as General Treussart asserted (CXIII), without external violence. Yet, I must say, that even in the driest situations, the only ones in which it ought to be used at all, chalk lime mortar opposes scarcely any resistance to the tools of the workman who attacks it, the slightest blow from which converts it into dust. I have particularly observed this extraordinary want of resistance recently, in looking at the workmen employed in pulling down the brickwork of the Royal Exchange, after the late fire, and also of some old houses in Pall Mall, which operations were accompanied with clouds of white dust, caused by the friable nature of the dry chalk lime. Thus chalk lime mortar, when wet is a pulp or paste, and when dry, it is little better than dust.

I have been induced to insert these remarks from having met with some respectable lime burners lately, who consider that chalk lime mortar may be improved by burning the chalk in a flame kiln, in preference to the common kiln, and as the former is a more expensive mode, they sell their flame burned chalk lime dearer than the common kiln burned lime. As this supposed superiority, of the former mode of burning over the latter, is not founded on any chemical property of chalk lime, it appeared worth while for me to expose the fallacy of such a supposition, and to repeat, that no precautions in the burning of the chalk or mixing of the mortar can possibly make this sort of lime good. Nothing in short but mixing it with trass, or with natural or artificial puzzolana of good quality, can improve chalk lime so far, as to render it fit either for concrete in wet soil, or for the construction of walls of any importance, whether exposed to air or to water.

END OF THE APPENDIX.

JAMES BURRILL, PRINTER, HIGH-STREET, CHATHAM.

ERRATA.

Page 46, *Second Note, Line* 3, *for* subcarbonate of soda, *read* carbonate of ammonia.

.... 78, *Line* 17 *from bottom, for* 1831 *read* 1830.

.... 113, *Line* 1, *for* beam *read* board.

.... 152, *Line* 15, *for* 29000 *read* 39000.

.... Do. *Line* 17, *for* 7000 *read* 8000.

.... 183, *Line* 20 *from bottom, for* slaked *read* quick.

APPENDIX, *Page* 86, *Line* 7, *for* Peyronet, *read* Perronet.

ERRATA.

Page 45. *Second Note*, line 3, *for* subcarbonate of soda *read* carbonate of ammonia.
,, 132, line 11 from bottom, *for* 1847 *read* 1839.
,, 148, line 1, *for* boat *read* board.
,, 152, line 1[illegible], *for* 23000 *read* [illegible].
,, Do. line 17, *for* 7000 *read* [illegible].
,, 153, line 20 from bottom, *for* slaked *read* quick.
APPENDIX, Page 60, line 7, *for* Peyronet *read* Parronet.

WORKS BY THE SAME AUTHOR.

Published by John Weale, at his Architectural Library, 59, High Holborn.

A Simple Course of PRACTICAL GEOMETRY AND PLAN DRAWING, founded on a principle of peculiar simplicity, and adapted for classes of [illegible] instruction. Second Edition, much enlarged. Cloth, bds, 10s.

Sold at Egerton's Military Library, near Whitehall.

Essay on the MILITARY POLICY, &c. of the British Empire. 4th Edition 12s.

Observations on the Expediency and Practicability of simplifying and improving the MEASURES, WEIGHTS & MONEY, used in this Country, without materially altering the present Standards, 5s. 6d.

Description of the UNIVERSAL TELEGRAPH for Day and Night Signals, 3s. 6d.

Printed in the United States
By Bookmasters